Electrostatics 1999

Other titles in the Series

Other Electrostatics titles in the Institute of Physics Conference Series:

143: Electrostatics 1995
Edited by S Cunningham

118: Electrostatics 1991
Edited by B C O O'Neill

85: Electrostatics 1987
Edited by J L Sproston

Other titles of interest include:

Electrostatics
J Cross   0 85274 589 3   1987

Contents: Historical background and elementary theory. Electrification of solids and liquids. Measurements and instrumentation. Electrometers and electrostatic voltmeters. Field meters and non-contracting voltmeters. Electrostatics in gas filtration. Appendices: Miscellaneous applications. Hazards and problems. Theory. Index.

Electrostatics 1999

Proceedings of the 10th International Conference, Cambridge, 28–31 March 1999

Edited by D M Taylor, University of Wales, Bangor, UK

CRC Press
Taylor & Francis Group
Boca Raton London New York

CRC Press is an imprint of the
Taylor & Francis Group, an **informa** business

CRC Press
Taylor & Francis Group
6000 Broken Sound Parkway NW, Suite 300
Boca Raton, FL 33487-2742

First issued in paperback 2019

© 1999 by Taylor & Francis Group, LLC
CRC Press is an imprint of Taylor & Francis Group, an Informa business

No claim to original U.S. Government works

ISBN-13: 978-0-7503-0638-6 (hbk)
ISBN-13: 978-0-367-40004-0 (pbk)

British Library Cataloguing in Publication Data

A catalogue record for this book is available from the British Library.

Library of Congress Cataloging-in-Publication Data are available

**Visit the Taylor & Francis Web site at
http://www.taylorandfrancis.com**

**and the CRC Press Web site at
http://www.crcpress.com**

Preface

The increasing impact of electrostatics in all spheres of today's complex world can be measured by the growth of this conference. In the first meeting in 1967 just over 20 papers were presented. In 1987 the number of contributions had risen to more than 50 while at this conference there were almost 90 presentations. A total of 149 participants from 18 countries came to absorb the Cambridge atmosphere, reflecting the truly international nature of the conference. Of the papers submitted, 80 passed through the thorough refereeing process and I take this opportunity to thank all the referees for their constructive criticism of manuscripts and in particular for their rapid response, which enabled these Proceedings to be delivered to the publisher on time.

Dr Keith Davies, that most famous son of Wales and baritone supreme, delivered the Bill Bright Memorial Lecture. In his inimitable style, he ranged from the historical to the futuristic, simultaneously delighting and inspiring his audience. A number of internationally renowned speakers were invited to present plenary lectures to 'set the scene' for each of the main conference topics. These included Professor Richard Friend (Cavendish Laboratory, Cambridge University, UK), Dr Martin Glor (Swiss Institute for the Promotion of Safety and Security, Switzerland), Dr Arnold Kelly (Charge Injection Corporation, USA), Dr Dave Swenson (ESD Association, USA) and Dr Kumar Wickramasinghe (IBM, USA).

The two 'hot topic' sessions generated considerable controversy and discussion respectively. From the discussion led by Dr Jeremy Smallwood, it seems that both measuring and interpreting the MIE of powders still pose a fundamental challenge. We were then treated to a most fascinating presentation by Dr Martin Buehler (Jet Propulsion Laboratories, USA) outlining the development of an electrometer for measuring the tribocharging characteristics of Martian soil. I am particularly grateful to Martin for agreeing, at very short notice, to write up his presentation as a paper for the Proceedings. It will be fascinating to learn at a later conference in the series how the electrometer performed!

It is a pleasure to thank the Advisory Committee for its guidance and suggestions, members of the Organizing and Programme Committees for their considerable efforts in ensuring the success of the conference and Belinda Hopley of the Institute of Physics for the support she provided to the Organizing Committee and for arranging the social programme. Once again this year a special thank you is are owed to Dr John Chubb for organizing and running a very successful pre-conference workshop.

I hope you will enjoy reading this record of Electrostatics '99. The subject has advanced considerably over the years. Bioelectrostatics and ESD/EOS have become an established part of the programme, adding to the recurring themes such as hazards and applications. Judging from the number of papers submitted, EHD seems to be an area attracting an escalating interest while scanning probe microscopy seems set to take the subject to a new dimension.

Once again, the friendly sociable atmosphere of the conference was a memorable feature. My sincere thanks go to everyone who attended and contributed to making this one of the most successful conferences yet.

Finally my thanks go to my secretary, Lynne Jones, for carrying most of the administrative burden associated with producing these Proceedings.

Professor Martin Taylor
Chair, Programme Committee

Bill Bright
Memorial Lecture

Inst. Phys. Conf. Ser. No 163
Paper presented at the 10th Int. Conf., Cambridge, 28–31 March 1999

3

Electrostatics - Then, Now and Maybe

D K Davies

Markab, Pear Trees, Bridge Road, Lymington, Hants, SO41 9BY

Abstract. Enormous development has taken place over the 30 year history of the quadrennial Institute of Physics conferences on Electrostatics, in terms of understanding, useful applications as well as solutions to grave industrial problems. Theories and experimental data - often originally presented at these conferences - form the background to a great deal of this development. Two projects; one to develop a sampling system for the capture and analysis of airborne particles; and the other, to establish the decisive factors controlling the explosive situation in aircraft fuel tanks are described to illustrate the facilities now available for electrostatic analysis and design. The former project involves design of nozzles to nucleate water droplets on the wanted species, measurement of the droplet size distribution, charging of the droplets, finite element analysis of the electrostatic lens focussing system and finally development of the complete sampling device. In the latter project, experiments showing the balance between charging and charge dissipation rates of liquids of variable conductivity illustrates the subtlety of modern problem solving.

It is concluded that the future in electrostatics based on new materials with 'designed' electrical properties (even at the single electron level), new devices providing microscopic control of materials and structures and, even, interaction with biological entities is limitless.

1. Introduction

Institute of Physics conferences on Electrostatics - or Static Electrification as it was known in an earlier epoch - have been held quadrennially for over 30 years. Enormous development has taken place in the subject over that period, in terms of understanding, useful applications as well as solutions to grave industrial problems. Indeed, over this period great industries have been established which are based entirely on electrostatic phenomena. The future for new electrostatics-based technology is equally promising.

The "Contents" pages of the Proceedings of this series of conferences provide an interesting insight into the subject. In the early days, the emphasis was on "fundamentals", individual sessions being given over to basic phenomena - charge generation and migration - in solids and liquids. Arguments

4

raged over the mechanism of contact charging - ions, electrons and even material transfer. How could an insulator accept very high surface charge and yet retain it, virtually, indefinitely? "Applications" were very definitely of secondary importance. Contrast the programme for this conference where by far the greater emphasis is in practical uses and discussion of the industrial hazards - in electronics, fires and explosions - that arise from electrostatic effects.

Some of the early work - often originally presented at these conferences - is reviewed briefly as a background to a presentation of examples of current research and some speculation about future development of electrostatics-based technology.

As an example of modern electrostatics, a project to develop a sampling system for the capture and analysis of airborne particles is described to illustrate the facilities now available for electrostatic analysis and design. The project involves experimental work and computational fluid dynamics to explore means for nucleating water droplets on the wanted species, measurement of the droplet size distribution, charging of the droplets, finite element analysis of the electrostatic lens focussing system and finally development of the complete sampling device. Similarly, experiments showing the balance between charging and charge dissipation rates of liquids of variable conductivity, which can be the decisive factor controlling the explosive situation in aircraft fuel tanks, illustrate the subtlety of modern problem solving.

Speculation on the future in electrostatics includes description of new materials with controlled electrical properties (even at the single electron level), new devices providing microscopic control of materials and structures and even novel biological diagnostic techniques.

2. Background

One of the earliest conferences in this series - in 1967 - included separate papers by Ion Inculet (1) and myself (2) on the relationship between contact charge transfer and material work function. The concept of plastics having active electronic characteristics was considered somewhat revolutionary but the prospective argument with Harper (3,4) - who presented the invited paper - a doyen of the field and an "ion" man, did not eventuate probably because of the weiLewis T J ght of experimental evidence. Similarly, the idea of mobility of charge carriers in these highly insulating materials (2), much less the possibility of modifying these electrical characteristics of plastics, was equally revolutionary. Contrast the present conference programme where papers on polymers with tailored electrical (and optical) properties are the norm.

The situation in liquids is not so clear cut, the fundamental models of charge generation and migration still being matters of debate. A number of original papers in this area have been presented in this series of conferences by Gibson and Lloyd (5), Gibbings et al (6) and Lewis (7). Krazucki (8) made the controversial suggestion by that the electrical characteristics of insulating liquids could be described by the motion of particulate impurities. Progress is such that, despite these uncertainties, liquids with controlled rheological properties have been described (9).

3. Modern Electrostatics

Two continuing projects are described as a means of both illustrating the familiar disparate aspects of electrostatics - the development of novel technology or problem solving - and the powerful experimental and modeling capabilities available.

3.1 An electrostatic aerosol sampler

Small airborne particles in the size range from $0.01\mu m$ to $10\mu m$, which present a particular health hazard since they can be readily inhaled and retained, are notoriously difficult to capture. An aerosol sampling method is being developed based on nucleation of a water droplet around each airborne entity to produce a larger particle, which can be captured, by means of electrostatic charging and deflection, into a small trap. The device provides an additional advantage in presenting the collected species in a minimal quantity of water to the analysis system.

The physical process comprises injection of the wanted species into a stream of water-saturated air just after minimal adiabatic expansion. A crucial factor in the device is, of course, the development of the novel, dual-concentric-channel nozzle, which can operate under normal (practical) atmospheric conditions and offer reasonable gas sampling flow rates. An important aspect in the evolution of the nozzle has been the theoretical analyses of the expansion of the water-saturated air stream and mixing of the two gas streams using the fluid-dynamic modeling programme "FLUENT". These analyses showed that efficient mixing of the two air streams is vital in order to establish the essential environment for growth of the droplets around the very small nuclei. The models demonstrated that effective mixing was best achieved by introducing a small barrier into the central, sampling stream just downstream of the nozzle exit.

The nozzle developed on the basis of these analyses and a parallel experimental programme was shown to be effective by means of a "Malvern", laser-scattering particle size spectrometer. Small, sub-micron carbon particles, introduced into the system by passing the sampling air stream through an arc discharge between carbon electrodes, could be seen to increase the droplet nucleation rate by several orders of magnitude to yield droplets of $3 \pm 1\mu m$ diameter.

It is important, of course, to minimize the baffle size to reduce particle loss by impaction. The barrier finally evolved was a very short thin wire. This provided the additional advantage in that it could also be used as an electrode to charge the water droplets, by means of a conventional D.C. corona discharge. An interesting beneficial, side-effect of the corona discharge was the demonstrable increased efficiency of mixing caused by the diverging corona wind. The charging of the droplets was confirmed by downstream capture in a Faraday pail.

The final stage in the development of the particle capture device is the design of the electrostatic deflection and focussing system for droplet capture. This has been greatly facilitated by using available "PC-OPERA", finite-element software. The system is essentially axi-symmetric and so the dimensions, location and potentials applied to the electrostatic lenses can be readily be modeled based on calculation of the tracks of particles of appropriate size and charge.

6

The whole system has been validated both theoretically and experimentally and it is evident that the development of such novel devices is greatly facilitated by the tools now available.

3.2 Evaluation of fuel ignition

There have been many recorded incidents of aircraft fuel tank ignition. The reason for many of these ignition events is still unknown and a matter of some controversy. While most of these have occurred during ground refuelling, some have been attributed to sparks produced by electrostatic charging of the fuel. Other conflicting theories include sparks from chafed wiring, overheating of on-board fuel pumps and incendive sparks from isolated objects charged by a charged fuel spray.

The electrostatic charging of jet of liquid ejected from a nozzle is well established. The well-known series of supertanker explosions in the 1970's, for example, were all attributed to sparks produced from slugs of water charged by falling through the highly charged mist generated by the high-pressure water jets used for removing residual hydrocarbon sludge from the empty tank walls. In this case it was necessary to postulate the charging of electrically-conductive objects of some capacitance - the slugs of water - so that sufficient energy to cause ignition could be released in subsequent spark to a grounded structure.

Clearly, it is important to establish whether incendive sparks can be obtained from isolated objects, which have acquired a charge by immersion in a charged, fuel mist. The objective of the investigation which is described here was, therefore, to determine the potential attained by a metallic objects - in the form of typical clamps used for fixing pipes in aircraft fuel tanks - while suspended in a mist of fuel spray and, subsequently, to establish the energy of sparks from the object to ground.

The spraying system used for this investigation was based on conventional, high-pressure nozzles fed from a reservoir which contained the fuel and pressurised by an air compressor producing pressures of up to 3 bar. The spraying characteristics of these nozzles are not greatly affected by pressure apart from throughput, of course. The modest pressure range used in the present experiments - with a maximum of about 2 bar which is about the average of the pressure range 25-42psi found in aircraft fuel lines - would not be expected to have any influence on the effects observed.

In all static electrification phenomena no unbalanced charge is produced. Contact charge separation which occurs in processes such as shearing of a liquid jet at a nozzle, the charge carried away by the liquid is balanced, both in magnitude and polarity by the charge left on the nozzle. Two techniques were, therefore, used to establish the electrostatic spray characteristics - measurement of the current to ground from the nozzle by means of an electrometer and the determination of the ratio of charge-to-mass of the charged spray droplets. The latter was obtained by capturing the spray in a Faraday pail and measuring both the charge accumulated and the mass increase.

The test object capacitance was measured using a proprietary hand-held instrument and the fuel conductivity determined using a conventional conductivity cell. The leakage resistance of the test object was determined by measuring the current to ground using the electrometer for known voltages. Thermocouples together with an electronic thermometer were used to make the various temperature measurements on sprayed fuel; nozzles; fuel tank; and, working chamber. The test object voltage was determined by connecting it to a metal plate, highly insulated from ground, and observing the voltage on the plate by means of a non-contacting electrostatic fieldmeter, the system being calibrated in-situ, using a high voltage power supply. Owing to the significance of the droplet size on the suspension characteristics of a droplet mist, the size distribution of droplets produced by the nozzles was examined using the "Malvern" particle size spectrometer described earlier.

The experimental results showed that the voltage produced on the test objects, with the supporting structure grounded, were not sufficiently high to cause an electric spark let alone have sufficient energy to cause ignition. To fix ideas, the minimum ignition energy of a stoichiometric mixture of flammable hydrocarbon fuel and air is about 0.2mJ. The capacitances of the test objects were of the order of 30pF and so the necessary minimum voltage to cause ignition ($1/2CV^2$) is 3651V. Given the measured leakage resistances of 2.3 x $10^{13}\Omega$ for a fuel conductivity of 570pS/m, the generation of such a voltage would require a charging spray current of 1.6 x 10^{-10}A. The experiments showed that while nozzle currents of this magnitude could be produced by these pressure-driven nozzles, owing to the divergence of the spray plume, only a fraction of the charged droplets actually collide with the test object itself.

The generation of the necessary voltages can only be achieved by almost total capture of the spray by the clamp at the present nozzle currents or, given the approximately 10% capture efficiency in some cases, an order of magnitude increase in nozzle charging current. Such an order of magnitude increase in nozzle was observed on increasing the fuel conductivity. However, the leakage current also increases by wetting of the test object insulation by the higher conductivity fuel but the evidence suggests that the decrease in leakage resistance is not proportional to increase in fuel conductivity. This may be due to a change in the thickness of the "wetting" fuel layer.

This investigation showed that while, an electrically isolated object can be charged by a fuel spray produced by grounded nozzle, the voltages produced under the present regime, which produces a divergent spray plume, were not sufficiently high to produce a spark let alone ignition. However, the dilemma of whether increasing the fuel conductivity solves or exacerbates the problem remains to be established.

8

4 The beginning of the future

The development of microstructure fabrication technology has resulted in materials investigative and manipulation methods such atomic force microscopy. Devices with spatial resolution of 20nm and temporal response time of only of 20ps have been described (11). Complete electrometers measuring only a few micrometres across have been fabricated (12) providing, potentially, the sensitivity of single-electron transistors. Microelectrode systems can now be produced which permit manipulation of biological materials at the single cell level with the possibility of single cell hybridization .

5 Conclusions

The conclusions to be drawn from this brief review of electrostatics are that the subject has indeed seen some remarkable changes of the past thirty years and that the future is very promising. In common with science, generally, this area of Physics is experiencing an increasing tempo of development. New methodology, which including computer-based modeling as well as novel experimental techniques offers great opportunity for the development of new products and processes and solutions to many serious industrial and environmental problems. New materials with tailored properties together with the modern micro-fabrication techniques will inevitably result in the development of novel electronic and optical devices and new methods for analyzing and modifying biological entities.

6 Acknowledgements

The financial support of DERA for the project on the development of the aerosol sampler and, in particular, the technical assistance of Don Clark, Steve Preston and Derek Shakeshaft of Porton Down is gratefully acknowledged. The investigation of the fuel charging problem was supported by the US Airforce - via an EOARD contract - and both the support and technical input of Steven Gerken of WPAFB is appreciated.

6 References

[1] Inculet I I and Wituschek E P 1967 *Proc IOP Conf. Ser. No.4* 37 - 43
[2] Davies D K 1967 *Proc IOP Conf. Ser. No.* **4** 29 - 36
[3] Harper W R 1967 *Proc IOP Conf. Ser. No.* **4** 3 - 10
[4] Harper W R 1967 *Contact and Frictional Electrification* (Oxford, Clarendon Press)
[5] Gibson N and Lloyd 1967 *Proc IOP Conf. Ser. No.* **4** 89 - 99
[6] Gibbings J C, Saluja G S and Machey A M 1975 *Proc IOP Conf. Ser. No.* **27** 16 - 33
[7] Lewis T J 1983 *Proc IOP Conf. Ser. No.* **66** 16 - 33
[8] Krazucki Z 1975 *Proc IOP Conf. Ser. No.* **27** 1 - 15
[9] Sproston J C 1983 *Proc IOP Conf. Ser. No.* **66** 53 - 58
[10] Steeves G M 1998 *Appl.Phys.Lett.* **72** *504*
[11] Cleland A N and Roukes M L 1998 *Nature* **392** *160 -162*

DISCUSSION - Bill Bright Memorial Lecture

Title: Electrostatics - Then, Now and Maybe
Speaker: D K Davies

Question: Martin Glor
              Is spraying a realistic charging mechanism during aircraft refuelling?

Reply: No. This project was concerned with spraying from a hole (fracture) in a fuel
              pipe <u>within</u> the fuel tank.

Question: M K Mazumnder
              What is the primary mechanism for charging particles smaller than 2 μm in
              diameter - field charging or diffusion charging? One would expect diffusion
              charging to be dominant. Your experimental data agreed with the Pauthenier
              Limit - which is based on field charging. Could you please comment.

Reply: We used the Pauthenier Limit simply to obtain an "order of magnitude"
              droplet charge. The focusing effect we are seeking is not particularly charge
              dependent in the sense that we can accommodate any droplet charge (within
              reason) by changing electrode position and applied potentials.

Section A

Fundamental Physics

Inst. Phys. Conf. Ser. No 163
Paper presented at the 10th Int. Conf., Cambridge, 28–31 March 1999
© *1999 IOP Publishing Ltd*

13

Spectroscopy and calorimetry of the spark discharge

Z Kucerovsky†[1], W D Greason† and M Wm Flatley‡

† Applied Electrostatics Research Centre, University of Western Ontario, London, N6A 5B9, Canada

‡ Suncor Energy, Sarnia, P.O. Box 307, Ontario, N7T 7J3, Canada

Abstract. Spectrally resolved optical energy of the spark discharge was measured in the visible and infrared regions, using a single-event spectrometer calibrated by a calorimeter. A spark gap generator was used, with an optical signal collecting system that provided further calibration and alignment by means of a set of light emitting diodes. The calorimeter detectivity was $D = 4.3 \times 10^7$ V.J^{-1} and its detection limit 2.3×10^{-13} J.

1. Introduction

The spark and the arc are used as electromagnetic energy sources when interactions of a system with electrostatic discharge are studied ([4], [5], [10]). Most of the spark discharge studies have dealt with the radio frequency emissions. Much of the current work in the optical region have been devoted to the arc formation on the switch contacts at low voltages [7], electric arc dynamics and the arc spectrum and its spatial distribution (*cf. e.g.* [13]). Our work focused on the spark's optical emission.

The spark pulses are relatively infrequent and inhomogeneous, making spark emission spectroscopy more challenging than *e.g.* corona spectroscopy [6]. The spark energy is comparable to that of a high power laser, and laser methods can be used for its measurement (*cf.* Radak *et al* [9]). A sophisticated calorimeter was used by Bednarz *et al* [2] for the determination of heat capacity of 20-mg samples, with millikelvin temperature resolution, and by Yao *et al* [14] for a nonadiabatic scanning calorimeter.

The calorimetric measurements require that a difference between the temperature of two bodies be determined, typically with a thermistor or themocouple. Thermocouples [9] have nearly linear response and small mass, yielding accuracy, sensitivity and fast response; unfortunately, they need a d.c.-coupled or a chopper type amplifier. Thermistors are relatively more nonlinear, but can provide a.c. signal and be linearized. The electronics used to process the calorimeter signal can be quite sophisticated *cf. e.g.* the amplifier of De Geronimo *et al* [3] and Murgatroyd's Blumlein bridge [8].

---

[1] E-mail xenon@julian.uwo.ca

2. Theoretical Part

As the goal of our work was to measure spectral energy output of the spark, calorimetry and spectroscopy methods were used. Theory of modern calorimetry has been discussed in literature (*cf.* [1], [11], [12]). If Q is energy deposited on a body of mass m and specific heat c, its temperature increases as

$$\Delta T = \frac{Q}{mc}.$$

The increment of temperature is converted *e.g.* by a thermistor into a corresponding change in resistance, which is mostly taken to be exponential, depending on the material and geometry. Having a relatively high resistance, thermistors generate Johnson noise. A typical thermistor with resistance $R_t = 1$ kΩ, at room temperature ($T \approx 300$ K), and its signal processed with a band-pass amplifier (*e.g.* the bandwidth $\Delta B = 50$ Hz), generates Johnson noise $4kTR\Delta B = 8.3{\times}10^{-16}$ V^2.s^{-1} (k is the Boltzmann constant). Without attempting linearization, we have used four thermistors in a Wheatstone bridge, which doubled the sensitivity and at low level signals caused but slight nonlinearity error.

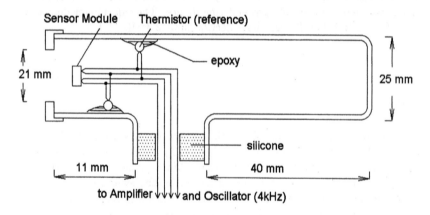

Figure 1. Calorimeter head (thermistors: sensing $R_{1,4}$; reference $R_{2,3}$).

3. Experimental Part

The apparatus consisted of a spark gap, an optical system for collecting the spark emission, a calorimeter with the calibrator and aligner, and a grating spectrometer. The spark gap comprised two stainless steel spheres (12.7 mm dia.), with an adjustable gap (up to 10 mm). A low inductance capacitor (Maxwell, 0.15 μF, 0.015 μH, 50 kV$_{max}$) was used to power the spark discharge. A light emitting calibrator diode was placed on the optical axis of the spark gap. Five diodes were used alternatively (blue, 473 nm; green, 556 nm; yellow, 587 nm; red, 640 nm; and infrared, 880 nm).

A two-lens, symmetrical condenser with an aperture of 10 cm and stereoangle $\Omega = 4.1 \times 10^{-2}$ sr was located at the condenser's first focus (12.4 cm). The gathered

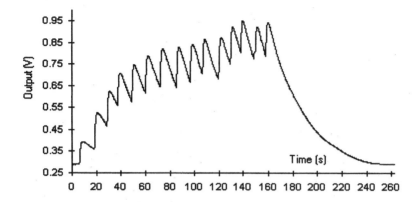

Figure 2. Calorimeter response to a sequence of 14 spark gap discharges.

radiation was focused by the second lens on either the calorimeter or the input fiber of the spectrometer.

The calorimeter body was of a thin-wall copper tubing (*cf.* figure 1). The entrance aperture was circular (21.4 mm dia), with the F–stop number of one. Two thermistors ($R_{1,4}$) were bonded to the black coated sensor, and two reference thermistors ($R_{2,3}$) to the copper body (Fenwall, 192-102DET-A01). The thermistor bridge was a.c.-powered (2.0 V, 4 kHz), and the bridge's differential voltage was processed by a signal conditioner, a second order band-pass filter (4 kHz), precision rectifier and integrator. A digital voltmeter (Keithley, M 2001) and an IBM PC-class computer (P6-II, LX) were then used.

The spark pulse was used to determine the time constant and sensitivity of the calorimeter, as shown in figure 2 for fourteen consecutive discharges (4-mm gap, 13.9 kV). The calorimeter signal was acquired simultaneously with the corresponding voltage on the spark gap capacitor. The visible and ultraviolet single pulse spectrum of the spark discharge in air is shown in figure 3.

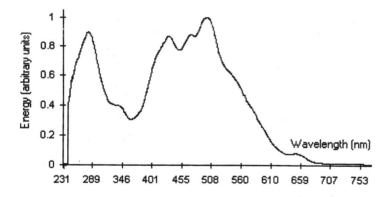

Figure 3. Pulse spectrum of the spark discharge in air.

4. Results and Discussion

The calorimeter response was obtained for a number of pulse events and for continuously emitting (c.w.) optical sources and its spectral sensitivity determined using a grating spectrometer (Ocean Optics, M 1000, 100 μm slit, fiber input, CCD). The calorimeter output was invariant with the spectral composition of the impinging signal and its range went beyond the i.r. limit of the spectrometer. The calorimeter system detectivity (D) at room temperature, in terms of output voltage to input energy, was determined to be $D = 4.3 \times 10^7$ V.J^{-1}. Its detection limit was measured to be 2.3×10^{-13} J, determined for the output noise level of 10 μV and the signal-to-noise level of one. The long term stability of the system was found to be acceptable.

The time constant and the detectivity of the calorimeter, can be improved by decreasing the mass or changing the material of the sensor. In measurements of the optical sources with a known spectral distribution, filtering of the impinging signal would improve the detection limit. The filter element would eliminate the extraneous light and limit the atmospheric interference, noticeable when the calorimeter is operated with open input aperture. An industrial class instrument could be designed, with signal processing assisted by a microprocessor or microcontroller.

In conclusion, the calorimeter system was found to be suitable for accurate and sensitive measurements of the emission energy of the spark. The combination of calorimetry and spectroscopy has worked well in non intrusive characterization of optical radiation, especially the studies of optical emission associated with electrostatic discharge.

Acknowledgments

This research was supported by grants from the National Science and Engineering Research Council of Canada and the Suncor Energy, Sarnia, Canada.

References

[1] Baloga J D and C.W. Garland C W 1977 *Rev. Sci. Instrum* **48** 105

[2] Bednarz G, Millier B and White M A 1992 *Rev. Sci. Instrum.* **63** 3944-3952

[3] De Geronimo G, Bertuccio G and Longoni A 1996 *Rev. Sci. Instrum.* **67** 2643-2647

[4] Greason W D, Kucerovsky Z, Bulach S and Flatley M W 1997 *IEEE Trans. Ind. Applicat.* **33** 435-443

[5] Greason W D, Kucerovsky Z, Bulach S and Flatley M W 1998 *Ibid.* **34** 867-873

[6] Kucerovsky Z, Inculet I I and Lee A K W 1985 *Ibid.* **IA-21** 17-22

[7] Lindmayer M and Paulke J 1998 *IEEE Trans. Comp. Packag. Man. Technol. A* **21** 33-39

[8] Murgatroyd P N and Belloufi M 1990 *Meas. Sci. Technol.* **1** 9-12

[9] Radak Br B and Radak Bo B 1991 *Rev. Sci. Instrum.* **62** 318-320

[10] Raizer Y P 1991 *Gas Discharge Physics* (Springer Verlag)

[11] Schwartz P 1971 *Phys. Rev.* **B 4** 920

[12] Sullivan P F and Seidel G 1968 *Phys. Rev.* 679

[13] Takeuchi M and Kubono T 1998 *IEEE Trans. Comp. Packag. Man. Technol. A* **21** 68-75

[14] Yao H, Ema K and Garland C W 1998 *Rev. Sci. Instrum.* **66** 172-178

Inst. Phys. Conf. Ser. No 163
Paper presented at the 10th Int. Conf., Cambridge, 28–31 March 1999

Surface-Wave Mechanism for the Phenomenon of Passive Insulator Charge Elimination

T S Lee

Electrical Engineering Department, University of Minnesota
Minneapolis, Minnesota, 55455 USA

Abstract. A grounded wire tip causes a dip in local surface potential. giving rise to two, and only two, possible but mutually exclusive field scenarios depending on geometrical and physical conditions. The first is conventional whereas the second has a field saddle-point in between point and insulator. If surface charge level is high enough, corona induction is indicated according to either scenario. Evolution dynamics predicates that the second kind of field description is inevitable and the saddle-point would play a central role in the ensuing wave phenomenon. It creates a circular central domain of protection with sharply defined boundary. As neutralization progresses, the saddle-point ascends with the associated protection domain in expansion. The latter constitutes the basis for the observed wave propagation.

1. Introduction

From the surface (horizontally placed) of a ground-backed insulator layer, charge can be passively eliminated, or reduced, by bringing a sharp grounded point near. Such is known as passive eliminator and has been reviewed[1,2]. Recent experimental evidence[3] reveals the surface-wave nature inherent in the intervening phenomenon through a time-evolving physical process of induced corona neutralization This work seeks to provide a descriptive physical explanation.

Let us assume that a uniform charge with density σ exists on surface of a dielectric layer of thickness d and permittivity ε. It is desired to bring in a passive sharply-tipped earthed object to cause elimination. We consider the positive charge case here. The negative charge case follows similar arguments with only sign changes.

Fig.1a illustrates the case of the layer backed by ground. Its top surface is at a potential $V_O = \sigma d/\varepsilon$. If a second grounded electrode is placed above at a height h, as portrayed in Fig.1b, a fraction of the Coulomb flux are drawn upward and the surface potential will be reduced to $V_1 = V_O [1+(\varepsilon_o d/\varepsilon h)]^{-1}$. Such a state corresponds to an upper electrode having an infinite radius of curvature. Now if the latter be made finite, i.e., in the

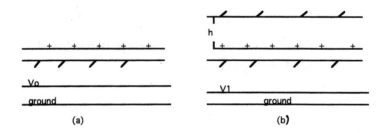

Fig.1 Potential distributions under the influence of an nearby object: (a)unperturbed surface. (b)effect of nearby plate electrode.

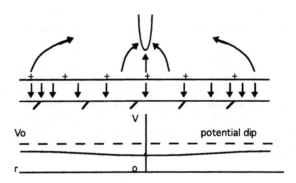

Fig.2 Local potential depression due to an nearby point

form of a probe tip as shown in Fig.2, diversion of Coulomb fluxes would be centered and the new two-dimensional potential pattern as shown would follow. In this case, the surface potential starts with V_0 at large r, where the effect of the probe has yet to be felt, and diminishes to a minimum at the center where the diverting effect would be most severe. Thus the surface will no longer be equipotential. It is this local *depression* in potential brought about by the probe presence which needs to be examined.

As separately portrayed in Fig.3a and Fig.3b are the potential patterns arising from a (+) sheet charge which has on a surface the potential decreases from left to right. Two possible near-field descriptions may apply. The first shows a case of local charge domination and has the normal field component reversing in sign in crossing the surface. The second corresponds to domination due to ambient field and the said normal component does not change sign.

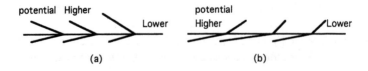

Fig.3   Two potential field patterns across charged surface; (a) charge domination
(b) distant field domination

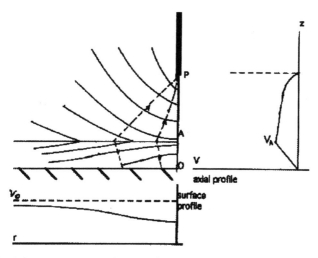

Fig 4. *Scenario I* suggesting pseudo-Laplacian back-corona generation .

Fig. 4 and Fig.5 globally sketch families of near field equipotential curves and their corresponding flux-lines, with *d* exaggerated to show clarity. For viewing potential distribution, we have found it helpful in drawing a potential profile along the central axis of symmetry *OP*. At both ground and the wire tip, the potential is zero. In between, a maximum is expected. By and large, the local field at the tip is strong by the large curvature assumption and is due to charges situated on the surface over a large domain. On the other hand, near the center of the surface it is more dependent on the local surface potential distribution. It is submitted that, depending on physical parameters *d, h, ε* as well as wire tip curvature given, one of two, and only two, possible field scenarios may be encountered.

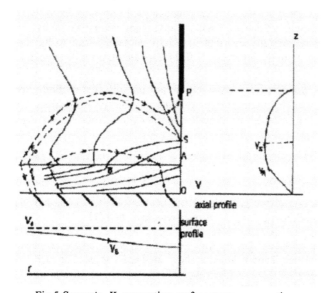

Fig 5 *Scenario II* suggesting surface wave propagation.

20

1.1. *Scenario I*

If the wire tip is remote and its sharpness is only moderate, then the near-tip field due to induction at the tip will be limited and the related *cathode drop* can only be minor. The field distribution is in a *pseudo-back corona* form. In that case, since the potential can be expected to climb more or less monotonically within the thin dielectric layer, an absolute maximum is at the surface, say point A. Thereafter the potential monotonically decreases to zero at wire tip P. It is noteworthy that, here the surface behavior prescription according to Fig.3a applies uniformly for the induced potential dip (Fig.2).

1.2. *Scenario II*

If the wire tip curvature is to be made severe, other things being equal, its depression effect at point A will be great, resulting in a reduction in V_A. On the other hand, the field at the tip will increase collaterally. Thus, the related *cathode drop* will be greater. On balance, a condition may come about where , on the profile drawn, a relative extremum will begin to emerge in the air region. (Similarly, reducing h can have the same effect.) Let us denote this point by S. The local potential will exceed V_A. Examination suggests that point S is a *saddle-point* of the general field, where the potential is the same as a surface point at an intermediate radius, say B. Zone of back corona generation is attached to the wire tip still. Ions of the opposite polarity created can only migrate toward the dielectric surface by following the flux lines. They are deflected on account of the presence of the saddle point. In umbrella fashion, a circular shaded domain of protection within B would be in place. Arriving ions can land only outside a sharply defined boundary. As they continue to do so, charge is removed from the surface progressively. As the tip field is weakened, S moves up in accompaniment. This picture serves to explain the appearance of an expanding *wave* front in motion. Also, in contrast to Scenario I, potential descriptions according to Figs. 3a and 3b would now apply to the outer and the inner surface domains separately.

There is a related possible event sequence yet to be considered. Even if initially the system should start with a Scenario I state. Subsequent neutralization by arriving ions, now with current density centrally distributed, would invariably lead to the emergence of the saddle point above A in the air region*. Therefore again, a wave expansion would be the outcome.

With electrostatic energy passively stored initially, the motion may be considered derived from an inherent system energy-releasing instability. Altogether, then, regardless of the initial state, it is inevitable for the system to gravitate to a surface wave generation mode.

2. Discussion

While the process considered projects charge neutralization, it does not preclude the well known possibility of charge *partial - or over-compensation* [1].

---

*A case of uniform σ distribution but with a central zero-value *cavity* assumed was tested. Finite-element simulation found it to occur above the barrier sheet. For such a worst, *albeit* special, case, the conclusion was confirmed .

For moderate initial charge level, ultimately the wave will be quenched resulting in a remnant disk. It is interesting that this agrees with the Lichtenberg-figure studies using wires[4].

Charge relaxation on oil films by a grounded wire generates a solitary expanding wave[5]. Subsequent investigation shows that triggering of the wave could be effected by a second scheme of h reduction likewise[6]. At the heart of this class of oil waves, both electrostatics and electrohydrodynamics play important roles. However, the latter is expected to affect fluid-dynamic details mainly. In essential field aspects, the oil wave problem and the passive charge neutralization problem are nearly identical. Thus the model proposed has relevance to both.

3. Summary

A physical theory has been proposed for the phenomenon of passive surface elimination: A grounded wire tip causes a local dip in surface potential, giving rise to two, and only two, possible but mutually exclusive field scenarios depending on geometrical and physical conditions. The first is conventional whereas the second has a field saddle-point in between point and insulator. When surface charge is strong enough, corona induction is indicated according to either scenario. Evolution dynamics predicates that the second kind of field description is inevitable and the saddle-point would play a central role in the ensuing wave phenomenon. It creates a circular central domain of protection with sharply defined boundary. As neutralization progresses, the saddle-point ascends with the associated protection domain in expansion. The latter constitutes the wave as observed.

References

[1]   Horvath T and Berta I 1982, *Static Elimination,* (Research Studies Press, Wiley) p.52

[2]   Noll C G 1995, chapter in *Handbook of Electrostatic Processes,* eds. Chang Jen-Shih, Kelly A, and Crowley J (Marcel Dekker, Inc.) pp.733-748

[3]   Lee T S 1998 *Proc. - Electrostatic Society of America- Institute of Electrostatics Japan Joint Symp.,* Palo Alto, Calif. USA, pp.121-126

[4]   Ota T and Ito Y 1990 *IEEE Tran. Ind. Appl.* Soc. **26,** 656-661

[5]   Lee T S and Phan T 1995 *Inst. of Phys. Conf.* Ser., **143,** 257-262

[6]   Lee T S 1997, IEEE Annual Report - *Conf. Electrical Insulation and Dielectric Phenomena* (Minneapolis, October 19-22) pp. 694-697

Inst. Phys. Conf. Ser. No 163
Paper presented at the 10th Int. Conf., Cambridge, 28–31 March 1999
© *1999 IOP Publishing Ltd*

23

Distortion of a water surface stressed by a perpendicular, alternating electric field

J A Robinson, M A Bergougnou, G S P Castle, I I Inculet

Applied Electrostatics Research Centre, Faculty of Engineering Science,
The University of Western Ontario, London, Ontario, N6A 5B9, Canada

Abstract. A conductive liquid subjected to an electric field can become distorted due to the electrostatic force on its surface. The distortion is often in the form of small projections that can become conical, e.g., the Taylor cone in a DC field [1]. This paper reports a study of the distortion of a water surface exposed to a perpendicular AC field, in the frequency range 20-200 Hz. The water was confined to a narrow channel. The wavelength of the observed (oscillating) projections was measured and found to agree closely with predictions using Mathieu and dispersion equations. The projections appear as a standing wave with the same frequency as the applied voltage.

1. Introduction

The authors' interest in the distortion of a water surface in a perpendicular AC field is spurred by a novel ozone generator, for use in water treatment, under development in the Applied Electrostatics Research Centre, The University of Western Ontario. This device uses a layer of water as its lower, grounded electrode [2], [3].

The instability caused by a strong electric field applied over the horizontal surface of a conductive liquid is similar to that caused by subjecting a horizontal free surface to vertical oscillations or by heating a fluid from below [4], [5]. Taylor and McEwen [5] investigated the case of a steady, initially uniform, vertical electric field applied to a horizontal interface between a conducting and non-conducting fluid. Using a linearized analysis, they determined a stability criterion by considering the balance between the electrostatic force, gravity, and surface tension.

In the case of a periodic vertical electric field, Yih [6] considered the hydrodynamics of the fluids, neglecting the effect of viscosity. This (linearized) analysis showed that the stability of the interface is governed by the equation

$$\frac{d^2a}{d\tau^2} + \left[p - 2q\cos(2\tau)\right]a = 0, \tag{1}$$

where a represents the time dependence of the interface displacement; $\tau = \omega t$; ω is the (angular) AC frequency; t is time; and p and q are parameters that depend on the wave number, surface tension, fluid densities and depths, permittivity of the upper (non-conducting) fluid, and gravitational acceleration. Equation (1) is the Mathieu equation;

solutions lie in stable or unstable (or marginally stable) regions (depending on parameters) of a stability chart. For given parameters, (1) can be evaluated numerically [7].

Jones [8] subjected the surface of a dielectric fluid to a high-intensity, periodic, primarily tangential electric field with a strong gradient. His analysis also led to the Mathieu equation. Although the Mathieu solution gives a band of unstable wave numbers for a given set of operating parameters, Jones predicted the wave number (k) of the observed instability by assuming the mode with the maximum growth rate to dominate, considering all possible values of k.

For the disturbances reported in this paper, the region of the stability chart in which the fastest growth rates occur is the unstable region closest to the origin. In this region, the instability has a frequency of one-half the driving frequency. In our case, the driving frequency is 2ω, since the electric stress is proportional to the square of the voltage.

The numerical calculations can be much simplified by noting that, to a very close approximation for the conditions here, the most unstable solutions occur when $p = 1$ [4].

Without consideration of the Mathieu equation, Melcher [9], also with a linearized analysis, derived the dispersion relation for waves on a fluid-fluid interface in the presence of a vertical electric field. He considered instability to occur when the phase velocity becomes imaginary (the disturbance grows exponentially in time). If the phase velocity is real, given the wave (disturbance) frequency, the dispersion relation determines the resonant k. It is interesting to note that if the disturbance frequency in the dispersion relation is set equal to the AC frequency (ω), the dispersion relation leads to essentially the same prediction of k as the Mathieu instability analysis (assuming $p = 1$).

2. Apparatus and procedure

Tap water was contained in an acrylic reservoir with a grounded stainless steel plate covering a portion of the bottom. Two brass plates were submerged in the water with their top (dry) surfaces flush with the water surface. The plates were separated from each other to form a channel between them, the width of which could be varied. Additional plates blocked the channel ends so its length was 179 mm. The plates rested on acrylic blocks; the water depth in the channel was about 25 mm. An Al disk electrode of diameter 150 mm was set at 9 or 10 mm above the water-brass surface. The top of the reservoir was open to the lab air.

Alternating voltage was supplied to the electrode by a high-voltage transformer fed by an audio amplifier. To limit the current in the event of sparks between the electrode and water or brass surfaces, a 50-MΩ resistor was connected between the transformer and electrode. Input to the amplifier was provided by a variable-frequency sine-wave generator.

Electrode voltages were measured using a high-voltage probe connected to an oscilloscope. Due to its very short electrical relaxation time, the tap water (resistivity of approx. 30 Ωm), was assumed to be a perfect conductor. Distance measurements were aided by a camcorder (12x zoom) connected to a video monitor.

3. Results and discussion

At low voltages, the water surface bulges slightly upward below the electrode. As the voltage is increased to a critical level, the damping effect of viscosity is overcome, and small rounded projections (instabilities) appear on the bulge. Further increases in voltage cause the projections to become higher, closer together, and more conical. At higher voltages, electric

Fig. 1. Photo of projections on channel. AC frequency: 20 Hz; channel width: 6.4 mm; exposure time: 1/60 s; projection wavelength: approx. 11.5 mm.

discharges occur at the cone tips. In the tests reported here, the narrow channel prevented a significant bulge from forming, and coerced the projections into an orderly, single-file row.

When viewed with a strobe light, there appears to be two sets of projections, each oscillating at the AC frequency, with the peaks of one set appearing midway between the peaks of the other set. One set lags the other by half the AC period. This phenomenon is a standing wave, with the upward peaks much more pronounced because of nonlinear growth due to the upward electric force. The frequency of the standing wave is equal to the AC frequency, as predicted by the Mathieu analysis. The projection wavelength $(2\pi/k)$ is taken to be the distance between two successive peaks of the standing wave.

To verify that the two sets of waves are not related to the voltage polarity, a full-wave rectifier was connected to the power supply; no difference in the disturbance was seen. (Since the electric stress is proportional to the square of the voltage, full-wave rectification does not change the stress (driving) frequency because the frequency of a fully-rectified sine wave is equal to that of a squared, unrectified sine wave.)

Repeatable results were recorded by setting the channel width to slightly greater than half the wavelength. Too wide a channel allows a two-dimensional arrangement of the projections, making it difficult to accurately measure the wavelength, while too narrow a channel artificially increases the projection curvature and hence the effect of surface tension.

Fig. 1 is a photo of a row of projections on the channel surface. Fig. 2 summarises the measurements, showing good agreement between the measured and calculated values of wavelength. The wavelength is very dependent on the AC frequency, and slightly dependent on the voltage at a given frequency. To measure the voltage dependence at a particular frequency, one set of measurements was taken at each of three different voltages: a voltage just above that needed to cause projections (labelled "low voltage" in Fig. 2), a voltage just below that causing sparking ("high voltage"), and a voltage midway between those two. Due to the very small differences in wavelength at different voltages, Fig. 2 shows the high- and low-voltage results only for the frequencies 20 and 40 Hz.

4. Conclusions

For a fluid interface in a vertical AC field, an instability analysis using the Mathieu equation leads to the same wavelength prediction as does use of the dispersion relation. The experimental results closely agree with this prediction for an air-water interface, even though the analyses are linearized, and the water surface was restricted to a narrow channel. The wavelength is highly dependent on the AC frequency (shorter at higher frequencies), and

26

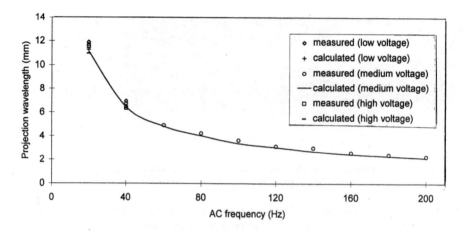

Fig. 2. Projection wavelength (measured and calculated) vs. AC frequency.

slightly dependent on the voltage magnitude (shorter at higher voltages). The disturbance appears as a standing wave.

Acknowledgments

The authors wish to express their thanks for the expert technical assistance provided by Dave Woytowich, and for helpful discussions with Joran Velikonja and Dr. T B Jones. NSERC and OGS scholarship support of the principal author (J A R) is sincerely appreciated.

References

[1]   Taylor G 1964 *Proc. R. S. Lond. A* **280** 383-397

[2]   Robinson J A, Bergougnou M A, Cairns W L, Castle G S P and Inculet I I 1998 *IEEE Trans. on Ind. Appl.* **34** 1218-1224

[3]   Robinson J A, Bergougnou M A, Cairns W L, Castle G S P and Inculet I I 1998 *Conf. Record IEEE-IAS Annual Meeting* **3** 1820-1827

[4]   Benjamin T B and Ursell F 1954 *Proc. R. S. Lond. A* **225** 505-515

[5]   Taylor G I and McEwen A D 1965 *J. Fluid Mech.* **22** part 1 1-15

[6]   Yih C-S 1968 *Phys. Fluids* **11** 1447-1449

[7]   McLachlan N W 1964 *Theory and Application of Mathieu Functions* (New York: Dover)

[8]   Jones T B 1972 *J. Appl. Phys.* **43** 4400-4404

[9]   Melcher J R 1963 *Field-Coupled Surface Waves* (Cambridge MA: MIT)

Inst. Phys. Conf. Ser. No 163
Paper presented at the 10th Int. Conf., Cambridge, 28–31 March 1999

Evidence of Bipolar Corona Emission from a DC Point

T S Lee[1,2] and G Touchard[2]

1) Electrical Engineering Department, University of Minnesota,
   Minneapolis, Minnesota, 55455 USA
2) Groupe Electrofluidodynamique L.E.A. (U.M.R. 6609 du C.N.R.S.)
   University de Poitiers, Poitiers, France

Abstract. Corona discharge in the presence of polymer barrier leads to unexpected behavior trend. Sudden switching to a lower magnitude produces one of three outcomes: 1) a modified Trichel-pulsing regime, 2) a quiescence regime, or 3) an inverse-pulsing regime. If the switching is only of a moderate magnitude, Trichel-like pulsing continues, but merely with its repetition rate somewhat reduced. If switching magnitude is severe enough, the presence of the surface charge will begin to offset and suppress the local Laplacian field   By sharply reducing the inter-electrode distance, regime of quiescence could be suppressed, resulting in the coexistence of both charging and inverse pulses. In effect, a basic DC-AC conversion mechanism is established. This report gives observed details of the first evidence from experiments conducted.

1. Introduction

Coronas arise from electron avalanche processes under inhomogeneous field conditions which bring the avalanche to cease after a short travel. Neutral molecule species are ionized by electron collision driven by the field[1]. Conventional laboratory studies have favored using a metallic point-plane arrangement in which the point provides the required intense field. The underlying physical mechanisms have been well reviewed[2-5]. Presence of an insulating cover over the plane electrode adds interesting ramifications in discharge behavior. In particular, it directly affects the well-known pulse ("Trichel") production: pulse-shape modification, pulse suppression, and inverse-pulse induction. Characterization of some of these was studied in the system of Fig.1 [6], in which a 0.06 mm Mylar dielectric sheet with radius dimension $D = 20$ cm. was used. Qualitative descriptions of corona change in response to a step-wise potential switching, in spite of the ever-present long-term drifts, are illustrated in Fig.2 for a point suspension h of a few cm's. For such a diagnostic figure, consider an initial corona charging tip potential V_C applied for a sufficiently long time.

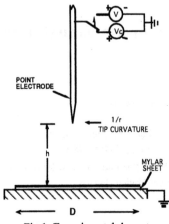

Fig.1 Experimental layout.

Sudden switching to a lower magnitude produces one of three outcomes: 1).a modified Trichel-pulsing regime, 2).a quiescence regime, or 3).an inverse-pulsing regime. If the switching is only of a moderate magnitude, Trichel-like pulsing continues, but merely with its repetition rate somewhat reduced. If switching magnitude is severe enough, the presence of the surface charge begins to offset and suppress the local Laplacian field. Here V_q represents the limiting potential of a quenched state in which quiescence in corona activity reigns. Finally, if the said magnitude is sufficiently great, a new state of inverse-corona pulsing will begin to set in. V_i denotes the threshold point potential in correspondence. In the present work, we explore means of making V_q and V_i approach each other, and even beyond, and investigate the physical consequence therefrom.

2. Experimental Observation

In order of importance to values of quenching potential V_q and inverse-corona threshold potential V_i, the following may be considered a list of relevant physical parameters: (1).point sharpness(radius of curvature r), (2).suspension height h , (3).dielectric sheet

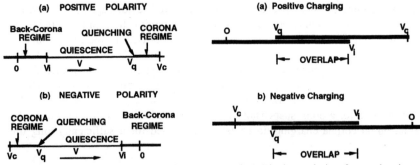

Fig.2 Illustration of corona regimes during potential switching with combined high h and moderate Vc values. (a) Positive polarity. (b) Negative polarity.

Fig.3 Bipolar emission from a barrier-influenced DC corona point at sharply reduced tip suspension: (a) Positive point (b) Negative point.

29

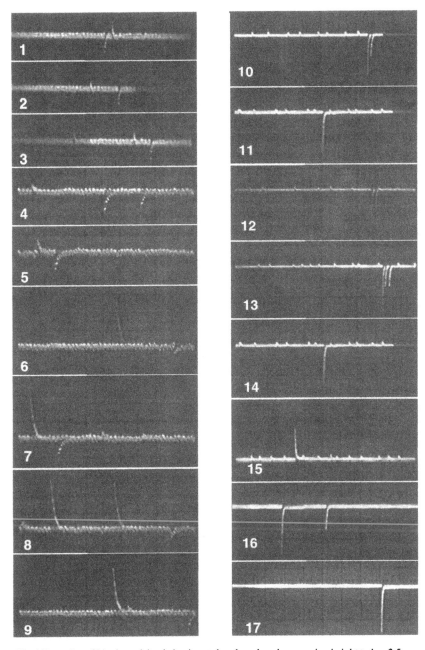

Fig.4 Examples of bipolar pulsing behavior at sharply reduced suspension height to h = 2.5 mm.
Vc = -24 kV. Parameters used are in Table 1.

radius D, (4).charging potential V_c, (5).sheet thickness, (6).sheet permittivity, (7).air density, and others. In our experiment, items (5) through (7) are either fixed, minor, or not subject to control. From experience[6], item (4) is of limited influence. In contrast, smaller r and h values would both contribute to an earlier corona onset and thus bring the two critical potentials in question closer together. Larger D values would lead to both easier and earlier corona quenching onset and inverse-corona onset. However, we have selected a sheet of great enough size and fixed $D = 20$ cm) for our experiment. Finally, a sufficiently thin wire is selected with a radius of curvature r (= 0.05 mm) fixed for sake of laboratory convenience.

The most important parameter adjustable is the point suspension height h . The range of 1 mm to 2 cm has been used in experimentation. We find that the drawing-together trend of V_q and V_i is always in evidence. With a sufficiently small h value used, the ranges of respectively Trichel-like pulsing behavior and inverse-pulsing behavior actually show a range of overlap, eliminating the corona-quiescence regime altogether. Figure 3 illustrates this circumstance according to either initial charging polarity. Within this overlap of potential ranges, pulses of both polarities can coexist. If the magnitude of V_c becomes great enough , at a small h, no switching is necessary for achieving the latter state, suggesting a wide overlap range.

An example of current trace is given in Fig.4, corresponding to V_c = -24 kV at h = 2.5 mm and switched to a new potential of -22 kV, -12 kV, and -9 kV respectively. Patterns of the two pulse trains are statistically scattered in both the repetition rate and the pulse magnitude. The oscilloscope traces have:

Table 1: Experimental Parameters

Trace No.	V (kV)	Horizontal (msec/unit)	Vertical (μA/unit)
1-6	-22	0.5	4.6
7-9	-22	0.1	4.6
10-15	-12	0.5	2.3
16-17	- 9	0.5	4.6

Inverse corona pulses being pointed downward, traces 1-9 are typically noisy. Corona is agitated. At lower voltage of -12 kV, traces 10-15 show that charging current pulses have two components: a high-frequency *steady* weak pulse train and a train of infrequent giant isolated pulses. The associated inverse pulses are equally isolated and statistical. Finally, at -9 kV, traces 16-17 show that only inverse pulsing remains. In all regimes seen, pulses are all distinct, never seen to occupy and share the same time intervals. In all regimes, tests show that none of the inverse-corona pulses coincides with any of the charging-corona pulses. All pulses, be they - or + , are distinct, suggesting that each takes place separately in time.

The observed - pulses have a mean repetition rate of < 0.5 per msec and that of the + pulses at < 0.08 per msec The mean amplitude for the - and + pulse trains are 1-20 μA and 5 -30 μA respectively. As the amplitude V decreases from 12 kV, the mean repetition rate and the mean amplitude change respectively for the - and + pulse trains. They decrease and increase collectively.

Basic corona has only received limited attention. Lama and Gallo [7] studied the point-plane negative Trichel pulsing phenomenon and pointed out that the between-pulse elapse in time is much shorter than the ionic transit time across the electrode gap. Thus, the space charge picture is one of distinct, successive ionic clouds, or puffs, emitted from tip. In the gap, the mutual Coulomb repulsion tend to cause each to progressively expand in lateral directions. Similar studies involving an inter-electrode charged barrier have not been pursued.

3. Summarizing Remarks

When a sharp-tipped conducting object, itself being maintained at a finite potential, comes near a highly charged ground-supported insulator surface, normally one of three possible scenarios, depending on the level of the potential being carried, among other factors, is the outcome: 1).modified Trichel-like corona pulsing, 2). quenched quiescence, or 3). induced inverse corona pulsing . By changing and manipulating geometrical parameters, ranges of the potential for the first and the third scenario regimes can be brought to overlap with each other, eliminating the possibility of the second type of occurrence. Typically in that case, ionic clouds, or puffs of both + and - polarities are intermittently and successively released from the point corona alternately. In this experimental work, such a phenomenon is created in a DC point-plane system incorporating the use of an intervening dielectric barrier of finite dimension. Evidence of the anticipated dual-pulsing behavior in the point current trace is clearly established, identifying a new basic mechanism of DC-AC conversion . In this system, the applied DC voltage supplies the Laplacian field for generating the Trichel-like pulses while the accumulated charges on the surface are responsible for generating pulses of the opposite polarity associated with the inverse corona. It is found that they all appear to have distinct existence and no pair of these pulses of opposite signs share time of injection at the point tip together.

References

[1]   Loeb L B 1965 *Electrical Corona,* University of California Press.

[2]   Trinh N Giao and Jordan J B 1968 *IEEE Trans.PAS-***87,** 1207-1211.

[3]   Trinh N Giao and Jordan J B 1970 *J. Appl. Physics,* **41** 3991-3999.

[4]   Trinh N G 1995 "Discharge in Air Part I: Physical Mechanisms" *Insulation Magazine* (IEEE) **11,** No.2, 23-29

[5]   Cobine J D 1956 *Gaseous Conductors-Theory and Engineering Application,* Dover Publications, New York, 1956.

[6].   Lee T S and Gasal J 1997 , *IEEE Trans.Ind. Appl.,* **33**, 692-696.

[7].   Lama W L and Gallo C F,1974 *J. Appl. Phys.,* **45**,103-113.

Inst. Phys. Conf. Ser. No 163
Paper presented at the 10th Int. Conf., Cambridge, 28–31 March 1999

Charge density of a high velocity electrified jet of diesel oil.

A Sélénou Ngomsi, H Romat, K Baudry

Laboratoire d'Etudes Aérodynamiques UMR 6609 du CNRS, Université de Poitiers
Boulevard 3 - Téléport 2 - BP 179 - 86960 FUTUROSCOPE Cedex - FRANCE

Abstract. Charge density measurements of a jet electrified by induction are reported. The liquid used is diesel oil doped with stadis450 (anti-static). The measurements of the charge density of the jet containing a 3% of additive are performed for potentials applied to the electrode varying, for different diameters of the orifice of the nozzle and also for velocities varying up to 100m/s.

1. Introduction

In the past, several authors tried to develop electrostatic atomisers for fuels in order to use them in combustion systems [1-2]. This electrostatic atomisation method presented some basic advantages because it prevented the agglomeration of small droplets. Nevertheless the different methods used were limited because the fuel had a very poor conductivity. Jones and Thong [3] were unable to charge kerosene (conductivity $\sigma=10^{-12}$ S/m) by the induction method. However, by adding a 3% of ASA3, the value of the conductivity of the liquid increased up to 10^{-6} S/m and allowed them to charge kerosene successfully.

In the experiments presented in this paper, we used diesel oil doped with an anti-static additive (stadis450). We study the parameters likely to enhance the charge density of the jet.

2. Theoretical basis

In our experiments the electrified jet is obtained by induction. An electric field applied to an electrode which is not in contact with the jet attracts the charges of a certain sign and repels the others. The efficiency of the charge induction method depends on the electrical relaxation time τ of the liquid given by $\tau = \dfrac{\varepsilon}{\sigma}$ which is the quotient of the electrical permittivity ε and the electrical conductivity σ of the liquid. Theoretically, if the relaxation time is small compared to the hydrodynamic time t_h which is the time necessary for a fluid particle to go from the injector to the electrode, the process of charging of the jet by induction is possible [4]. In our experiment, we chose to work with diesel oil and because we knew that its relaxation time was high, we added an anti-static (Stadis450), to enhance the conductivity of the liquid, this addition of Stadis450 changed the composition of the bulk of the liquid and its electric properties. The value of the conductivity of the solution with 3 % Stadis450 is 3.3 10^{-6} S/m.

34

3. Experimental set-up

The experimental equipment used for this study has been extensively documented in [5]. However we are going to describe the three main parts of the device:

Injection system : The important element of this system is the piston pump which pressurizes the liquid. The maxima for the flow-rate and the pressure are respectively 21.1 cm³/s and 138 bars. We use nozzles whose diameter are 220 μm, 300 μm and 400 μm.

Spray Electrification system : As shown in Fig. 1, with switches I_1 and I_2 we can connect the electrode to the generator, the ammeter then giving the current detected in the high potential branch. The electrode is a 3 mm thick annulus made of stainless steel, its distance to the nozzle is 3 mm. The electric potentials can vary up to 3 kV.

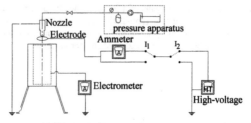

Fig. 1. Experimental apparatus.

Reception spray system : The system is composed of a tank and of a current measuring system. The current created by the discharge of the droplets at the contact with the wall is measured by the Keithley electrometer 610C. The current i_s divided by the volume flow rate $\bar{\dot{Q}}$ gives the mean charge density ρ : $\rho = i_s/\dot{Q}$

4. Results and discussion

4.1 Influence of the concentration of the additive

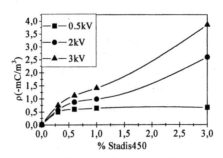

$\phi_j = 0.3$ mm, $U_0 = 68$ m/s
Figure 2 : Influence of the concentration on the charge density

In Fig. 2 we plot the charge density versus the concentration of additive Stadis450 for various values of the potential on the electrode. The first thing that we can notice is that the charge density is relatively constant for 500 V. Then, when the potential of the electrode is greater than 500 V the charge density increases with the concentration of the additive. If the presence of this kind of thershold (which is in fact at 1 kV) is not really understood, the influence of the concentration above 1 kV is clear. When the concentration increases, the number of electrical

charges carriers increases too and so the charge density. In fact, above 1 kV we know that the electric field may modified the energy of interaction between the molecules. The dissociation constant K_d of the species giving rise to ions increases when the electric field is greater than 10^4 V/m, which is the case when the applied voltage is 1 kV. The dependence of the reaction of dissociation on the electric field (called reinforced dissociation by electric field) which has been substantially studied by Onsager [6] is therefore to be taken into account in ours results.
We finally can conclude saying that :
Under 1 kV, the influences of the concentration of the additive and the electric field are not really understood. Neither the first, nor the second seems to influence the charge density of the liquid. Above 1 kV, the presence of the additive enhances the charge density of the jet and the effect of the electric field, can be fully understood if we take into account the mecanisms in the bulk of the liquid

4.2. Influence of the potential and the electrode diameter.

Fig. 3 shows the charge density versus the applied voltage for a fixed velocity (68 m/s), for a diameter of the nozzle of 0.3 mm and for three diameters of electrode. We first observe a relative constant charge density until 1 kV and then that the charge density is quite the same for the three diameters (until 1 kV). Below 1 kV, the action of the electric field is not high enough to change the equilibrium of the molecules of the liquid. Above 1 kV and for a fixed diameter, the charge density increases with the potential, a bit for a large diameter and a lot for a small one. The fact that whatever the diameter the charge density increases (above 1 kV, which corresponds to an electric field greater than 10^4 V/m) is quite normal if we consider that the system acts as a capacitor whose charge increases when the potential increases. But once again we must take into account the reinforced dissociation if we want to fully explain the phenomenon.

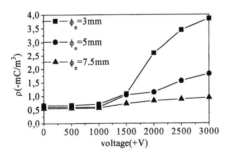

ϕ_j=0.3 mm, U_o=68 m/s
Figure 3 : Influence of the electrode potentiel and electrode diameter (Gazole + 3% Stadis450)

In fig. 3 we can also see that for a fixed potential the charge density increases when the diameter of the electrode diminishes. This is due to the capacitor of the system. When the diameter of the electrode is large, the capacitance of the system diminishes and therefore so the charge density.

4.3. Influence of the velocity

Fig. 4a and 4b show that the charge density is independent of the velocity whatever the potential. This suggests that the relaxation time is small compared to the hydrodynamic time.

Indeed, when it is the case the capacitor is fully charged very quickly and the velocity has no influence on the charge process.

The figures also show that for a null potential the charge density depends on the diameter of the orifice of the nozzle. The charge density increases when the diameter decreases. This is a problem which has been widely investigated [7]. It is due to the diffuse layer inside the nozzle. The thickness of the diffuse layer is very small for non conductive liquids but the charges can be convected if the velocity profile is significant near the wall of the nozzle where the diffuse layer exists. It is the case here, the velocity can perturb the diffuse layer with small radius but cannot perturb it with a large one.

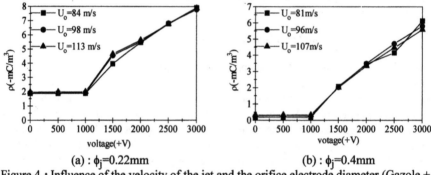

(a) : ϕ_j=0.22mm   (b) : ϕ_j=0.4mm

Figure 4 : Influence of the velocity of the jet and the orifice electrode diameter (Gazole + 3% Stadis450)

5. Summary and conclusion

It is possible to atomise insulating liquids by charge induction systems provided that they contain additives. The aim of this work was to study the parameters likely to enhance the charge density of diesel oil. It has been shown that it :

    - depends on the properties of the fluids
    - increases with the voltage
    - is independent of the jet velocity
    - increases as the electrode diameter decreases
    - increases as the nozzle orifice diameter decreases

Recently we did other experiments on the parameters enhancing the charge density of diesel oil droplets. The trend of the results plotted in this paper is confirmed

References

[1]   Kelly A 1984 *IEEE Trans IA-18* 2 267-274
[2]   Bailey A 1986 *Atomizer and Spray Techn.* 2 95
[3]   Jones A R and Thong K C 1971 *J. Phys.D: Appl. Phys.* 4 1159-1166
[4]   Law S E and Cooper S C 1987 *Transaction of the ASAE* 30 75-79
[5]   Sélénou Ngomsi A, Romat H and Artana G 1997 *J of Electrostatic* 40&41 609-614
[6]   Onsager L 1934 *J. Chem. Phys.* 2 599-615
[7]   Romat H 1980 Thesis

Inst. Phys. Conf. Ser. No 163
Paper presented at the 10th Int. Conf., Cambridge, 28–31 March 1999
© *1999 IOP Publishing Ltd*

Pulsed streamer discharge in water through pin hole of insulating plate.

Masayuki Sato, Yukio Yamada and Bing Sun

Department of Biological and Chemical Engineering
Gunma University, Kiryu, Gunma 376-8515, Japan

Abstract. Experimental investigation was carried out on high voltage pulsed discharge through pin hole of insulating plate in water. High voltage was applied between parallel plate electrode, and insulating plate was placed between them. Pulsed electric field was concentrated around the pin hole, as a result, streamer discharge was formed. Streamer length in the high voltage electrode side was shorter than that of earth electrode side, which was independent of the electrode gap length. Phenol degradation was tested by alternating the path way of the liquid flow through the reactor. The longer streamer (earth electrode side) degraded phenol much faster than the case of shorter streamer (high voltage electrode side).

1. Introduction

Some researches have been reported on a high voltage pulsed discharge in water for a purpose to break up chemical bonds of organic materials containing water (Clements, et al. 1987; Sharma, et al. 1993; Joshi, et al. 1995). By a spectrum analysis of a discharge light, many active species are found such as H, O, OH radicals and hydrogen peroxide (Sato, et al. 1996; Sun, et al. 1997). They react with a trace amount of organic compound in water to degrade to the other materials with less molar weight

A needle-plate electrode configuration has been used to generate the pulsed discharge in water, because a highly convergent electric field is formed at the tip of the needle, which leads to electrical breakdown of water. Streamer discharge appears by a high intensity electric field and also spark discharge occurs with changing the electrical operating conditions. The convergent electric field was generated by using pin hole of insulating plate which was placed between two parallel plate electrodes (Sato, et al. 1997; Yamada, et al. 1998). In the present study, streamer formation and phenol degradation were investigated experimentally.

2. Experiment

Schematic diagram of the experimental apparatus is shown in Fig. 1. Two plate electrodes (30 mm diameter) were faced with each other in a Plexiglas cylinder (45 mm inside diameter). The electrodes were separated by a thin insulating material (Polyvinyl chloride with 1 mm thickness) having a small hole (1 mm diameter). The electrode separation distance was 30 mm, and insulating plate placed from 5 to 25 mm separation from the electrode surface.

38

Water was fed to the reactor with the following manner: (a) no water flowing (no pumping); (b) water flowing into the right side reactor, and through the pin hole, then out from the left reactor (in the case of the figure); (c) water flowing into the right side reactor, and out from the right hand side (*i.e.*, reactions occur separately), and *vice versa.*; (d) water flowing in and out both reactors simultaneously

Aqueous phenol solution was used for the sample liquid, and was analyzed by high performance liquid chromatography (LC-9A, Shimadzu) with the column packed with polymethacrylate gel (RS pak, Shodex-Showa Denko). The high voltage positive pulse (controlled from 0 to 25 kV and 10 to 50 Hz) was used. The pulse forming capacitor was 6,000 pF. The discharge light was recorded using video camera and recorder, and the streamer length was determined by averaging the longest streamers that appeared in a period of five seconds on TV monitor.

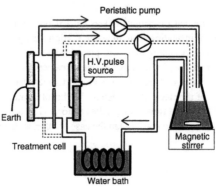

Fig. 1 Schematic diagram of experimental apparatus.

3. Results and discussion

Intense electric field strength was formed around the pin hole when high voltage pulse was applied between electrodes. Magenta colored streamer discharge developed from the pin hole to each electrode. The streamer length forming at the left side reactor (earth electrode side) was longer than that of right hand side. It was reported that the streamer length was longer with applying positive pulse to the needle electrode than that with negative pulse using needle-plate

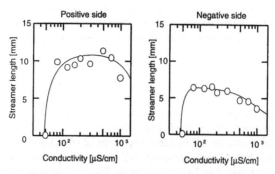

Fig. 2 Variation of streamer length with changing electrical conductivity of water, where applied voltage: 25 kV, pulse frequency: 50 Hz, electrode gap length: 30 mm.

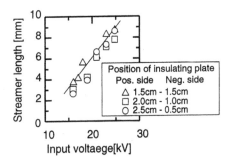

Fig. 3 Variation of streamer length at positive side with changing distance between insulating plate and electrode, where pulse frequency: 25 Hz, electrical conductivity of water: 2.0×10^{-4} S/cm.

electrode configuration (Clements, et al. 1987). In the present experiment, the ground electrode side (left, hereafter referred to positive side) was considered to be applying positive pulse to the pin hole, and vice versa, therefore the left side streamer was longer than the right.

As shown in Fig. 2, the streamer length was varied by changing electrical conductivity of water. In the case less than 5×10^{-5} S/cm, no discharge was observed at the applied voltage up to 25 kV. From the figure, streamer length at the positive side was longer than the negative side in conductivity between 10^{-3} and 10^{-4} S/cm.

Fig. 3 shows the results of the streamer length with changing distance between electrode and insulating plate from 5 to 25 mm (electrode distance was kept constant). As shown in the figure, large difference in streamer length was not observed. It was because high intensity nonuniform electric field is formed at the pin hole, that is independent from the electrode gap. Because of the difference between streamer length in each side of the reactor, it was suggested that the amount of generated active species was also different in each case.

Fig. 4 shows the degradation rates of phenol with changing path ways of the sample

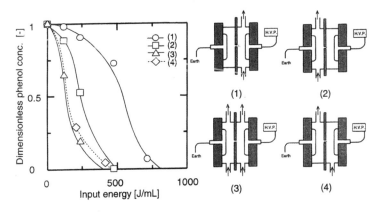

Fig. 4 Phenol degradation with alternating flow path ways, where phenol initial concentration: 50 ppm, pulse voltage: 23 kV, pulse frequency: 25 Hz, flow rate: 1.07 mL/min, electrical conductivity of solution: 0.2 mS/cm, electrode gap length: 30 mm.

40

liquid in the reactor. As illustrated in the figures (1) to (4), four kinds of flow path were tried, i.e., (1) negative side only, (2) positive side only, (3) two path ways through both reactor, (4) flowing through pin hole. A long time was needed to degrade phenol in the case of (1), negative side, on the other hand, flowing through pin hole (4) and flowing both reactor (3) showed higher degradation efficiency. The result shows longer streamer has higher electron energy for generating active species to react with phenol molecules.

References

Clements, J. S., M. Sato and R. H. Davis: "Preliminary investigation of prebreakdown phenomena and chemical reactions using a pulsed high voltage discharge in liquid water," IEEE Trans. Ind. Appl., IA-23, 224 - 235 (1987).

Joshi, A. A., B. R. Locke, P. Arce and W. C. Finney: "Formation of hydroxyl radicals, hydrogen peroxide and aqueous electrons by pulsed streamer corona discharge in aqueous solution," J. Hazardous Materials, 41, 3 - 30 (1995).

Sato, M., T. Ohgiyama and J. S. Clements: "Formation of chemical species and their effects on microorganisms using a pulsed high voltage discharge in water," IEEE Trans. Ind. Appl., 32, 106 - 112 (1996).

Sato, M., Yamada, Y., Sun, B., Nakane, T.: "Preliminary study on pulsed streamer discharge in water through pin hole of insulating plate," Int. Symp. on Non-Thermal Discharge Plasma Tech. for Air Pollution Control, pp. 120 - 121 (1997).

Sharma, A. K., B. R. Locke and W. C. Finney: "A preliminary study of pulsed streamer corona discharge for the degradation of phenol in aqueous Solutions," Hazardous Waste & Hazardous Materials, 10, 209 - 219 (1993).

Sun, B., M. Sato and J. S. Clements: "Optical study of active species produced by a pulsed streamer corona discharge in water," J. Electrostatics, 39, 189 - 202 (1997).

Yamada, Y. Sun, B., Ohshima, T., and Sato, M.: "Pulsed discharge in water through pin-hole," Proc. of Asia-Pacific Workshop on Water and Air Treatment by Adv. Tech.: pp. 168 - 169 (1998).

Inst. Phys. Conf. Ser. No 163
Paper presented at the 10th Int. Conf., Cambridge, 28–31 March 1999

Modelling the Electrical Conductivity of Two-Phase Composite Materials

Nicola Harfield†, John R. Bowler† and Gary C. Stevens‡[1]

† Department of Physics and ‡ Polymer Research Centre,
University of Surrey, Guildford, Surrey GU2 5XH, UK

Abstract. A model has been developed to calculate the bulk electrical conductivity of a two-phase material composed of spheroidal particles arranged on a simple-cubic lattice and embedded in a matrix with different conductivity. Results are compared with independent theoretical calculations and experimental data and good agreement is observed. The method can be generalised to model quasi-random microstructure.

1. Introduction

Conducting two-phase polymeric materials may be tailored for specific applications by adjusting the characteristics of the phases. Commonly, non-conducting polymers are loaded with particles of a conductive material to form a composite which retains the flexibility of the polymer and also conducts electrically. For example, conducting polymer composites are used in packaging and casings to provide electrostatic shielding of electronic components. Another important application is electrical stress grading and relief in high-voltage devices and power plant.

Historically, these materials have been designed largely by trial and error but optimisation via modelling is attractive. We have developed an analytical model to calculate the electrical conductivity of a material composed of particles dispersed in a matrix with different electrical properties. The model is based on spheroidal particles arranged on a simple-cubic lattice (see figure 1). The lattice structure enables a large ensemble of particles to be considered through the application of periodic boundary conditions.

2. Model

The model is a development of a calculation for a simple-cubic lattice of spheres [1]. In the presence of a uniform applied electric field, the electric potential in the material obeys Laplace's equation. The solution can be written as a series expansion in spheroidal

[1] E-mail: g.stevens@surrey.ac.uk

harmonics in each of the two phases, and the coefficients determined by applying the field continuity conditions at the surface of the particles. The bulk conductivity of the material is then determined from one of the low-order coefficients [2, 3]. The method can be generalised to model quasi-random microstructure [4].

The adjustable parameters of the model are

1. The particle volume fraction, f. This is the ratio of the volume occupied by the particles to the total volume of the material.

2. The particle aspect ratio, $\epsilon = a/b$, figure 1. As ϵ becomes large, the particles become fibre-like. As $\epsilon \to 0$, the particles become disc-like, and $\epsilon = 1$ describes a sphere.

3. The ratio of the particle conductivity to the conductivity of the matrix, α.

3. Results

In figures 2 and 3, the bulk conductivity is shown as a function of f for particles with aspect ratio 1 (spheres) and 0.5 (oblate spheroids). Comparison is made with results of an independent calculation [5] and experimental data [6] and good agreement is obtained. Figure 2 shows results for highly-conducting particles in a low-conductivity matrix ($\alpha \to \infty$). The maximum value of f for $\epsilon = 1$ is $\pi/6$ and for $\epsilon = 0.5$ is $\pi/12$. For the material containing oblate spheroids, the conductivity is greater in the direction in which the particle dimension is greatest, as you would expect. In figure 3, results are given for insulating particles in a conducting matrix ($\alpha = 0$).

Figures 4 and 5 reveal the effect of particle shape on the bulk conductiivty. The horizontal (xx) and vertical (zz) components of the conductivity tensor are shown as a function of the particle aspect ratio for high-conductivity particles with $f = 0.1$. Both of these figures show that the bulk conductivity increases significantly as the particles are elongated in the direction of the applied field.

The effect of varying the conductivity ratio of the two phases, α, is shown in figure 6 for prolate spheroidal particles with $\epsilon = 5$ and $f = 0.02$. The anisotropy of the material is clearly visible. This figure shows that it is possible to increase the conductivity of a material considerably even with a very low volume fraction of particles. The way in which the value of the bulk conductivity saturates as α increases is also clear.

Acknowledgment

This work was supported by an EPSRC ROPA.

References

[1] Zuzovsky M and Brenner H 1977 *J. Appl. Math. Phys. (ZAMP)* **28** 979–992

[2] Harfield N 1999 *Eur. Phys. J. Appl. Phys.* in print.

[3] Harfield N 1999 *J. Phys. D: Appl. Phys.* submitted.

[4] Sangani A S and Yao C 1988 *J. Appl. Phys.* **63** 1334–1341

[5] Kushch V I 1997 *Proc. R. Soc. Lond.* A **453** 65–76

[6] Meredith R E and Tobias C W 1960 *J. Appl. Phys.* **31** 1270–1273

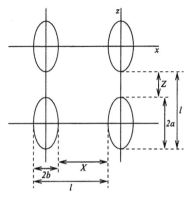

Figure 1. 2D schematic diagram of a simple-cubic lattice of spheroidal particles. The polar axes of the spheroids are parallel to z.

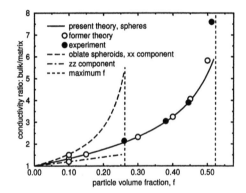

Figure 2. Bulk conductivity as a function of volume fraction, f, for highly-conducting spheres ($\epsilon = 1$) and oblate spheroids ($\epsilon = 0.5$) in a low-conductivity matrix. In the calculation, $\alpha = 10^{10}$. The results of former theory are taken from [5] and the experimental data is taken from [6].

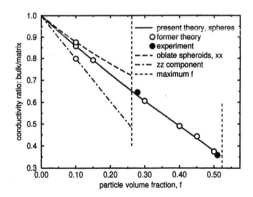

Figure 3. As for figure 2 but with insulating particles in a conducting matrix ($\alpha = 0$).

44

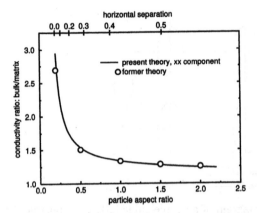

Figure 4. The xx-component of the bulk conductivity as a function of particle aspect ratio, for $f = 0.1$. The horizontal separation X between the particles is also shown, normalised with the side length of the cubic cell. In the calculation, $\alpha = 10^{10}$. Results of former theory are taken from [5].

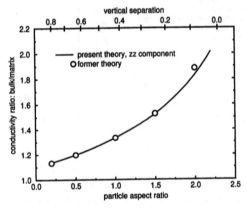

Figure 5. As for figure 4 but now showing the zz-component of the bulk conductivity and the vertical separation Z between the particles.

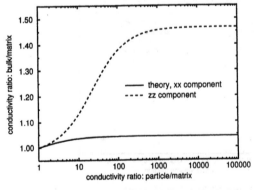

Figure 6. Bulk conductivity as a function of the particle/matrix conductivity ratio α for prolate spheroids with aspect ratio 5 and $f = 0.02$.

Inst. Phys. Conf. Ser. No 163
Paper presented at the 10th Int. Conf., Cambridge, 28–31 March 1999

About a thermodynamical approach for capacitor relaxation current

L C Bassani, L D Di Lillo

Centro de Investigacion y Desarrollo en Fisica - Instituto Nacional de Tecnologia Industrial–C C 157 - (1650) San Martin - Argentina

Abstract. Several relaxation processes show a similar temporal behaviour, suggesting that a thermodynamical approach is possible. Linear irreversible thermodynamics as was developed by Onsager and Biot allows to define the variation of the entropy when coupled fluxes of different variables act on the system. In this paper we apply the above results assuming that the coupled variables are the polarizations in the n cells of an homogeneous perfect dielectric when a step electric field is applied. If in the ensemble of n relaxation normal modes obtained, all the frequencies are different, the electric current is the superposition of n exponential terms with amplitudes proportional to the frequencies. In that case, for the continuum limit, the resulting temporal behaviour is of the form $i(t) = Kt^{-1}$. Many dielectrics show a law $i(t) = Kt^{-m}$ with $m < 1$, and approaching 1 when they have a small loss factor.

1. Introduction

When a capacitor is subject to a constant step voltage, or when it is discharged from a charged state, the measured electric current is a function of time, often known as the Curie-Von Schweidler law, $i(t) = Kt^{-m}$. Many important papers deal with experimental and theoretical researches in this field [1, 2, 3, 4]. P. Debye [5] treated the relaxation of polar molecules diluted in a non polar solvent, after the removal of the impressed electric field, arriving at an expression with a single relaxation time. A generalized Debye equation for an isotropic material, using the formalism of so-called extended irreversible thermodynamics [6], determines the time rate of change of polarization when there is a coupled thermal flux. Two different characteristic times are involved in this treatment. For a solid, it seems reasonable that a distribution of many times could appear in the relaxation process. When a perfect dielectric is subject to an electric field, in an isothermal quasi-static way, the polarization P is the new thermodynamic state variable. Thermodynamics allows to evaluate the entropy variation; in other words, the amount of heat absorbed or liberated isothermally by the system [7].

In an isolated dielectric, the temperature variation depends on $\left(\frac{\partial \epsilon}{\partial T}\right)_p$, ρ, C_p, T and E values, where ϵ, ρ, C_p, T and E, are the permittivity, density, specific heat, temperature and the electric field, respectively. For typical values of the above mentioned quantities, the temperature variation is less than $10^{-5}\,^\circ C$. For this reason we do not take into account the coupled thermal effects. Neither is electrostriction considered in this approach. This analysis for the quasi-static process is the basis for our attempt to describe a relaxation process taking as a background the concepts of linear irreversible thermodynamics developed by Onsager and Biot [8, 9].

2. Linear irreversible thermodynamics and relaxation

Following Onsager and Biot, an ensemble of a system and a large reservoir at temperature T is defined. For the equilibrium state (when there are no applied external forces), the total entropy is a maximum and its variation can be defined as

$$TS = -\frac{1}{2}\sum_{i,j=0}^{\infty} a_{ij}q_i q_j \qquad (1)$$

where $a_{ij} = a_{ji}$, S is the total entropy variation, and q_i, q_j are the generalized coordinates such as electric charge, polarization charge, chemical concentration, etc. When there are applied forces and the system is displaced from equilibrium , the derivatives $\frac{dS}{dq_i}$ are not zero. Onsager assumed that these derivatives are the "forces" producing the time variation of q_i ($\dot{q}_i$) through the law

$$T\frac{dS}{dq_i} = \sum_{j=0}^{n} b_{ij}\dot{q}_j \qquad (2)$$

where $b_{ij} = b_{ji}$. The time derivative of the entropy variation results in

$$T\dot{S} = \frac{1}{2}\sum_{i,j=0}^{n} b_{ij}\dot{q}_i \dot{q}_j. \qquad (3)$$

When there are generalized applied forces Q_i, the expression for irreversible processes is

$$\sum_{j=0}^{n} a_{ij}q_j + \sum_{j=0}^{n} b_{ij}\dot{q}_j = Q_i. \qquad (4)$$

According to Biot [9] this equation can be written in an operational form

$$\sum_{j=0}^{n} A_{ij}q_j = Q_i \qquad (5)$$

where $A_{ij} = a_{ij} + pb_{ij}$, and p is the operator $\frac{d}{dt}$. Biot has shown that the modes are orthogonal and the roots are real and negative. The solution for eq.(5) is

$$q_i = \sum_{j=1}^{k}\left[\sum_{s}\frac{C_{ij}^{(s)}}{p+\lambda_s}+C_{ij}\right]Q_i \qquad (6)$$

where $C_{ij}^{(s)} = \Phi_i^{(s)}\Phi_j^{(s)}$ and $C_{ij}^{(s)} = C_{ji}^{(s)}$.

There is an ensemble of relaxation modes (index s). If there are no infinite roots, $C_{ij}=0$. Zero roots are discarded because they correspond to a stationary state. If the forces Q_i are applied suddenly, at $t=0$, it can be written in terms of the unit step function $\Theta(t)$

$$Q_i = Q_i^\star \times \Theta(t). \tag{7}$$

Taking into account the operational relation

$$\frac{1}{p+\lambda_s}\Theta(t) = \frac{1}{\lambda_s}\left(1 - e^{-\lambda_s t}\right) \tag{8}$$

equation (6) becomes

$$q_i = \sum_{j=1}^{k}\sum_s \frac{C_{ij}^{(s)}}{\lambda_s}Q_j^*\left(1 - e^{-\lambda_s t}\right) \tag{9}$$

where

$$\lambda_s = \frac{\sum_{ij}a_{ij}\Phi_i^{(s)}\Phi_j^{(s)}}{\sum_{ij}b_{ij}\Phi_i^{(s)}\Phi_j^{(s)}} = \sum_{ij}a_{ij}C_{ij}^{(s)} \tag{10}$$

because the normalization for the finite modes is $\sum_{ij}b_{ij}\Phi_i^{(s)}\Phi_j^{(s)} = 1$.

We make the assumption that our system is a plane plate perfect capacitor subject to a constant external voltage V. We considered coupled relations between polarizations in different cells, that is, variables of the same nature in different coordinates of the system. It is an extension of the usual treatment that considers coupled variables of different nature, such as the electrical, thermal, chemical, etc. The q_i terms are the polarization p_i in the direction of the electric field E in the n equal cells into which the system can be divided. The current density is $j = j_p = \frac{\partial P}{\partial t} = \frac{n}{V_d}\sum_i \dot{p}_i$, where P is the polarization and V_d is the dielectric volume. When $Q_j^* = E_j = 1$, then

$$\dot{p}_i = \sum_{j=1}^{k}\sum_s C_{ij}^{(s)}e^{-\lambda_s t}. \tag{11}$$

Summing over all cells i,

$$\dot{P} = k_1\sum_i \dot{p}_i = k_1\sum_s e^{-\lambda_s t}\sum_{ij}C_{ij}^{(s)} \tag{12}$$

and taking into account eq.(10), we can write

$$\sum_{ij}a_{ij}C_{ij}^{(s)} = \lambda_s = k_2\overline{a_{ij}C_{ij}^{(s)}} \tag{13}$$

where k_2 is a constant factor. The a_{ij} values depends on the material structure but are independent of the amplitudes $C_{ij}^{(s)}$. Then it is possible to put $\overline{C_{ij}^{(s)}} = K\lambda_s$, the expression for $\dot{P}$ results

$$\dot{P} = \sum_s K\lambda_s e^{-\lambda_s t}. \tag{14}$$

In this discrete superposition of modes, if there are no multiple roots, the amplitudes are proportional to the relaxation frequencies. When n is great, in the continuum limit, the distribution density function of the form $f(\lambda)\,\alpha\,\lambda$ leads to $I = \dot{P} = At^{-1}$.The time $t = 0$ is excluded due to the fact that the setting up of the electric field is not instantaneous. In this paper an ideal capacitor is considered, not taking into account all kinds of conduction currents. In a real capacitor, the law is not valid after a time t_1 when conduction current overcomes polarization current.

3. Conclusion

With the assumptions we have made here, a linear irreversible thermodynamical treatment shows that for an ensemble of relaxation modes of different frequencies, the Curie-von Schweidler expression with exponent $m=1$ is obtained. In this case, a principle of minimum rate of entropy production can be postulated [9]. It seems that $m = 1$ is a limit value for the fractional laws actually encountered in many experiments. Two values of m ($m < 1$ and $m > 1$) are sometimes observed in two different temporal scales. Fractional power laws are also obtained in other fields, such as mechanical relaxation and chemical reaction kinetics [10].

The model analysed in [3] allows the value of the exponent m to be related to $tg(\delta)$. For many dielectrics with a small loss factor, values of m approaching 1 are obtained.

Some works deal with time-scale invariance in photo conduction in disordered systems using a model with multiple traps. It is interesting to remark that the expression for the so-called waiting time distribution function that determines in the model the transient current is of the same form as eq.(14) when the capture rate of the traps (w) is constant and there is a distribution of release rates (r_i) [11, 12].

References

[1] Jonscher A K 1977 *Nature* **267** 673-679

[2] Jonscher A K August 1987 *IEEE Trans. on Electrical Insulation,* **EI-22** 4 361-364

[3] Westerlund S, Ekstam L October 1994 *IEEE Trans. on Dielectric and Electrical Insulation* **1** 5 826-839

[4] Das-Gupta D K April 1997 *IEEE Trans. on Electrical Insulation* **4** 2 149-156

[5] Debye P 1929 *Polar Molecules* (The Chemical Catalog Co., Inc, New York)

[6] Castillo L F, Garcia Collin L S April 1986 *Phys. Rev. B* **33** 7 4944-4951

[7] Landau L, Lifchitz E 1969 *Electrodynamique des milleux continus, Chapitre II* (Moscou:Editions Mir)

[8] Onsager L, Machlup S September 1953 *Phys. Rev.* **91** 6 1505-1515

[9] Biot M A March 1955 *Phys. Rev.* **97** 6 1463-1469

[10] Jonscher A K, Bozdemir S March/April 1995 *IEEE Electrical Insulation Magazine* **11** 2 30-33

[11] Noolandi J November 1977 *Phys. Rev. B* **16** 10 4466-4473

[12] Scher H , Shlesinger M F and Bendler J T January 1991 *Physics Today 26-34*

Inst. Phys. Conf. Ser. No 163
Paper presented at the 10th Int. Conf., Cambridge, 28–31 March 1999
© *1999 IOP Publishing Ltd*

Streamer properties in air and in the presence of insulators

N L Allen and P N Mikropoulos

Department of Electrical Engineering and Electronics, UMIST, UK

Abstract. Under uniform electric field, the propagation of a single streamer along insulating surfaces was studied. The dynamics of streamer growth were investigated as influenced by the amplitude of the voltage, and therefore energy, used for initiation and the nature of the insulating material; in this case fibreglass and resin. It is shown that the stable propagation of the streamer can be described in terms of an intrinsic propagation field together with an associated velocity. These properties are characteristic of the medium employed and were used to formulate the relation between the propagation velocity and electric field. Along an insulating surface a streamer system propagates with a surface and an air component.

1. Introduction

The detailed investigation of the discharges along insulating surfaces is of considerable fundamental and practical interest. Previous work has discussed the influence of an insulating surface on the time-dependent growth of an avalanche [1], and on the branched streamer corona development in non-uniform field [2]; other valuable information can be found in reference [3]. Work, carried in a uniform electric field geometry, has shown that the stable propagation of a single streamer in air alone [4] or along insulating surfaces [5] is characterised by an *intrinsic* propagation field with an associated propagation velocity which are typical of the dielectric medium employed. These properties together with the relation between the propagation velocity and electric field have been shown to be influenced by the energy used for streamer initiation, the air density and the nature of the insulating material; thus, the propagation of a single streamer can now be modelled [4, 5]. The objective of this paper is, in the light of new measurements, to extend the knowledge on the influence of the insulating material on the streamer properties by studying the propagation of a single streamer along fibreglass and epoxy resin insulators.

2. The experimental arrangement

The uniform electric field geometry used has been described elsewhere [4, 5]. Cylindrical insulators were inserted perpendicularly between two parallel plane electrodes. A sharp point was located in a small aperture at the centre of the lower earthed plane, at the same level with the latter and in contact with the insulator. A positive pulse voltage, with 20 ns rise time, 135 ns duration and variable in amplitude, was applied at the point. Positive streamers were thus initiated from the point and propagated towards the upper plane, which was at a steady negative potential. The streamer growth was monitored at various levels by three identical photomultipliers, each with a vertical field of 6 mm; their outputs were displayed simultaneously on the screen of a 400 MHz digital storage oscilloscope.

50

Two different specimens, made of fibreglass and epoxy resin, were tested; they were 12 cm long and 3 cm in diameter and each time before the experiments they were cleaned with ethanol so as to eliminate any deposited surface charges. All the field measurements were adjusted to standard air density δ according to the IEC correction procedures [6].

3. Results and discussion
3.1. *Streamer propagation field*

Propagation probability distributions were obtained through tests analogous to those used for sparkover measurements, Class 1 (multiple-level) tests [6]. The frequency of propagation, up to the cathode and to various distances of traverse from the point of origin, was measured by applying 20 voltage pulses at the point, at gradually increasing field levels until traverse occurred with a probability of 1. This measurement procedure was applied for several values of pulse voltage amplitude, u. As all the distributions followed the normal Gaussian distribution, the required electric field for propagation corresponding to 0.975 propagation probability, termed *stability* field E_{st} was computed.

The stability fields corresponding to several distances of streamer traverse from the point, obtained when using a pulse voltage amplitude of 4 kV, are shown in figure 1. For the full streamer traverse up to the cathode, the relation between the stability field and the pulse voltage amplitude can be derived from figure 2. Both figures indicate that higher fields are required for streamers to propagate along insulators than in air alone. Also, the stability field increases with distance of traverse tending towards asymptotes at a traverse of 12 cm (figure 1) and decreases linearly with increasing pulse voltage amplitude (figure 2). These results are consistent with previous ones reported in [5], where it was suggested that the higher stability field when streamers propagate along insulators rather than in air alone results from greater energy losses for propagation over an insulator than in air. Also, the linear decrease of the stability field with pulse voltage amplitude was attributed to the decreasing energy imparted to the streamer at the point of origin as the pulse voltage amplitude decreases [5].

As figures 1 and 2 indicate, both the stability field and the rate of its decrease with pulse voltage amplitude vary significantly with material. Thus, both the energy losses for stable propagation and the way that the streamer consumes the energy supplied at initiation are functions of the nature of the dielectric medium. As indicative of the energy losses may be considered the *intrinsic* propagation field E_{in} i.e. the stability field corresponding to propagation with the minimum possible energy supplied at initiation; this can be calculated from the linear regression lines in figure 2 for $u = 0$. Accordingly, the slope of each line in figure 2 may indicate the way that the streamer consumes the energy supplied at initiation.

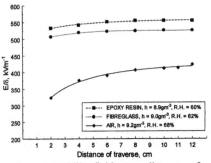

Figure 1: Stability fields up to distances of traverse, $u = 4$ kV

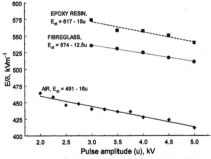

Figure 2: Linear regression of stability field over pulse amplitude

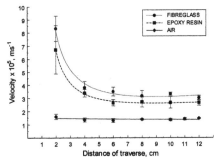

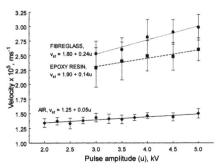

Figure 3: Propagation velocities up to distances of traverse, $u = 4$ kV, vertical bars represent 2σ

Figure 4: Linear regression of the propagation velocity over pulse amplitude, 12 cm traverse distance, vertical bars represent 2σ

3.2. Streamer propagation velocity

At each applied pulse voltage the streamer transit times were measured, as the delay between the start of the rise of each light signal, detected at the point and the cathode respectively, hence, the velocity of propagation was calculated. The propagation velocities associated with the stability fields, shown in figures 1 and 2, are displayed in figures 3 and 4 as a function of the distance of traverse and the pulse amplitude respectively. Streamers, propagate stably with higher velocity along insulating surfaces than in air alone; they decelerate at the beginning of their travel, but after a short distance from the point their velocity tends eventually to a constant value for the main part of the gap (figure 3). Also, the velocity of propagation increases linearly with the pulse voltage amplitude (figure 4) and shows a significant dependence upon the insulating material.

In the presence of an insulator, a streamer system was found to reach the cathode, consisting of a *surface* and an *air* component; the same phenomenon was also observed in previous studies [5, 7]. Figure 5 displays the velocity of both components as a function of electric field. The velocity of the *surface* component is higher than that for propagation in air alone; it increases according to a power law with electric field and depends on the nature of the insulating material. However, these parameters exert negligible influence on the velocity of the *air* component, which is smaller than that for propagation in air alone. The higher propagation velocity along an insulating surface rather than in air supports the argument that there is an enhancement of the ionisation close to the insulator caused by photo-electron emission from the surface [1, 2, 5]. However, the dependence of the propagation velocity on the material suggests that this mechanism is a function of the insulating material's nature.

As in reference [5], the following empirical equation can be used to express accurately the relation between the streamer propagation velocity and electric field under any value of δ and for any pulse voltage amplitude and dielectric medium employed. The curves drawn alongside the experimental points in figure 5 result from this equation, where E_{in} and v_{in} are the constants of the linear regression lines of the stability field and velocity over pulse amplitude respectively, while β and α are the corresponding slopes (figures 2 and 4); the coefficients γ and n are chosen to match the experimental values. Thus, the coefficients α and γ may be taken to indicate the extent of the effects of the energy used for streamer initiation and of the electric field on the streamer propagation velocity respectively.

$$v_{str}(p,t) = \left(v_{in} + \frac{\alpha u}{100} \right) \left(\frac{E(p,t)}{(E_{in} - \beta u)\delta} \left(1 + \frac{\gamma}{100} \right) \right)^n \times 10^5, \text{ ms}^{-1}.$$

52

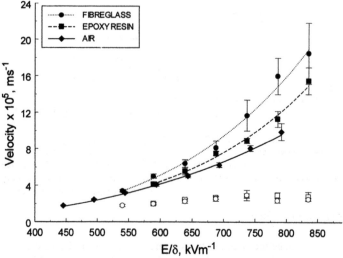

Figure 5: Propagation velocities as a function of electric field,
$h = 9 \text{ gm}^{-3}$, R.H. = 64%, vertical bars represent 2σ

5. Conclusions

The stable propagation of a single streamer along an insulating surface is characterised by an *intrinsic* propagation field and an associated velocity both higher than for propagation in air. These properties are characteristic of the dielectric medium employed and depend on the energy supplied for streamer initiation.

   In the presence of an insulator and for electric fields greater than the stability field, a streamer system propagates with a *surface* component that travels faster than in air and an *air* component with propagation velocity smaller than that in air alone. An empirical formula accurately predicts the relation between the streamer propagation velocity and electric field for any insulating material.

Acknowledgements

The authors acknowledge the support of an EPSRC Grant.

References

[1]   Verhaart H F A Tom J Verhage A J L and Vos C S 1987 *5th I.S.H.* Braunschweig paper 13.01

[2]   Gallimberti I Marchesi G and Niemeyer L 1991 *7th I.S.H.* Dresden paper 41.10

[3]   Sudarshan T S and Dougal R A 1986 *IEEE Trans. Elect. Insul.* **EI-2** (5) 727-746

[4]   Allen N L and Mikropoulos P N 1999 *J. Phys. D: Appl. Phys.* accepted for publication

[5]   Allen N L and Mikropoulos P N 1999 *IEEE Trans. on Dielectrics and Elect. Insul.* accepted for publication

[6]   IEC publication 60-1 1989

[7]   Allen N L and Ghaffar A 1995 *Annual Report of CEIDP* 447-450

Inst. Phys. Conf. Ser. No 163
Paper presented at the 10th Int. Conf., Cambridge, 28–31 March 1999

Corona discharge from a metal filamentary particle and its motion within parallel plate electrodes

K Asano, T Yamaki and K Yatsuzuka

Electrical and Information Engineering, Yamagata University, Yonezawa, Japan

Abstract. In order to avoid the serious damage triggered by the discharge from a metal particle within a GIS, the fundamental study of the discharge and the motion of a metal filamentary particle was performed. When a metal filamentary particle is laid on the lower electrode, it shows several different patterns of motion with the increase of applied voltage; such as still-standing, precession motion, jumping up-and-down, and so forth. Those motions depends on many different parameters. The interesting result is that the threshold voltage of bouncing motion is lower than the standing motion. The key factor for the change of motion could be discharge current.

1. Introduction

Particle motion and DC corona discharge from a metal filamentary particle within parallel plate electrodes have been extensively investigated for the protection of particle-initiated breakdown of a gas-insulated switchgear (GIS).[1~4] According to the previous observations, the particle shows several different motions depending on the condition; namely, standing, precession, up-and-down motion, hanging-down from the upper electrode, and so on. Although these motions of a metal particle are due to the Coulombic force within an electric field, it was also clearly shown that the corona discharge from a particle has significant influence on the motion.

The measurement of discharge from a fixed particle on the electrode suggested that there is a relation between the particle motion and discharge current. During the standing motions or spinning motion, there is stable corona discharge from the edge of a particle. However, when a particle is moving around on the electrode, the current of corona discharge differs from the fixed particle. Thus, this relation was carefully investigated.

2. Experimental apparatus

The experimental apparatus is basically the same as the previous one and illustrated in Fig. 1. Two brass disc electrodes of 80 mm diameter forms a parallel plate electrode system. They

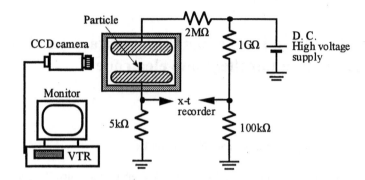

Fig. 1 Experimental apparatus

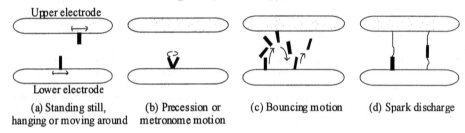

| (a) Standing still, hanging or moving around | (b) Precession or metronome motion | (c) Bouncing motion | (d) Spark discharge |

Fig. 2 Schematic diagram of mode of particle motion

are settled in a plastic box to keep the gap constant up to 20 mm. The most experiments were performed with 20 mm gap. A piece of copper and aluminum wire of 0.25 mmϕ cut of various length was used as a metal filamentary particle.

DC high voltage is applied to the upper electrode through 2 MΩ resistor to prevent the damage of instruments from flashover. The current which flows through the filamentary particle is measured by using a series resistor and a volt meter. The motion of a metal particle is monitored by a CCD camera or a high speed camera.

3. Experimental results

3.1. Particle motion

When the applied voltage to the upper electrode is gradually raised, a metal particle which was laid on the lower electrode will raise itself at one end. Further increase of applied voltage results in different mode of particle motion; still-standing, precession motion, metronome motion, moving around, and bouncing motion between two electrodes as shown in Fig. 2. In many cases, a particle motion changes its mode from one to the other.

The occurrence frequency of each mode is counted for each particle length. One of the results is shown in Fig. 3. The mode of particle motion differs with the particle material, polarity and particle length. Generally speaking, bouncing motion is predominant when the particle length of copper is 2 mm or less. For the case of copper particle, still-standing or precession motion is mainly observed with positive voltage to the upper electrode. Polarity

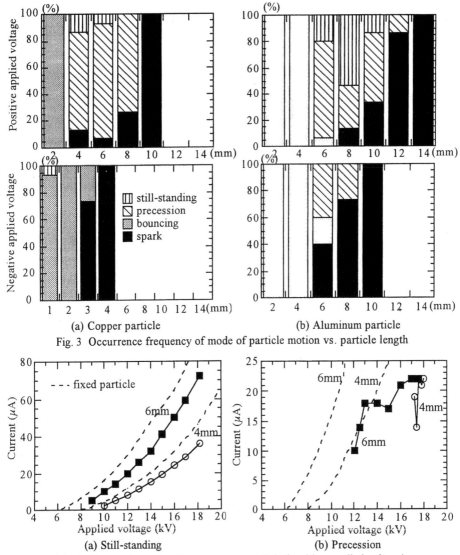

Fig. 3 Occurrence frequency of mode of particle motion vs. particle length

Fig. 4 Current in motion from a copper particle (positive applied voltage)

dependency for spark discharge is quite high. For both copper and aluminum particles, spark discharge is likely to occur with negative polarity comparing with positive polarity. Hanging motion is more frequently seen with negative polarity of aluminum particles. However, when a copper particle is used, hanging motion is only observed with 1 mm particle.

3.2. Corona current

When the positive voltage is applied to the upper electrode, negative corona starts from the edge of a standing or moving particle on the lower electrode. The corona current with a fixed particle shows smooth increasing curve as shown in the previous report[5].

56

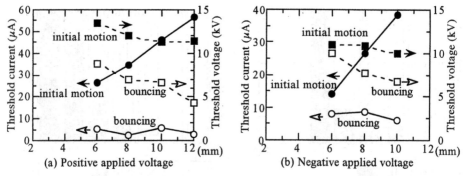

Fig. 5 Threshold voltage and current of initial and bouncing motion (Al)

Experimental results shown in Fig. 4 are obtained as follows. Firstly the applied voltage is increased up to a certain mode of motion to occur. Then the voltage is gradually decreased up to the motion to stop. The dashed curves in the figures are from a fixed particle, which were obtained from previous experiments[5]. The results indicate that the current from an unfixed particle is less than the one with a fixed particle. The current from a still-standing particle shows smooth decreasing curve as shown in Fig. 4(a). However, the current curve of precession motion shown in Fig. 4(b) is not as smooth as still-standing. When the applied voltage is high, the solid angle of precession motion is large. Then the angle gradually becomes small with decreasing voltage and reaches to standing motion.

3.3. Threshold voltage and current

When the applied voltage is gradually decreased, starting from still-standing the particle becomes suddenly to jump up for bouncing motion. The threshold voltage and current for each length of aluminum particle were carefully measured. The results are shown in Fig. 5. It is interesting to note that the threshold voltage of bouncing motion is lower than the standing motion. Therefore, the key factor for the change of motion could be discharge current.

4. Conclusion

Many different modes of particle motion were observed. Each mode depends on the particle material, length and polarity of applied voltage. The current of still-standing or precession motion of a particle is lower than the fixed particle. Current curve of still-standing motion is smoother than the one of precession motion. It was found that the threshold voltage of bouncing motion is lower than the standing motion.

References

[1] K. Asano et at, 1993 J. Electrostat., **30** 65
[2] K. Asano, et al, 1995 Inst. Phys. Conf. Ser. No. 143, 17
[3] K. Asano, T. Kitamura and Y. Higashiyama, Proc. 1996 Ann. Meet. Inst. Electrostat. Japan, 263 (in Japanese)
[4] K. Asano, K. Anno and Y. Highashiyama, 1997 IEEE Trans. IAS, **33** 679
[5] K. Asano, T.Yamaki and K. Yatsuzuka, 1998 IEEE IAS meeting, St. Louis, p1814

DISCUSSION - Section A - Fundamental Physics

Title: Spectroscopy and Calorimetry of Spark Discharges
Speaker: Z Kucerovsky

Question: D K Davies
The non-intrusive characterisation of sparks is important for prediction of ignition as well as electronic damage. Is this method applicable to the very fast, ~ ns, timescale of real ESD?

Reply: We believe that the method is suitable for nanosecond signals, but so far we have not attempted measurements in that region. Our belief is based on some of our work with pulsed lasers and pyroelectric detectors. We intend to extend the range of our measurements into the nanosecond region.

Question: N L Allen
(a) What currents were measured? (b) Were filters used to distinguish UV from visible.

Reply: (a) The spark current was measured with a ferrite core pulse transformer as shown in fig.(a). The cut-off frequency of the transformer was 3MHz. Since the circuit was self-triggered, the gap voltage depended on the electrode separation. the pulse current ranged from 500 to 1500 A pk-pk. The current was given by $i(t) = (V/R) \exp -t/RC$ (where $R=9.6 \ \Omega$, $C=0.15 \ \mu F$). The parasitic inductance of the capacitor was $0.015 \ \mu H$.

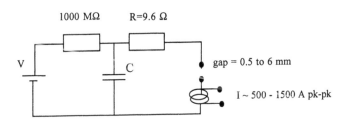

Figure (a)

(b) Absorption filters were not used, but rather the wavelength separation was achieved with a spectrometer provided with a diffraction grating of the echellete type. The grating blaze angle was 23.5°. The grating was illuminated with an optical fibre and the diffracted signal was detected by a silicon charge-coupled device of the linear array type. The nonlinearity of the spectral response of the CCD detector was rectified at the signal processing stage using a look-up table. The spectrometer made it possible to work without absorption filters.

Title: Surface-wave Mechanism for the Phenomenon of Passive Insulator Charge Elimination

Speaker: T S Lee

Question: D K Davies
Can the expanding wave velocity be predicted from the field change and ion mobilities?

Reply: In the model proposed, the expanding wave velocity is intimately related to the saddle-point ascent motions as well as to associated field changes. In turn, the latter depends on ion production of induced inverse corona which again is field-dependent. The angular spread in corona output at the point, which determines the extent of charge neutralisation outside the observed wave front on the surface, the ability of the resultant time-dependent surface charge pattern to induce corona and the medium properties all have an influence. Although the wave velocity can be determined precisely, its prediction is not possible owing to the lack of methodologies for assessing all these.

Title: Distortion of a Water Surface Stressed by a Perpendicular Alternating Electric Field

Speaker: J A Robinson

Question: Tom Jones
It is nonlinearity which presumably limits the amplitude of these standing waves, but the linear theory seems to predict the initial wavelength.

Reply: Yes. This paper makes no attempt to analyse the nonlinear growth of the waves; the linear theory is used only to predict the wavelength. Our results indicate that even for large-amplitude waves, the linear theory is useful for predicting the wavelength. Thank you Professor Jones for the clarification.

Question: Zayed Huneiti
Did you observe any relationship between the applied AC frequency and the cone angle?

Reply: We have not yet done a thorough study of the angles of the so-called "cones". Both the wavelength and amplitude of the cones decrease when the AC frequency is increased, so it may be that the cone angle is independent of the AC frequency.

Question: G Touchard
Do you have an explanation for the shorter wavelength at higher potentials?

Reply: According to the dispersion relation for waves on a water surface (see reference (9) of the paper), an increase in the electric field perpendicular to the surface causes a decrease in the phase velocity. The standing wave observed in our experiments, with wavelength $2\pi/k$ (where k is the wavenumber), can be considered to be the result of two travelling waves, each moving with phase velocity ω/k (where ω is the angular frequency), but in opposite directions. A decrease in the phase velocity of the travelling waves, with ω constant, will result in an increase in k, and therefore a decrease in wavelength.

Section B

Bioelectrostatics

Inst. Phys. Conf. Ser. No 163
Paper presented at the 10th Int. Conf., Cambridge, 28–31 March 1999

Positioning, Levitation and Separation of Biological Cells

W. M. Arnold (m.arnold@irl.cri.nz)

Sensors and Electronics Group, Industrial Research Ltd.
Gracefield Research Centre, Lower Hutt, New Zealand

Abstract. Manipulation of microparticles and biological cells by means of alternating electric fields applied between arrays of interdigitated planar microelectrodes is described. Negative dielectrophoresis enables positioning of particles in multiple traps, as well as levitation and sorting from other particles or cells having different dielectric properties. Yeast cells levitated in growth medium can be grown to form isolated clones. Concentration of cells from dilute or very dilute suspension, as well as sorting using long-range transport are demonstrated by use of lev-vection: the combination of levitation and convection. Scale-up of lev-vection is straightforward.

1. Introduction

The demonstration of electric-field driven short-range motion, or else local immobilisation, of cells and other microparticles by the dielectrophoretic force [1,2] is readily carried out with interdigitated arrays of planar microelectrodes [3-5]. The arrays require single-phase, low-voltage excitation, and so are simple to fabricate and operate. There have been many reports of their use to accomplish *short-range* separations. However, the use of such arrays to separate cells into two or more *bulk* collections, or to electro-filter them, is more difficult. Such methods as do exist use externally-induced flow [5-7]. The use of electrical travelling waves [2] offers another solution [8-10], but only at the cost of very significant increases in electrical and fabrication complexity (poly-phase excitation and multilayer metallisation respectively). A simpler method of causing long-range motion is desirable. Such a method, and also results that show that cell growth is not inhibited by the electrical fields used, is described.

2. The Dielectrophoretic Force and Modified Media

The time-averaged dielectrophoretic (DEP) force exerted by a non-uniform field of peak strength E on a particle of radius a is given by:

$$\overline{F} = \pi\, a^3 \varepsilon_m \varepsilon_0\ U'_{(\omega)}\ \nabla E^2 \tag{1}$$

64

The DEP force may attract or repel particles from the regions of higher field ($\overline{F}$ positive or negative respectively), and this is determined by the sign of $U'_{(\omega)}$, the real part of the complex Clausius-Mossotti factor [2,11]:

$$U^*_{(\omega)} = \frac{\varepsilon^*_p - \varepsilon^*_m}{\varepsilon^*_p + 2\varepsilon^*_m} \qquad \text{where all} \qquad \varepsilon^* = \varepsilon' - j\sigma/\omega\varepsilon_{,0} \qquad (2)$$

and symbols have the following meanings: p, particle; m, medium; ε, relative permittivity; σ, conductivity; ε_0, vacuum absolute permittivity; $j = \sqrt{-1}$; $\omega = 2\pi f$.

Negative DEP is a necessity if cells or particles are to be passively and stably trapped or levitated: its production requires the polarizability of the medium to exceed that of the cells. This implies either a high medium permittivity, and/or a high conductivity [12-14]. In accordance with Eq. 2, the conductivity difference determines the sign of $U'_{(\omega)}$ at very low frequencies, whilst the permittivity difference dominates at very high frequencies. In addition, cell polarizability decreases as the frequency is raised [15]. Therefore, cellular DEP commonly changes from negative to positive and back again as the frequency is raised through the practical and useful range (1kHz to 100MHz).

At a given field strength, raising the permittivity of the medium will generally increase the magnitude of the DEP force (positive or negative) because of the term $\varepsilon_m \varepsilon_0$ in Eq. 1. Raising the conductivity does not have this advantage: rather, the increased current flow may cause too much heating.

3. Positioning

When particles or cells were allowed to settle from suspension onto an energised interdigitated electrode array, they could often be made to position themselves.

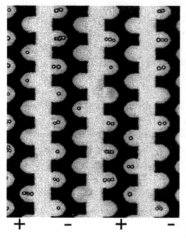

Fig. 1. Polystyrene latex particles (5.2μm diameter) positioned, by means of a field of frequency 9MHz and 1.0V peak, in the minimum field regions close to the electrode plane of "headlands and bays" pattern interdigitated electrodes. Alternate electrodes are energised by opposite phases ("+" and "-") of the RF voltage. When the relative permittivity of the suspension was raised from 78 (water) to 155 by use of 1M EACA (the zwitterion ε-aminocaproic acid [12,16]), only 0.25V peak amplitude was necessary. The particles settled from a suspension in 10mM KCl solution (130mS/m conductivity). The gap between adjacent electrode stripes, and the width of a "bay", are 22μm; the repeat distance is 70μm. Microelectrode patterns were generated in-house and fabricated (Au on Ti on glass) as before [16].

Using the "headlands and bays" electrode pattern (Fig. 1), distinct stable positions for particles showing either negative (in the bays) or positive DEP (on the headlands) are available. Polystyrene particles (Fig. 1) show only negative DEP in conductive solutions: all particles centralise themselves, or their centre of mass in the case of particle aggregates, in the lowest-field regions (the bays).

4. Microseparation

Attempts to produce results as orderly as Fig. 1 with yeast cells failed: it was not possible to obtain trapped cells exclusively, because there were always at least a few cells that showed positive DEP. This was initially interpreted as being due to biological variability, but the use of staining methods allowed the results to be more satisfactorily interpreted in terms of which cells were viable. Non-viable cells were detectable in every culture, and not added intentionally.

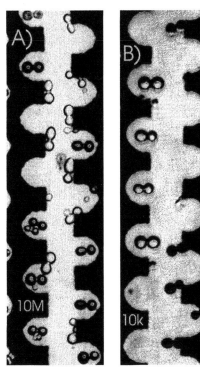

Fig. 2. Yeast cells positioned and micro-separated in the maximum and minimum field regions of the "headlands and bays" array, dimensions as in Fig. 1. The applied voltage was 0.6V peak. Dark-appearing cells had taken up methylene blue dye, a criterion of non-viability. In A), the applied frequency was 10MHz, whereas in B) it was 10kHz. The change of frequency reversed the micro-separation. The aqueous medium (total conductivity 19mS/m) contained methylene blue (2.4µg/ml) HEPES buffer (2.5mM, 50% as Na salt) and 1M EACA to increase the relative permittivity to 155. EACA did not influence cell staining by the dye. *Saccharomyces cerevisiae* were obtained freeze-dried as a commercial supply for baking, and used to seed cultures (25°C for 16-20h, in an oscillating shaker) in YEG (1% yeast extract, 2% glucose). Colour images were cyan-separated to better contrast blue and undyed cells.

In Fig. 2, the dark cells (dyed vivid blue in the original) indicate themselves to be non-viable, or at least to have a poorly functioning plasma membrane. At 10MHz (Fig. 2A), the dyed cells showed negative DEP and partitioned themselves into the bays, whereas the undyed cells showed positive DEP and attached themselves to the high-field (headland) regions of the electrodes. At 10kHz the opposite was true: the dyed cells are in maximum-field regions, mostly on headlands, whilst the negative-DEP traps now hold only undyed cells.

The above results indicate that a separation, *at least at the micro-scale*, of viable from non-viable cells can be carried out and visualised. The non-viable cells appeared similar in size and shape to the viable ones, not broken or visibly shrunken as is the case with heat-killed cells. Yet there was a dielectric difference. To minimise the possibility that the methylene blue itself had caused a change in cells, the separation was repeated using a very different type of dye: the cyanine dye FUN-1 [17] (6µM for 1h at 33°C in 2% glucose) which causes red fluorescence only in metabolically active cells (not shown).

66

5. Lev-vection: combined levitation and convection

The fluid flow necessary to cause *long-range* separation of cells exhibiting different DEP behaviours can be generated as a by-product of the heating that results when intense electric fields are applied to biological media.

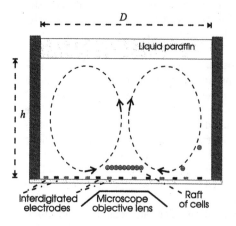

Fig. 3. The production of a convection cell by the heat developed just above an electrically energised array of planar microelectrodes. The circulation concentrates suspended (subsequently levitated) micro-particles to a predictable region just above the centre of the electrode array. Cells follow the combined forces due to the circulation and gravity, except close to the array, where DEP dominates. Dimensions h and D are approximately equal and run from several mm to several cm in present examples. The liquid paraffin is required for sterile or long-term work.

To take advantage of this heating and consequent tendency to convection, it is necessary to employ not a thin layer of liquid above the electrodes, as is often done to permit observation with a conventional microscope, but instead a much deeper chamber (Fig. 3) which implies the use of an inverse microscope.

A single convection cell can now develop, which has some very useful consequences. Cells which are sufficiently levitated to have horizontal freedom of movement will be swept into the region just above the centre of the electrode array. Unless the convection cell is very powerful, they must remain there at the levitation height, under the influence of gravity.

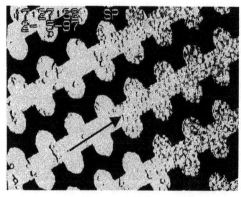

Fig. 4. Production of a raft of yeast cells in a device as shown in Fig. 3, using a microelectrode array as in Fig. 1 and cells suspended in 25 mM KCl solution (0.4S/m conductivity). The higher magnification view (left) shows the edge of the cellular raft: the arrow indicates the direction of motion of incoming levitated cells. The lower magnification view shows the approximately circular accumulation of yeast cells, which formed during several minutes field-induced lev-vection (2V peak at 2.5MHz).

The convection cell sweeps out the whole chamber, thus all cells that levitate under the given field conditions will be concentrated into one place, and are observed to form an almost planar raft whose outline depends on the cross-section of the container. Depending upon the chamber size, the concentration factor can be many orders of magnitude. Cells which do not levitate, i.e. show zero or positive DEP, will deposit on the electrodes, mostly at the rim of the electrode array. In this way a very useful separation of cells of different dielectric properties can be obtained.

Fig. 4 shows the above effects. In the right hand side the raft of cells contained the cells from 30μl of yeast suspension after concentration by lev-vection. The left-hand image shows levitated cells (both individual and pearl-chained) being swept into the central mass. The darkening of the raft towards the RHS is evidence of a greater thickness (multiple cells) there, which can be observed directly by focussing through it.

6. Cell growth and division whilst levitated

The ability to levitate cells in normal growth medium raises the question of whether the cells can grow and divide whilst exposed to the levitation field. The use of the lev-vection device allows a seed colony of cells to be brought to a pre-determined spot for observation, and studied there (Fig. 5). Despite earlier work on growth of fibroblasts on energised microelectrodes [18], there has been little if anything reported on microbial growth at field strengths of 10 kV/cm or more.

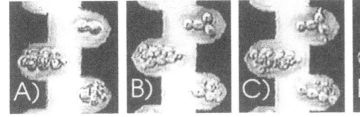

Fig. 5. Growth and division of yeast cells whilst levitated and trapped, using growth medium (YEG, conductivity 115mS/m) in a pre-sterilised device as in Figs. 1 & 3. Cells from a very dilute suspension were accumulated (2.0V peak at 0.3MHz) and then held in traps (0.6V peak). At 0.6V, yeast cells are still free to show Brownian motion within the potential wells from which they cannot escape (except by growth and division). Frames A-D, extracted from a time-lapse video, show the budding and repeated division of a yeast doublet which grew to a field-trapped clone, as well as growth of other cells.

Fig. 5A (at time 0) features a doublet cell (post-division) in a trap at top right;
Fig. 5B (at 1hr 27min) the doublet has now become a quadruplet, still trapped;
Fig. 5C (at 2hr 0min) further buds are appearing on two of the quadruplets;
Fig. 5D (after 4hr 29min) shows further division, the group is still trapped.

In this and parallel experiments, growth continued over 2 days to form a levitated raft of close-packed cells covering one half or more of the culture vessel diameter (3.5mm). At its centre, the raft was observed to be several cells thick, but became a monolayer towards the edge. The growth rate was rapid (appreciably faster than typical generation times for yeast, which are 2-3 hours).

7. Conclusion

Lev-vection, i.e. concentration and separation based on levitation and thermoconvection, is useful for screening liquids for small numbers of cells, and also as a mass cell-sorting technique. Further work with plain interdigitated electrodes has shown that these simpler structure also give lev-vection, but that the "cliff and headland" or castellated [3,4] geometries improve the retention and separation of those particles showing positive DEP. In view of the simplicity of these structures, scale-up is less of a problem than with the poly-phase electrodes required for the travelling-wave techniques of transporting cells.

Positioning of cells in individual cages will facilitate studies of growth and development, and their dependence on cell-surface and cell-cell interactions. Likewise the culture of cells in levitation is a most interesting phenomenon for those interested in the influence of electric fields on biological organisms.

Acknowledgement: this work was supported by the New Zealand PGSF.

References

[1] Pohl H A 1978 *Dielectrophoresis* (Cambridge, Cambridge University)

[2] Jones T B 1995 *Electromechanics of Particles* (Cambridge, Cambridge University)

[3] Huang Y and Pethig R 1991 *J. Meas. Sci. Technol.* **2** 1142-1146

[4] Pethig R, Huang Y, Wang X-B and Burt J P H 1992 J. Phys. D: Appl. Phys. **25** 881-888

[5] Markx G H, Talary M and Pethig R 1994 *Biotechnology* **32** 29-37

[6] Docoslis A, Kalogerakis C, Behie L A, and Kaler K V I S 1995 *US Patent 5,626,734*

[7] Markx G H and Pethig R 1995 *Biotechnol. Bioeng.* **45** 337-343

[8] Masuda S, Washizu M and Kawabata I 1988 *IEEE Trans. Ind. Appl.* **24** 217-222

[9] Fuhr G, Hagedorn R, Müller T, Benecke W, Wagner B and Gimsa J 1991 *Studia Biophys.* **140** 79-102

[10] Morgan H, Green N G, Hughes M P, Monaghan W, and Tan T C 1997 *J. Micromech. Microeng.* **7** 65-70

[11] Foster K R, Sauer F A and Schwan H P 1992 Biophys. J. **63** 180-190

[12] Arnold W M 1992 *Biochem. Soc. Trans.* **20** 119S

[13] Fuhr G, Arnold W M, Hagedorn R, Müller T, Benecke W, Wagner B and Zimmermann U 1992 *Biochim. Biophys. Acta* **1108** 215-223

[14] Arnold W M and Fuhr G 1994 *Conference Record, IEEE/IAS Annual Meeting, Oct. 1994 Denver USA* 1470-1476

[15] Schwan H P 1985 in: *Interactions Between Electromagnetic Fields and Cells*, Eds: Chiabrera A, Nicolini C and Schwan H P (New York, Plenum) pp. 75-97

[16] Arnold W M and Turner G C 1998 *IEEE Ann. Rep. Conf. Elec. Ins. Diel. Phenom. Atlanta* 360-363

[17] Millard P J, Roth B L, Thi H-P T, Yue S T and Haugland R P 1997 *Appl. Environ. Microbiol.* **63** 2897-2905

[18] Fuhr G, Müller T, Schnelle T, Hagedorn R, Voigt A, Fiedler S, Arnold W M, Wagner B, Heuberger A and Zimmermann U 1994 *Naturwissenschaften* **81** 528-535

Inst. Phys. Conf. Ser. No 163
Paper presented at the 10th Int. Conf., Cambridge, 28–31 March 1999
© 1999 IOP Publishing Ltd

Single *Cryptosporidium* oocyst Isolation and Capture using a Travelling-Wave Dielectrophoresis Device.

Goater, A.D.[1], Pethig, R.[1], Paton, C.A.[2] & Smith, H.V.[2]

[1]Institute of Molecular and Biomolecular Electronics, University of Wales, Dean Street, Bangor, Gwynedd, LL57 1UT, UK

[2]Scottish Parasite Diagnostic Laboratory, Stobhill NHS Trust, Springburn, Glasgow G21 3UW, UK

Abstract. A micro-electrode device is described for manoeuvring single particles from a suspension for capture onto a membrane filter. Fabricated on a thin polyimide film by photolithography, the device is demonstrated here using 5 µm oocysts of *Cryptosporidium parvum*, a waterborne pathogen of man. This repeatable process can also reveal information about the physiological state of the oocyst through its electrorotational response. The methods described have potential for sample handling prior to Polymerase Chain Reaction (PCR) analysis of DNA or infectivity studies where precise numbers and viability information are important.

1. Introduction

The development of microelectrode devices for the electrokinetic-characterisation and selective manipulation of cells and other bioparticles is currently an active area of research. One important electrokinetic phenomenon is dielectrophoresis (DEP) [1] whereby translational motion of particles is induced by exposing them to non-uniform, stationary, AC electric fields. The magnitude and sense of the dielectrophoretic force depends on the nature of the dipole moments induced in the particles, and this in turn is a function of the dielectric properties of each particle and its surrounding medium. If, instead of a stationary electric field, a particle is subjected to a moving field, then under appropriate conditions it can be induced to move by the effect of travelling wave dielectrophoresis (TWD) [2]. A third technique, electrorotation (ROT) [3], involves monitoring the induced spin of particles subjected to a rotating electric field. It has been shown to be a sensitive method for monitoring the physiological viability of cells [3].

As both rotating and travelling wave electric fields can be induced using similar electric signals their integration into a single microelectrode structure is possible [4]. An application of such a device is described here with the transmission stage of the parasitic protozoan *Cryptosporidium parvum*. Normally a self-limiting illness in immunocompetent people the clinical signs of cryptosporidiosis include 'flu-like' symptoms, diarrhoea and abdominal pain, and as yet there are no effective treatments or vaccines. The largest outbreak of water-borne disease in recorded history involved this parasite in 1993, affecting over 400,000 individuals in Milwaukee, Wisconsin [5].

Detection of oocysts from environmental samples by the Polymerase Chain Reaction (PCR) is problematic due to contamination and a lack of efficient concentration methods. There is also a requirement for a method of isolating low numbers of particles, for example to test the sensitivity of the PCR protocols.

2. Circular-Spiral Microelectrode Array

The 20 µm width circular spiral microelectrode elements were fabricated onto a 50µm thick polyimide film (Kapton HN®) substrate using standard photolithographic techniques. A total area of ~3 mm² was created with just 5 turns of the spiral. When the four spiral electrode elements are energised with sinusoidal voltages of relative phase difference shown (Figure 1), a travelling electric field is generated that travels radially from the centre towards the periphery of the spiral array. At the centre of the spiral an electrode-free region creates a ROT chamber of diameter 140 µm. A hole of diameter 100 µm was created in the centre (Figure 2) to allow particle capture onto a wetted polycarbonate membrane positioned below the film (Figure 3).

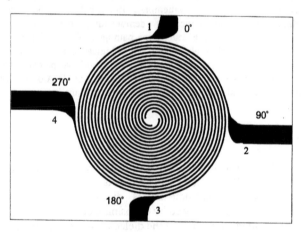

Figure 1 The Spiral Microelectrode design and phases of applied sinusoidal voltages.

Figure 2 The 100µm hole in the centre of the spiral electrode array.

A rubber 'o' ring spacer with hypodermic needle inlet and outlets allowed sealed sample handling (Figure 3).

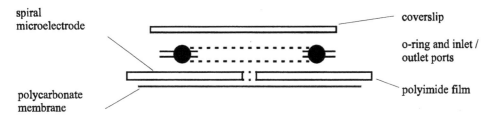

Figure 3 Schematic of the oocyst isolating chamber.

3. Collection / Isolation Procedure

The *C. parvum* oocysts (Iowa isolate) were prepared by repeated washes in ultra pure water to reduce the suspending medium conductivity, and then resuspended in a dilute solution of standard Phosphate Buffered Saline solution of conductivity 4.5 mS m^{-1}. From previous DEP, TWD and ROT data obtained for this particle the following procedure (Table 1) was determined for their sequential collection, viability determination and isolation.

The electrode elements were energised with phase-quadrature sinusoidal signals of frequencies between 60 kHz and 1.5 MHz. The direction of the travelling wave was determined by the phase relationship between adjacent energised electrodes.

Table 1. Reproducible procedure of the isolation, viability determination and capture of single oocysts.

Action	Field conditions	Result
Cell injection	Off	Oocysts suspended over electrodes
Assisted sedimentation	30 s at 1 MHz 6 V pk-pk	Oocysts attracted to electrode edges by positive DEP
Concentration of oocysts	TWD wave from centre to periphery 160 kHz 1 V pk-pk	Oocysts levitate and move toward centre of spiral
Remove TWD force	Off	First oocyst drops into ROT chamber
Check viability	1.5 MHz 1 V pk-pk	Co-field ROT = nonviable Anti-field ROT = viable
Negative DEP	60 kHz 6 V pk-pk	Isolated oocyst centred over 100µm diameter hole
Removal of other oocysts to periphery	TWD wave from periphery to centre 160 kHz 1 V pk-pk	Oocysts travel to periphery of spiral, ~900µm distance
Holding of peripheral oocysts	1 MHz 6 V pk–pk	Oocysts held by positive DEP

While holding the unwanted peripheral oocysts the membrane is removed and examined for successful capture (Figure 4).

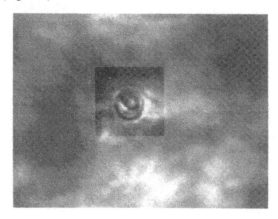

Figure 4 Highlighted region showing the 5μm oocyst on a polycarbonate membrane fibre.

5. Conclusions

We have demonstrated that the combined TWD/ROT device can be used to selectively manipulate single particles. In this case a parasite oocyst of diameter 5 μm, could be selectively characterised for its viability and then isolated for further analysis. Importantly, for sample handling prior to PCR analysis, debris is not concentrated by TWD. Other applications envisaged for this rapid micro-electrode device include the preparation of precise doses for infectivity studies or as a controlled input module for laboratory-on-a-chip type devices.

Acknowledgements

We gratefully acknowledge the support of the BBSRC (Grant 5/E10206) and NFCR, and thank M S Talary, C J Hayden and J A Tame for technical assistance.

References

[1]   Pohl H A 1978 *Dielectrophoresis* (Cambridge: Cambridge University Press)

[2]   Fuhr, G, Hagedorn, R, Muller, T, Benecke, W, Wagner, B and Gimsa, J 1991 *Stud. Biophys.* **140** 79-102

[3]   Arnold W M and Zimmerman U 1988 *J. Electrostatics* **21** 151-191

[4]   Goater A D, Burt, J P H and Pethig R 1997 *J. Phys. D : Appl. Phys.* **30** L65-L69

[5]   MacKenzie, W R, Hoxie, N J, Proctor, M E, Gradus, M S, Blair, K A, Peterson, D E, Kazmierczak, J J, Addiss, D G, Fox, K R, Rose, J B and Davis, J P 1994 *New Eng. J. of Medicine* **331** 161-167

Inst. Phys. Conf. Ser. No 163
Paper presented at the 10th Int. Conf., Cambridge, 28–31 March 1999

Characterisation of the dielectrophoretic movement of DNA in micro-fabricated structures

D. Bakewell, H. Morgan[*] **and J. J. Milner**[+]

Department of Electrical Engineering, University of Glasgow, Glasgow G12 8LT, UK
[+]Institute of Biomedical & Life Sciences, University of Glasgow, Glasgow G12 8QQ, UK

Abstract. The controlled micro-scale movement of DNA is possible using DC and AC electrokinetic techniques. This paper describes the frequency dependent dielectrophoretic trapping of DNA by positive DEP forces. Experiments are described where 12 kilobase pair DNA plasmids, are trapped on the edges of 10µm width interdigitated electrodes in a range of medium conductivities and frequencies. The collection of DNA is imaged using fluorescence microscopy, the data is image processed and analysed semi-quantitatively. The results point to a DEP collection rate which has an exponential time dependence. The measured collection rates concur with measurements of the frequency-dependent polarisability of the DNA made using dielectric spectroscopy. A method for analysing the data to give values of the polarisability of the DNA is discussed.

1. Introduction

Techniques for non-contact manipulation and separation of biological macromolecules are widely used in many areas of biotechnology and pharmaceutics. Although these methods are very successful, new methods for handling and separating small numbers of molecules (such as DNA) are necessary for the new generation of micro-analytical devices. Small molecular weight fragments of DNA can be efficiently separated using electrophoresis methods, but larger fragments (Mbp – 10Mbp) are more difficult since the mobility becomes independent of chain length for fragments greater than 50kbp [1]. In order to circumvent this problem, pulse field gel electrophoresis methods were developed, where separation occurs over long time periods (one day). As an alternative to DC field techniques, methods for the non-contact transportation of DNA using dielectrophoretic forces have been presented [2] and even "molecular surgery" of DNA has been demonstrated [3]. As a method for improving the efficiency of electrophoresis methods, a theoretical model of an enhanced DNA separation system was described by Ajdari and Prost in 1991 [4]. They described how the use of periodic dielectrophoretic traps (giving an a-periodic sequence of energy traps) together with continuous field free-flow electrophoresis could potentially increase the separation efficiency for large molecular weight DNA fragments by a factor of 100. However, despite the promise and potential of AC electrokinetic manipulation methods, little experimental or theoretical work has been reported for DNA.

DNA is a member of a class of polyelectrolytes; macromolecules which in solution carry a high charge, and which possess high polarisabilities. In a DC field the mobility of the

---

[*] Author to whom correspondence should be addressed

molecule is a function of the net charge (which depends on its molecular weight). In a non-uniform electric field, the dielectrophoretic force acting on the molecule is a function of the polarisability given by

$$\mathbf{F}_{DEP}(\mathbf{x}) = \tfrac{1}{2}\alpha v \nabla |\mathbf{E}(\mathbf{x})|^2 \qquad (1)$$

where v is the volume, $\mathbf{E}(\mathbf{x})$ the electric field at position $\mathbf{x} = (x, y, z)$ and the polarisability, α, depends on the length of the molecule. Polyelectrolytes are molecules that have many ionisable groups along the length of the molecule and are consequently surrounded by a cloud of counter-ions. Application of an electric field produces distortion of this counter-ion cloud leading to large values of polarisability. This high value means that, in general, the dielectrophoretic force can be very high (for the size of particle) and also the direction is always towards regions of high electric field strength (positive DEP).

The dielectric properties of DNA have been studied by a number of groups [5 - 8] and the mechanism responsible for the polarisation is attributed to fluctuations in counter-ion movement along the major axis, although other mechanisms also considered include Maxwell-Wagner polarisation. It is generally accepted that the counter-ion cloud surrounding the DNA undergoes a relaxation in the frequency range between 100kHz and 1MHz so that the magnitude of the DEP force is expected to change in this frequency window.

This paper describes the AC electrokinetic manipulation of DNA together with a method for the semi-quantitative measurement of the DEP force acting on the molecules. Preliminary data is presented and compared with dielectric measurements of the same DNA molecules.

2. Materials and methods

A 1 cm x 0.5 cm array of parallel wires, (10μm wide, 10μm separation) was used for the DEP experiments. The entire area was divided into 8 separate electrodes sets, each of which could be addressed with a different frequency as shown in figure 1. The electrodes were manufactured on glass microscope slides using standard photolithographic techniques. They were powered using an AC voltage up to 20V peak to peak over the frequency range 1Hz-20MHz. The signal from the generator was buffered using a power amplifier which ensured a uniform voltage and current into a range of termination impedances. Measurements were made at frequencies from 50 kHz to 20MHz.

Supercoiled plasmid DNA, pTA250, was prepared according to standard protocols [10] and purified using Qiagen Megaprep 2500. The 12 kilobase pair double-stranded DNA had an effective size of less than 1 μm, thus being very suitable for DEP experiments with electrode sizes and gaps of the order of 10 μm. The DNA was mixed with anti-fade agent 2-Mercaptoethanol (2 % v/v) and DAPI (Molecular Probes, D-1306) in ultra pure water. Samples were micro-pipetted onto the electrode structures, and enclosed with cover-slips to reduce evaporation.

With a different frequency for each one of the 8 arrays, movement of the DNA was observed using an epifluorescent microscope Nikon Microphot SA (X20 objective), recorded with a JVC digital camera and S-VHS video. The video recordings were frame-captured (40 msec between frames) and processed with MATLAB 5.2 (Math Works, Inc.) to provide accumulation characteristics of fluorescently labelled DNA on the electrode edges.

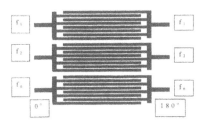

Figure 1. Diagram showing the separately addressable interdigitated electrode arrays (10 μm width and 10 μm gap)

Figure 2. Captured video image showing DNA plasmid accumulation on the electrode edges (f = 1 MHz, V = 18 V_{pk-pk})

3. Results

Figure 2 shows the accumulation of DNA on an electrode edge, 10 seconds after the application of a constant voltage (18 V_{pk-pk}, 1 MHz). The trapping of DNA under positive DEP forces could be achieved over a wide range of frequencies, from 50 kHz to 20MHz. Measurements were made in medium conductivities from 0.25mS/m (ultra pure water) to 0.2 S/m SSPE (Sodium chloride Sodium Phosphate EDTA buffer) and with the interdigitated electrodes, only positive DEP of the DNA was observed, although the magnitude of the DEP force varied. Collection at the electrode edges occurred within msecs of applying the electric field and the increase in fluorescence continued for up to a 1 minute.

Quantification of the rate of arrival of DNA was performed using image processing. In general, the rectangular symmetry and spatial periodicity of the electrodes was exploited so that an average accumulation of DNA around a representative electrode edge could be computed. An example characteristic of DNA fluorescent intensity (a. u.), as a function of time, is shown in figure 3. One of the interesting features is the near exponential rise of intensity, which in this example is approximately 3 seconds. Experiments were conducted over a range of frequencies from 100 kHz to 10MHz, and the results indicated that the rise time increased with increasing frequency. This indicates the rate of arrival of DNA, and hence, the DEP force on each of the DNA macromolecules, decreases with increasing frequency. This tendency concurs with the decrease in polarisability measured for the same DNA using dielectric spectroscopy [5-8]. This data is summarised in table 1 and shows the relaxation frequencies f_R, and dielectric decrements, $\Delta\varepsilon'$ for a solution of the DNA with an estimated volume fraction 1 %.

Table 1

f_R (MHz)	$\Delta\varepsilon'$
0.14	19.1
1.7	5.7
10.4	0.3

DNA solution dielectric decrement $\Delta\varepsilon'$ and relaxation frequencies f_R

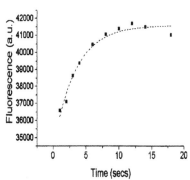

Figure 3. Graph of the time evolution of DNA collection at electrode edges.

76

4. Discussion

The exponential characteristic of DNA accumulation, shown in figure 3, is likely to be the due to a combination of diffusion limited kinetics, and possibly electrostatic shielding which occurs as the macromolecules accumulate on the electrode edges. Considering only diffusion, the time constant yields a measure of the DNA 'particle' accumulation near the electrode surface, which in turn is dependent on the frequency dependent polarisability α. This is evident by considering the spatio-temporal evolution of concentration $c(\mathbf{x}, t)$ under the influence of force field $F(\mathbf{x}, t)$. For non-interacting particles [11], $c(\mathbf{x}, t)$ is described by the joint probability density $P(\mathbf{x}, t)$, normalisation constant N, and particle flux $\mathbf{J}(\mathbf{x},t)$

$$\frac{\partial c(\mathbf{x},t)}{\partial t} = \frac{N \partial P(\mathbf{x},t)}{\partial t} = - div \, \mathbf{J}(\mathbf{x},t) = -\frac{N}{\zeta} \nabla.\left(P(\mathbf{x},t)\mathbf{F}(\mathbf{x},t)\right) + N \, D \, \nabla.\left(\nabla P(\mathbf{x},t)\right) \tag{2}$$

where ζ and D are the Stokes' friction and diffusion constants respectively. Since buoyancy and gravitational forces nearly cancel, then $\mathbf{F}(\mathbf{x}) \approx \mathbf{F}_{DEP}(\mathbf{x})$ as expressed in (1). Rectangular electrode symmetry and suitable force approximations enable (2) to be simplified and solved. Using the time dependent collection characteristics together with appropriate boundary conditions, the frequency dependence of α can be estimated and compared with values obtained from dielectric measurements. Further work is in progress to develop these models.

5. Conclusion

DNA in solution is capable of being dielectrophoretically trapped onto electrode edges under a wide range of conductivities and radio frequencies. This paper has outlined the basic concepts underlying dielectrophoresis and described DNA collections on micro-fabricated interdigitated electrode arrays imaged using fluorescent microscopy. Subsequent analyses of DNA collections on electrode edges have revealed an exponential time dependence, and further, that collection rates decrease with increasing frequency. These observations concur with values of permittivity; motivating further analytical and experimental investigations.

Acknowledgements

The authors acknowledge Dr Y. Feldman and Dr I. Ermolina for the dielectric measurements.

References

[1] Olivera B M, Baine P and Davidson N 1964 *Biopolymers* **2**, 245-257
[2] Morishima K, Fukuda T, Arai F, Matsuura H and Yoshikawa K 1996 *Proc. IEEE Int. Conf. Robotics and Automation* 2214-9
[3] Yamamoto T, Washizu M, Kurosawa O and Shimamoto N 1998 *Proc. SPIE Int. Soc. Opt. Engineering USA* **3202** 228-36
[4] Adjari A and Prost J 1991 *Proc. Nat. Acad. Sci. USA* **88** 4468-71
[5] Bone S and Small C A 1995 *Biochim. Biophys. Acta* **1260** 85-93
[6] Mandel M 1977 *Ann. N. Y. Ac. Sci.* **303** 74-87
[7] Pedone F and Bonincontro A 1991 *Biochim. Biophys. Acta* **1073** 580-4
[8] Saif B, Mohr R K, Montrose C J and Litovitz T A 1991 *Biopolymers* **31** 1171-80
[9] Diekmann S Hillen W Jung M, Wells R D and Pörschke D 1982 *Biophys. Chem.* **15** 157-67
[10] Sambrook J, Fritsch E F and Maniatis T 1989 *Molecular cloning – a laboratory manual* (Cold Spring Harbor Lab. Press)
[11] Doi M and Edwards S F 1986 *The theory of polymer dynamics* (OUP)

Inst. Phys. Conf. Ser. No 163
Paper presented at the 10th Int. Conf., Cambridge, 28–31 March 1999
© *1999 IOP Publishing Ltd*

Electrostatic Properties of Metered Dose Inhalers.

Joanne Peart and Peter R. Byron.

Aerosol Research Group, Virginia Commonwealth University, Richmond, VA, USA.

Abstract. Electrostatic properties of the aerosols generated by chlorofluorocarbon (CFC) and hydrofluoroalkane (HFA) metered dose inhalers (MDIs) were characterized using an aerosol electrometer. Both Ventolin® and Airomir® MDIs conferred net electronegative charges on their fine particle clouds of the order of -160 pC, despite their different propellant systems, drug salt forms, drug concentrations and metering volumes. Furthermore, estimated electronic charges per aerosol drug particle ≈ 300 e; a value sufficient to influence regional lung deposition.

1. Introduction

Pressurized metered dose inhalers (pMDIs) are the most widely used inhalation device in the treatment of asthma [1]. pMDIs use a propellant to aerosolize a drug formulation (micronized drug suspended in the propellant together with a suitable surfactant, e.g. oleic acid) into a cloud of respirable particles [2]. Until 1995, all pMDIs contained chlorofluorocarbon (CFC) propellants, when the phase-out of CFCs challenged the pharmaceutical industry to reformulate and design replacement products with alternative propellants – the hydrofluoralkanes (HFAs) [3].

Aerosolisation is known to induce static charges by triboelectrification on individual aerosol particles or droplets [4], and as a consequence, aerosol electrostatics represents an important and emerging property of pharmaceutical aerosols. It has long been recognized that electrostatic charge can influence aerosol deposition in the respiratory tract [5-9]. Aerosol particles with diameters 0.3, 0.6 and 1.0 μm charged to between 50 and 100 electrons per particle demonstrated 15-30% increased upper airways deposition [5]. Other literature confirms that low particle charges are essential for enhanced pulmonary deposition [8]. Previous studies have indicated that the electrostatic properties of aerosols generated by dry powder inhalers were dependent upon both the inhaler and the formulation [10], and in some instances the charges were of sufficient magnitude to influence regional lung deposition, according to data reported in the literature [6-8]. This paper describes the quantification of electrostatic charges of the aerosols generated by CFC and HFA containing metered dose inhalers. More importantly, this study enables us to determine whether such charged particles would be expected to influence regional drug deposition in the respiratory tract.

2. Methods

Electrostatic properties of fine particles generated by chlorofluorocarbon (Ventolin® CFC) and hydrofluoroalkane (Airomir® HFA) containing metered dose inhalers (pMDIs) were reviewed. Ventolin® contains albuterol base suspended in CFC 11 and 12, with oleic acid present as a suspending agent, while Airomir® contains albuterol sulfate suspended in HFA 134a, with oleic acid and ethanol acting as suspending agents.

Electrostatic charges on the "fine particle" aerosol output were characterized using the apparatus shown diagrammatically in Figure 1. The MDI was inserted horizontally into the aerosol sampling apparatus, and actuated with an airflow passing through the apparatus $= 45$ l min^{-1}. The electrical currents induced by the collection of fine particles (aerodynamic diameters < 5.8 μm) in the aerosol electrometer (Figure 1: Model 3068, TSI Inc., St. Paul, MN) were measured as functions of time. The coarse aerosol fraction from each puff was collected in the throat and the coated impaction stage, identical in cut-off characteristics to the Marple Miller 5 μm aerodynamic cut-off stage (Model 160, cascade impactor, MSP Corporation, St. Paul, MN). At 45 l min^{-1}, the 50% cut-off diameter was calculated to be 5.8 μm.

The doses from actuation numbers 1 through 10 were drawn sequentially from each MDI. Following actuation 10, the apparatus was disassembled, drug was washed from the apparatus and assayed. The area under each individual aerosol current (pA) versus time (seconds) profile gave the "fine particle charge" in picocoulombs. The method of aerosol capture and charge determination precluded the direct analytical determination of the fine particle dose in all cases; where this was determined, it was calculated by subtracting drug collected in the throat and impaction stage from the "delivered dose". Five Ventolin® and five Airomir® inhalers were tested. All testing was performed in a controlled environment with temperatures between 20 and 23 °C and relative humidities ranging 40.4 to 48.3 %.

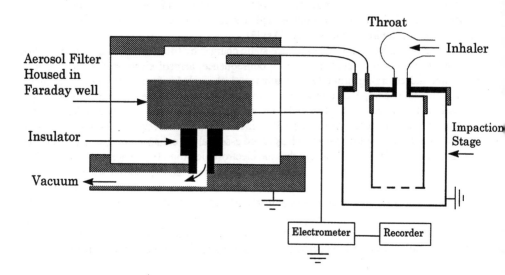

Figure 1: Aerosol electrometer and sampling apparatus.

3. Results and Discussion

Fine particle doses (as albuterol) from Ventolin® and Airomir® were determined to be 61.9 (±2.74) μg for Ventolin®, and 40.2 (±7.19) μg for Airomir®, respectively. These values appeared to be consistent with other literature reports [11].

Figure 2 shows the results for the mean aerosol current versus time profiles for aerosols generated by Ventolin® and Airomir® MDIs. The fine particle dose in both cases was highly, and net negatively charged (Ventolin® = -161.8 (± 1.8) pC, Airomir® = -167.7 (± 6.9) pC). It is important to recognize that these recorded charges result from the summation of the charges on individual particles, and may actually underestimate the total charges held by individual particles in the aerosol cloud. Notwithstanding this observation, the magnitude of these measured charges was significant. For example, if the albuterol fine particle aerosol generated from Ventolin® was assumed to be spherical and monodisperse, with an MMAD = 3 μm and density = 1.3 gcm^{-3}, a net charge of approximately 160 pC of a fine particle mass of 60 μg would correspond to approximately 300 electronic charges per single particle (electronic charge = -1.6 x 10^{-19}C). For Airomir, the estimated electronic charges per particle were higher, ≈ 490 e. According to the literature, charges per unit drug mass in the case of both Ventolin® and Airomir® were potentially capable of influencing regional deposition in the lung [6-8]. Interestingly, there was only a small, but significant difference between the fine particle dose charges of these aerosols (p=0.02), despite the major formulation differences between the products. A systematic investigation of the formulation components' contribution to the measured charge showed significant differences between the CFC and HFA propellant systems [12]. There is also compelling empirical evidence in the pharmaceutical aerosol literature to show the importance of spacer static charge on drug retention in plastic aerosol devices, used with metered dose inhalers [13-14]. But, it is also likely that the charged aerosol particles will themselves readily adhere to oppositely charged surfaces, or electrophoretically migrate towards surfaces.

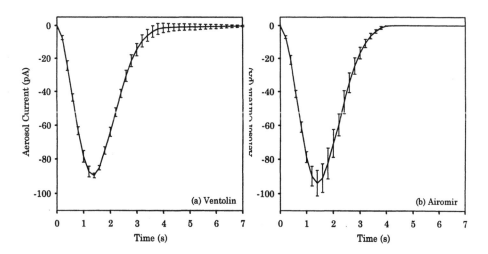

Figure 2: Mean aerosol current versus time profiles following actuation of (a) Ventolin® and (b) Airomir® metered dose inhalers (n=50; 10 actuations from each of 5 inhalers). Error bars represent sample standard deviations.

4. Conclusions

A method for characterizing the electrostatic charge associated with metered dose inhalers has been developed. Using this methodology, electrostatic properties of the fine particle clouds generated from marketed CFC and HFA albuterol inhalers were found to differ insignificantly. Highly charged electronegative aerosol clouds were developed following actuation, which could influence their *in-vivo* deposition pattern in the lung, according to data in the literature. Most of the classic models of aerosol deposition in the human lung are based on data collected following inhalation of "charge neutralized" aerosol particles [15]. As a consequence, many of the common assumptions concerning drug deposition from inhalers require that the particles are uncharged. Clearly, the aerosols, which are generated by some medical inhalers, are highly charged, and electrostatic charge cannot be ignored. Electrostatics is emerging as an important feature of pharmaceutical aerosols, with research encompassing the following areas: bulk handling procedures for raw materials, formulation and inhaler design issues, deposition in the respiratory tract, *in-vitro* testing aspects of inhalers and the clinical consequences of static charge associated with drug retention on spacer devices used with MDIs.

References

1. Clark, A.R. 1995. Aerosol Science and Technology, **22**, 374-391.
2. Byron, P.R. 1990. In: Byron, P.R. (Ed.), Respiratory Drug Delivery, CRC Press: Boca Raton, FL, pp 167 - 205.
3. Tansey, I.P. 1997. The Pharmaceutical Journal, **259**, 896-898.
4. Hinds, W.C. 1982. Wiley Interscience: New York.
5. Longley, M.Y. 1960. American Industrial Hygiene Association Journal, **21**, 187-194.
6. Melandri, C., Prodi, V., Tarroni, G., Formignani, M., De Zaiacomo, T., Bompane, G.F. and Maestri, G. 1977. In: W.H. Walton (Ed.). Inhaled particles IV. Pergammon Press: Oxford, pp 193-201.
7. Melandri, C., Tarroni, G., De Zaiacomo, T., Formignani, M. and Lombardi, C.C. 1983. Journal of Aerosol Science, **14**, 657-669.
8. Balachandran, W., Ahmad, C.N. and Barton, S.A. 1991. Institute of Physics Conference Series, 118, 57-62.
9. Cohen, B.S., Xiong, J.Q., Fang, C and Lei, W. 1998. Health Physics, **74**, 554-560.
10. Byron, P.R., Peart, J. and Staniforth, J.N. 1997. Pharmaceutical Research, **14**, 698-705.
11. Leach, C. 1996. In: Dalby, R.N., Byron, P.R. and Farr, S.J. (Eds.), Respiratory Drug Delivery V, Interpharm Press: Buffalo Grove, IL, pp 133 - 144.
12. Peart, J., Magyar, C. and Byron, P.R. 1998. In: Dalby, R.N., Byron, P.R. and Farr, S.J. (Eds.), Respiratory Drug Delivery VI, Interpharm Press: Buffalo Grove, IL, pp 227-233.
13. Barry, P.W. and O'Callaghan, C. 1995. British Journal of Clinical Pharmacology, **40**, 76-78.
14. N.J. Dewsbury, C.J. Kenyon and S.P. Newman 1996. International Journal of Pharmaceutics, **137**, 261-264.
15. Task Group on Lung Dynamics 1966. Health Physics, **12**, 173-207.

Inst. Phys. Conf. Ser. No 163
Paper presented at the 10th Int. Conf., Cambridge, 28–31 March 1999

Dielectrophoretic measurements of sub-micrometre latex particles following surface modification

Michael P. Hughes[+], Mary F. Flynn and Hywel Morgan[*]

Bioelectronics Research Centre, University of Glasgow, Glasgow G12 8QQ, UK

Abstract. Dielectrophoresis is the term given to the force induced on a polarisable particle when subjected to a non-uniform AC electric field. Dielectrophoresis has been used as an analytical technique to determine the properties of particles such as cells and bacteria, but only recently has it been applied to particles of significantly less than one micrometre in size. The dielectrophoretic force can be towards (positive) or away from (negative) high field regions. The sign and magnitude of the force on the particle varies with medium conductivity and frequency. For a particular frequency and medium conductivity the force can go to zero. For latex spheres it has been shown that measurements of the variation of the zero-force frequency as a function of medium conductivity can be used to characterise the dielectric properties of single particles.

The dielectrophoretic properties of sub-micrometre latex spheres are dominated by surface charge effects. In this paper the dielectrophoretic properties of sub-micrometre latex spheres before and after chemical modification of the surface are presented. It is shown that immobilising an antibody onto the bead surface gives rise to a reduction in the surface charge density of the beads. This change translates into a modified DEP zero-force spectrum. For the case of 216nm-diameter carboxylate-modified latex spheres, surface charge density decreases by approximately 50% due to surface modification by 1-Ethyl-3-(3-dimethylaminopropyl)carbodiimide, and in total by 80% following the binding of mouse polyclonal IgG. Binding of a secondary antibody (goat anti-mouse IgG) to this surface caused a further reduction in surface charge density of between 0% and 100%.

1. Introduction

Dielectrophoresis (DEP), [1] has been demonstrated to be an effective means of manipulating and characterising polarisable particles in solution. Recent advances in fabrication methods, such as the use of electron beam lithography, means that DEP can now be used to manipulate sub-micrometre particles such as viruses [2-4] and latex spheres [5].

The magnitude and direction of the dielectrophoretic force is controlled by the gradient of the electric field and the polarisability of the particle and suspending medium, which in turn depends on the frequency of the applied field. Thus the magnitude and/or direction of

---

*Author to whom correspondence should be addressed
[+] Current address: European Institute of Health and Medical Sciences, University of Surrey, Guildford, Surrey GU2 5RF, UK

force exerted on a particle can vary with the frequency so that the particle is attracted to regions of high electric field (positive DEP), or repelled from them (negative DEP).

For a solid homogeneous particle undergoing a single interfacial relaxation process, the DEP force changes sign at a particular characteristic *crossover* frequency. This can be measured and analysed to estimate the dielectrophoretic properties of a particle.

In this paper it is shown that the dielectrophoretic behaviour of 216nm diameter fluorescent latex spheres is largely governed by the chemical composition of the surface. Similar effects have been noted following electrorotation measurements of 0.8μm diameter latex beads [6]. The dielectrophoretic properties of the spheres were measured for the following cases; (i) beads as supplied by the manufacturer; (ii) after surface activation (for protein attachment) with EDAC (1-ethyl-3-(3-dimethylaminopropyl) carbodiimide), (iii) after immobilisation of a primary antibody and (iv) after binding a secondary antibody. From simple analysis it is shown that the surface charge density on the particle varies with surface functionality.

2. Materials and Methods

2.1 Experimental

Electrodes of a "castellated" design with square dimensions of 10μm along all faces were manufactured on glass using a layer of 100nm Au sandwiched between two layers of 10nm Ti. Electrodes were powered using a Hewlett-Packard signal generator providing 5V peak to peak sinusoidal signals over the frequency range 1Hz-20MHz. Experiments were observed using a Nikon Microphot. An aliquot of particles was suspended in KCl solution of appropriate conductivity at a concentration of 10^8 particles/ml. Approximately 25μl of the particle solution was placed into the electrode chamber and the assembly sealed with a coverslip.

2.2 Latex spheres: Protein Coupling

Fluorescent, carboxylate-modified latex spheres, 216nm diameter (Molecular Probes, USA), were activated for protein coupling according to the procedure outlined in [7]. Nanosphere concentration was estimated using light scattering and the protein content was estimated spectrophotometrically [7] and found to be 30 ±10 μg/ml, equivalent to approximately 1200 IgG molecules per sphere, indicative of monolayer coverage. A fluorescently labelled (TRITC) secondary antibody was coupled to the primary immobilised antibody by suspending the spheres in a solution of goat anti-mouse IgG in ultra pure water. These steps are outlined in figure 1. Subsequent to binding the secondary antibody, the red fluorescence of the beads was measured as an indication of binding efficiency.

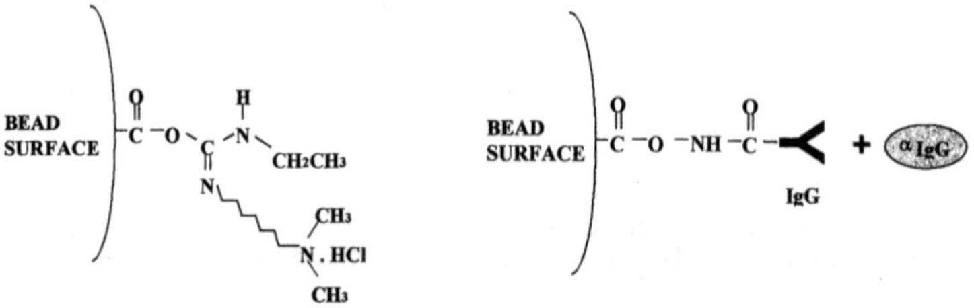

Figure 1. Diagram illustrating the sequence of antibody binding to the surface of a carboxylate-rich latex sphere. Left-hand diagram shows EDAC coupled to the carbonyl. Right hand diagram shows amide bond between protein lysine group and surface carboxyl group.

3. Results and Discussion

The dielectrophoretic force, F_{DEP}, acting on a homogeneous, isotropic dielectric sphere, is given by:

$$F_{DEP} = 2\pi r^3 \varepsilon_m Re[K(\omega)]\nabla E^2 \qquad (1)$$

where r is the particle radius, ε_m is the permittivity of the suspending medium, E the local *rms* electric field and $Re[K(\omega)]$ the real part of the Clausius-Mossotti factor, given by:

$$K(\omega) = \frac{\varepsilon_p^* - \varepsilon_m^*}{\varepsilon_p^* + 2\varepsilon_m^*} \qquad (2)$$

where ε_m^* and ε_p^* are the complex permittivities of the medium and particle respectively.

The effective conductivity of a latex particle can be written as the sum of the conductivity *through* the particle (the bulk conductivity σ_{pbulk}) and the conductivity *around* the particle due to the surface conductance K_s, [8] so that the total conductivity, σ_p is:

$$\sigma_p = \sigma_{pbulk} + \frac{2K_s}{r} \qquad (3)$$

For latex spheres, the bulk conductivity is negligible, so that the effective conductivity of the microsphere is dominated by surface conductance. The dielectric parameters of solid particles can thus be obtained through measurements of the cross-over frequency as a function of medium conductivity and analysed using equations (1) to (3).

The cross-over frequencies for the four different types of spheres, plotted as a function of electrolyte conductivity, are shown in figure 2. The vertical lines indicate the ranges over which crossovers were observed for a population of many thousands of particles. It can be seen that at low conductivities, the cross-over frequency increases slightly with medium conductivity but at a particular threshold conductivity there is step reduction in cross-over frequency where theoretical predictions indicate that the particle should only experience negative DEP. This is not observed experimentally due to the onset of the relaxation of the double-layer.

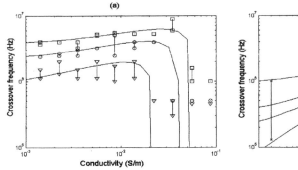
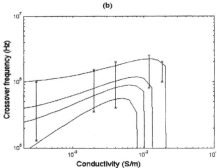

Figure 2 (a). Experimental crossover spectra for untreated beads (□), EDAC-activated beads (○) and beads functionalised with IgG (▽). Best-fit lines were calculated using equations (2) and (4) with K_s = 1.2nS, 0.7nS and 0.25nS respectively. **(b).** Experimentally observed crossover frequency for beads complexed with a secondary antibody. Best-fit lines are generated using equations (2) and (4) with values of (from top) 0.3nS, 0.1nS, 0.03nS, 0nS. The wide variation in binding is reflected in the distribution in cross-over frequencies.

Equations (2) and (3) predict a constant crossover frequency vs. conductivity relationship. However, the results indicate that the particle's net conductivity rises with medium conductivity. An approximate fit to the data can be made by adding an additional conductivity component to the surface charge which can be written as follows:

$$\sigma_p = \sigma_{pbulk} + \frac{2K_s}{r} + \frac{2\sigma_m}{\sqrt{\kappa r}} \tag{4}$$

where κ^{-1} is the Debye length. This equation can be successfully used to predict the DEP responses of latex spheres with diameters in the range 93-557nm, and also Herpes viruses [9] suspended in KCl solutions.

The experimental results were modelled by means of equations (2) and (4). For each type of sphere the best fits are shown in figure 2 with the following surface conductances: untreated beads 1.2nS, the EDAC-activated beads 0.7nS and primary-labelled beads, 0.25nS. This last value is approximate, as the magnitude of force on the labelled beads is relatively small compared with the unlabelled beads.

Following secondary antibody coupling, a broad range of crossover frequencies was observed. (e.g. 125kHz-2MHz in pure water), indicative of a wide range of surface conductances, summarised in figure 2(b). The maximum estimated value was 0.3nS (presumably no secondary binding) to approximately zero (maximum surface coverage of secondary antibody).

4. Conclusions

The dielectrophoretic behaviour of 216nm-diameter latex microspheres has been characterised by measuring the crossover frequency of the particles as a function of medium conductivity. It is shown that the surface conductance is reduced following surface modifications. For untreated beads, the surface conductance is shown to be 1.2nS; with EDAC activation this drops to 0.7nS and with IgG coverage to 0.25nS. Secondary antibody labelling causes the surface conductance to drop to between 0.25nS to close to zero depending on the coupling efficiency.

Acknowledgements

This work is supported by the Biotechnology and Biological Sciences Research Council (UK) grant no. 17/T05315

References

[1] Jones, TB 1995 Electromechanics of particles (Cambridge University Press)
[2] Müller T, Gerardino A, Schnelle T, Shirley SG, Bordoni F, De Gasperis G, Leoni R, Fuhr G 1996 *J. Phys. D: Appl. Phys.***29** 340-349
[3] Morgan H and Green NG 1997 *J. Electrostatics,* **42** 279-293
[4] Hughes MP, Morgan H, Rixon FJ, Burt JPH and Pethig R 1998 *Biochim. Biophys. Acta.***1425** 119-126
[5] Green NG and Morgan H 1999 *J. Phys. Chem B.* **103** 41-50
[6] Burt JPH, Chan KL Dawson D, Parton A and Pethig R. 1996 *Ann. Biol. Clin.* **54** 253-257
[7] Hughes MP and Morgan H 1999 *Anal. Chem.* Submitted
[8] Arnold WM, Schwan HP and Zimmermann U 1987 *J. Phys. Chem.* **91** 5093-5098
[9] Hughes MP, Green NG and Morgan H Unpublished work

Inst. Phys. Conf. Ser. No 163
Paper presented at the 10th Int. Conf., Cambridge, 28–31 March 1999
© *1999 IOP Publishing Ltd*

Electrorotation of oocysts of *Cyclospora cayetanensis*

C Dalton[1], R Pethig[1], C A Paton[2] and H V Smith[2]

[1]Institute of Molecular and Biomolecular Electronics, University of Wales, Dean Street, Bangor, Gwynedd, LL57 1UT, UK

[2]Scottish Parasite Diagnostic Laboratory, Stobhill NHS Trust Hospital, Springburn, Glasgow G21 3UW, UK

Abstract. Electrorotation is sensitive to the dielectric properties of microbiological organisms. Using a microfabricated planar gold electrode structure, we investigated the electrorotational properties of oocysts of the emerging coccidian pathogen, *Cyclospora cayetanensis*. We present differences in the electrorotational response of sporulated and unsporulated oocysts. Variations in the responses of morphologically similar oocysts, which are characteristic of a difference in oocyst viability, are also described. This rapid, non-invasive technique has potential applications in diagnostics.

1. Introduction

Cyclospora cayetanensis is a coccidian pathogen which causes prolonged diarrhoeal disease in both immunocompetant and immunocompromised hosts. Waterborne and foodborne routes have been implicated as likely routes of transmission. *Cyclospora* was first reported in humans by Ashford in 1979 [1] and is resilient to disinfection regimes used in water treatment. Viable, sporulated oocysts are responsible for the transmission of infection, and are not readily detected by current methods used to assure the safety of water supplies. Thus, distinguishing between non-viable and viable oocysts is of great significance to a variety of professionals, including epidemiologists, public health officials, and water industry personnel. The oocysts are remarkably uniform in size and shape (spherical, 8-10 μm diameter), making them ideal for electrorotational studies.

2. Electrorotation

As described by Pohl in 1978 [2], non-uniform (inhomogeneous) fields can induce a torque on a particle suspended in an electrolyte, causing it to spin. Arnold and Zimmerman [3] and Mischel *et al.* [4] first reported methods to induce controlled cellular spin by using rotating electric fields. A uniform rotating electric field can be generated by energising four electrodes with sinusoidal voltages, with 90° phase difference between adjacent electrodes (*figure 1*). The rate and direction of the induced electrorotation is determined by the imaginary component of the dipole moment *m* of the particle, which is itself dependant on the dielectric properties of the system as a whole. The rotational torque $\Gamma(\omega)$ exerted on a particle by the rotating electric field is given by :

$$\Gamma(\omega) = -\text{Im}\{m(\omega)\}E$$

where *Im* indicates the imaginary component of the dipole moment *m* and *E* is the electric field strength. If the imaginary component of *m* is positive, then the torque exerted will be negative causing the particle to rotate in a direction that *opposes* that of the rotating field (*figure 1b*).

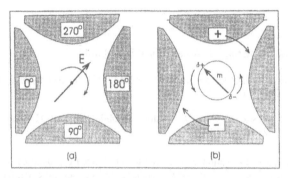

Figure 1 - (a) Rotating field generated by applying sinusoidal voltages to 4 electrodes, with phases spaced 90° apart. (b) The rotational torque acting on the particle will be either co- or anti- field, depending on the angle of the induced dipole moment. In the case represented here, the particle is rotating counter to the clockwise rotating field.

For biological particles, such as oocysts of *C. cayetanensis*, structures such as the oocyst wall and plasma membrane, along with the cytoplasm and its contents, each contribute to the overall electrical permittivity and conductivity of the particle.

3. Equipment and Materials

The electrodes were deposited by photolithography onto glass microscope slides, with a chromium adhesion layer of 5 nm and a gold layer of 70 nm. The inter-electrode analysis chamber was 2 mm in diameter. The electric fields were generated by a four phase sinusoidal signal generator, and applied to the electrodes in the frequency range 100 Hz to 10 MHz. Oocysts of *C. cayetanensis* were supplied by the Scottish Parasite Diagnostic Laboratory (SPDL), in either 2% potassium dichromate solution, or water. Oocysts were washed in ultra pure water and re-suspended in dilute solutions of normal phosphate buffered saline (Sigma Chemical Co.), which typically had conductivites of 0.6-1.0 mS/m, to reduce electrolysis and heat production at the electrodes. Sample results were visualised through a Nikon Optiphot 2 microscope using phase contrast microscopy with a magnification of x40 and a zoom of x2.25. Results were recorded to video tape for subsequent analysis using a JVC colour video camera head (model TK-1280E) attached to the microscope.

4. Results and Observations

Over 100 electrorotation spectra were obtained which facilitated the differentiation of sporulated and unsporulated oocysts of *C. cayetanensis*. Unsporulated oocysts rotate at a higher rate than sporulated oocysts (*figure 2*). The frequency at which the direction of rotation reverses from anti-field to co-field at high frequency (the so called cross-over frequency), is significantly different between sporulated and unsporulated oocysts. Thus, at 1MHz the unsporulated oocysts will rotate in the opposite direction to the sporulated oocysts, allowing rapid discrimination between the two types of oocyst.

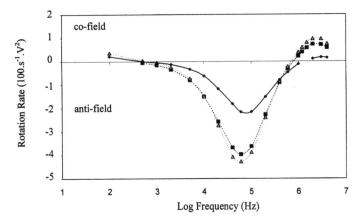

Figure 2 – Rotation spectra of unsporulated oocysts (dashed lines), which rotate at a higher rate than sporulated oocysts (solid line).

It is not possible to distinguish morphologically between viable and non-viable oocysts containing similar morula. Non-viable and viable oocysts can be either sporulated or unsporulated. The membrane of a viable oocyst is semi-permeable (selectively) to ions and non-lipid, soluble molecules. Upon cell death, membrane integrity is lost, and the cell freely exchanges material with the external suspending medium. This greatly affects the rotation spectra produced, as less induced charge differences occur between the inside of the oocyst, and the exterior medium. The peak rotation frequency was found to vary considerably for visibly similar, particulate free, unsporulated oocysts of *C. cayetanensis* (*figure 3*).

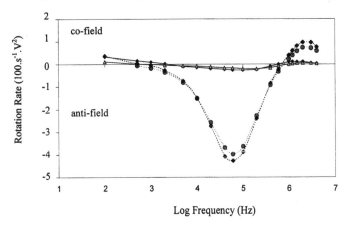

Figure 3 – Rotation spectra of unsporulated *C. cayetanensis* oocysts showing the difference between viable (dashed lines) and possibly non-viable (solid lines) oocysts.

This difference in the magnitude of rotation rate is similar to that reported by Goater *et al.* [5] for *Cryptosporidium parvum*, a smaller, but similar pathogen to *C. cayetanensis*. Goater confirmed that this observation was due to differences in oocyst viability by fluorogenic vital dye assay [6]. The viability of *C. cayetanensis* oocysts cannot be independently verified by such methods, because as yet none have been developed. Excystation of the oocysts at SPDL confirmed that 75% of the sporulated oocysts from the sample were viable.

88

The effect of small amounts of debris on the oocyst wall was investigated. The peak rotation frequency, and the rotation cross-over frequency values, for the debris-coated and uncoated oocysts, all remained the same (*figure 4*). The only difference was a slower rotation rate for the debris-coated oocysts. This can be attributed to viscous drag caused by the debris.

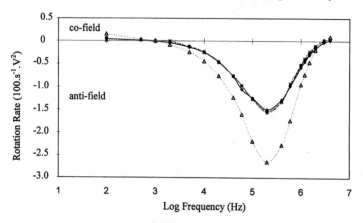

Figure 4 – Rotation spectra of uncoated (dashed line) and debris-coated (solid lines) sporulated oocysts of *C. cayetanensis*.

5. Conclusions

In this work, the sensitivity of the electrorotation assay in determining characteristic differences between morphologically indistinguishable oocysts of the pathogen *Cyclospora cayetanensis* has been shown. An important result has been the apparent ability of the assay to distinguish between viable and non-viable oocysts, previously not possible. As electrorotation is a non-invasive process, the undamaged oocysts can be tested further by other techniques in the hope of determining the life-cycle and modes of transmission.

6. Acknowledgements

We gratefully acknowledge the support of the BBSRC and NFCR, and thank A D Goater for valuable discussions and J Tame for technical assistance.

References

[1]   Ashford R W 1979 *Annals of Tropical Medicine and Parasitology* **73** 497-500

[2]   Pohl H A 1978 *Dielectrophoresis* (Cambridge: Cambridge University Press)

[3]   Arnold W M and Zimmerman U 1988 *J. Electrostatics* **21** 151-191

[4]   Mischel M, Voss A and Pohl H A 1982 *J. Biol. Phys.* **10** 223-226

[5]   Goater A D, Burt, J P H and Pethig R 1997 *J. Phys. D : Appl. Phys.* **30** L65-L69

[6]   Goater A D *personal communication*

Inst. Phys. Conf. Ser. No 163
Paper presented at the 10th Int. Conf., Cambridge, 28–31 March 1999

Sub-micrometre ac electrokinetics: particle dynamics under the influence of dielectrophoresis and electrohydrodynamics

Nicolas G Green[1], Antonio Ramos[1], Hywel Morgan[2], Antonio Castellanos[1]

[1] Departamento de Electronica y Electromagnetismo, Facultad de Fisica, Universidad de Sevilla, Avda Reina Mercedes s/n, 41012 Sevilla, Spain
[2] Department of Electronics and Electrical Engineering, University of Glasgow, Rankine Building, Oakfield Avenue, Glasgow G12 8LT, Scotland, UK

Abstract. The term AC Electrokinetics refers to the use of AC electric fields for the movement of particle. Dielectrophoresis is the movement of a polarisable particle in a non-uniform field due to the interaction of the field and the induce dipole. Electrohydrodynamics is the study of the interaction between electric fields and fluid media. This paper discusses dielectrophoresis as it applies to sub-micrometre particles suspended in the ionic media required for biological material. The presence of the double layer in such media complicates the dielectric properties of the particle and can also lead to fluid motion around the electrodes. The dynamics of sub-micrometre particles are also discussed with particular attention being paid to Brownian motion and an *observable force*. The characterisation, manipulation and separation of particles is briefly reviewed and discussed.

1. Introduction

AC Electrokinetics refers to the movement of particles under the influence of AC electric fields. The subject covers a wide range of direct effects such as electrophoresis (which occurs at low frequencies), dielectrophoresis and electrorotation. In addition the effects of the field on the fluid and indirectly on the particle through viscous drag have recently been examined.

Recently dielectrophoretic (DEP) characterisation and separation methods [1-4] have conclusively been demonstrated for sub-micrometre particles [5-10]. In addition, electric field driven fluid flow has been used in combination with dielectrophoresis for the separation of sub-micrometre particles [11]. Electrokinetic techniques have found particular application to biological and medical fields where non-contact and non-destructive methods are useful. The potential for the application of these techniques to biological sub-micrometre particles such as viruses, DNA and protein macromolecules has been demonstrated [7,8,12] and successful practical technologies would be of tremendous benefit.

2. Experimental

AC Electrokinetic experiments are generally performed on microelectrode arrays manufactured on planar substrates using semi-conductor manufacturing techniques [13]. The electric fields necessary for sub-micrometre particle manipulation must be of the order 10^5 –

10^6 Vm^{-1}, fields which can be generated from low potentials using microelectrodes. The fields also generally have quite complicated geometries, which are generated using precisely defined shapes of electrodes. Using electron beam lithography electrodes can be fabricated with feature sizes down to a micrometre and resolutions of down to 50nm [13]. The high strength electric fields have the unavoidable effect of heating the fluid which can result in fluid flow.

3. Forces on sub-micrometre particles

When an electric field is applied to a dielectric such as a particle and the fluid medium it is suspended in, the dielectric polarises [1]. In general the dielectric properties are frequency dependent. For the particular case of a dielectric particle, the polarisation of the particle and the medium give rise to an induced dipole at the interface between the two dielectrics. The interaction of this dipole and a non-uniform electric field gives rise to movement, which is referred to as dielectrophoresis [1]. For a spherical particle, the dielectrophoretic force is given by:

$$\langle \mathbf{F}_{DEP} \rangle = \pi a^3 \varepsilon_m \left(\frac{\varepsilon_p^* - \varepsilon_m^*}{\varepsilon_p^* + 2\varepsilon_m^*} \right) \nabla |\mathbf{E}|^2$$

where a is the radius, ε_m is the permittivity of the medium and $\mathbf{E}$ is an electric field of single frequency ω. The subscripts m and p refer to the medium and the particle respectively and ε^* is the complex permittivity given by $\varepsilon^* = \varepsilon - i\sigma/\omega$, where ε is the permittivity and σ is the conductivity. The term in brackets is referred to as the Clausius-Mossotti factor and describes the frequency variation of the dielectrophoretic force. For a homogeneous charged sphere, the Clausius-Mossotti factor is positive at low frequencies and the dielectrophoretic force acts towards high field regions. As the frequency is increased, the dipole relaxes, the Clausius-Mossotti factor becomes negative and the force acts towards low field regions. For biological particles, where the physical structure can be modelled using multiple concentric shells, this behaviour is more complicated and unique for a given particle type. As a result, particle types can be identified and mixtures of particle separated by controlling the frequency of the applied field [1-4].

For charged particles, the charge on the surface results in the particle being surrounded by an electrical double layer [14]. The double layer also polarises under the influence of the electric field resulting in another relaxation term in the dipole moment of the particle. In general, the effects of this component of the dipole moment are larger than for the polarisation of the interface and the frequency of the relaxation is lower [14,15]. The effects of the double layer polarisation on the dielectrophoretic properties of particles are currently being examined, with several experimental papers already published [6,9].

In addition to the electrical forces, a sub-micrometre particle is heavily affected by thermal effects and the medium it is suspended in. A concentration of particles experiences a force acting outwards (diffusion) but this effect can be ignored if the particles are sufficiently isolated. Here, only single particles are considered and the effects of random Brownian motion [14,16]. For a single particle moving in one dimension, the expected behaviour of the particle over a time interval dt was given by Einstein [16] in terms of a Gaussian probability distribution in position around mean zero with:

$$\langle |\mathbf{x}|^2 \rangle = 2Ddt$$

where $D = kT/f$ is the diffusion coefficient of the particle, k Boltzman's constant, T is the temperature and f is the friction factor of the particle. If the particle experiences a constant force **F**, it will accelerate up to a terminal velocity **v** given by Stokes as $v = F/f$. For biological particles like cells and viruses and sub-micrometre particles in general, the acceleration phase lasts less than 10^{-6} seconds. Such particles can be assumed to always move at the terminal velocity. An *observable force* can be defined in terms of the threshold value required to *see* the effects of the force against the background noise of Brownian motion. In other words, if the particle is observed to move a certain distance, what is the probability that the force rather than Brownian motion is producing the movement? The standard deviation of the Brownian motion distribution is $\sigma_x = \sqrt{(2Ddt)}$ in one dimension. The threshold observable force is defined in terms of the force required to move the particle three standard deviations in time dt. This is shown schematically in figure 1.

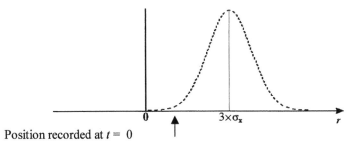

Position recorded at $t = 0$

Figure 1. A displacement of dx $= 3 \times \sigma_x$ along the direction of the force against the Gaussian probability distribution produced by Brownian motion.

Comparing $dx = 3 \times \sigma_x$ and $dx = |F| \, dt/f$, the relationship between force and time of observation is:

$$|F|\sqrt{dt} = \sqrt{18kTf}$$

This gives the threshold force for an observer to be 99.7% certain that the movement of a particle over a single time interval dt is due to the applied force. As the length of the time interval is increased the threshold force decreases and, since the friction factor is proportional to size in some manner, the threshold force decreases as the size decreases. This idea is only to calculate the order of magnitude of the force required. Previous publications have demonstrated that the force required to dielectrophoretically manipulate sub-micrometre particles is much smaller than has previously been thought [8]. The particle being manipulated in this case was a plant virus and the experimentally determined threshold force was $\sim 6 \times 10^{-16}$ N. Using an ellipsoid model for the virus to calculate the friction factor, the threshold force using the above equation was 1.75×10^{-15} N $- 6 \times 10^{-15}$ N for an observation time of $1 - 10$ seconds.

In addition to the effect of Brownian motion, electrically induced fluid flow (Electrohydrodynamics) is common in microelectrode structures used for dielectrophoresis. There are two important types of fluid flow observed in the non-uniform fields: Electrothermal and Electroosmotic. The electrothermal force arises from the localised electrical heating of the medium, which results in local changes in permittivity and conductivity [17]. The electric field then acts on those gradients producing forces on the fluid and in turn a drag force on the particle [17]. Electroosmosis arises from the interaction of the double layer at an electrode surface and a tangential DC electric field [18]. As discussed in other publications [17] and in another paper in these proceedings [19], in non-uniform AC fields, the field generating the double layer has a tangential component above the electrode surface. As a result, there is a fluid flow with a continuous direction acting across the electrode surface from the edge [19]. This is referred to as AC electroosmosis and is significantly stronger than electrothermal effects. It is also frequency dependent and only observed below the charge relaxation frequency [17]. This effect is has been used in combination with dielectrophoresis for the separation of sub-micrometre latex spheres [14].

4. Conclusions and prospects

AC electrokinetic techniques have been demonstrated to be extremely effective for the characterisation, manipulation and separation of particles, particularly biological particles. The techniques have been successfully applied to the sub-micrometre scale despite the effects of Brownian motion and electrically induced fluid flow. The electrically induced fluid flow has also been used for practical purposes, in combination with dielectrophoresis, for the separation of sub-micrometre particles. Since the techniques are non-contact and generally non-destructive, there are many potential applications in fine chemical and medical areas and there are good indications that they will become successful Biotechnological tools.

Acknowledgements

The authors would like to acknowledge the European Union for the award of a Marie Curie fellowship to N.G. Green (contract no. BIO4-CT98-5010 (DG12-SSMI)) and the DGES (Spain) (contract no. PB96-1375).

References

[1]   Pohl H A 1978 *Dielectrophoresis* (New York: Cambridge University Press)
[2]   Pohl H A and Crane J S 1971 *Biophys. J* **11** 711-727
[3]   Pethig R, Huang Y, Wang X-B and Burt J P H 1992 *J Phys D: Appl Phys* **24** 881-888
[4]   Gascoyne P R C, Huang Y, Pethig R, Vykoukal J and Becker F F 1992 *Meas. Sci. Tech* **3** 439-445
[5]   Green N G and Morgan H 1997 *J. Phys. D: Appl. Phys* **30** L41-L44
[6]   Green N G and Morgan H 1997 *J. Phys. D: Appl. Phys* **30** 2626-2633
[7]   Green N G, Morgan H and Milner J J 1997 *J. Biochem. Biophys. Methods* **35** 89-102
[8]   Morgan H and Green N G 1997 *J. Electrostatics* **42** 279-293
[9]   Green N G and Morgan H 1999 *J. Phys. Chem.* B **103** 41-50
[10] Hughes M P and Morgan H 1998 *J. Phys. D: Appl. Phys* **31** 2205-2210
[11] Green N G and Morgan H 1998 *J. Phys. D: Appl. Phys* **31** L25-L30
[12] Washizu M, Suzuki S, Kurosawa O, Nishizaka T and Shinohara T 1994 *IEEE Trans. Ind. Appl.* **30** 835-843
[13] Green N G and Morgan H 1998 *J.Electrostatics* (submitted)
[14] Lyklema J 1991 *Fundamentals of Interface and Colloid Science* (London: Academic Press)
[15] Schwan H P, Schwarz G, Macjuk J and Pauly H 1962 *J. Phys. Chem.* **66** 2626-2635
[16] Einstein A 1905 *Ann. Phys. Lpz.* **17** 279-293
[17] Ramos A, Morgan H, Green N G and Castellanos A 1998 *J. Phys. D: Appl. Phys* **31** 2338-2353
[18] Grossman P D and Colburn J C (eds.) 1992 *Capillary Electrophoresis* (New York: Academic Press)
[19] Ramos A, Castellanos A, Gonzalez A, Morgan H and Green N G 1999 *Electrostatics '99*

DISCUSSION - Section B - Bioelectrostatics

Title: Positioning, Levitation and Separation of Biological Cells
Speaker: W M Arnold

Question: Martin Taylor
Is there any significance in the fact that the yeast cells appear to collect in pairs?

Reply: In a rapidly-dividing yeast culture, a majority of the cells normally remain as doublets after division. Under growth during levitation, cells remain together as much larger groups, presumably because there are no shearing forces.

Question: Mary-Frances Flynn
Could the effect be used to coagulate highly colloidal particles? Have there been any numerical indications of a lower mass limit for which this effect is useful?

Reply: Yes it could, subject to the particles showing negative DEP and fast enough sedimentation to remain in stable levitation.

Question: D K Davies
Hybrid generation by fusion of cells in contact under high fields has been described at a previous conference. Is this possible under your conditions?

Reply: Certainly, and the ability to trap single cells would be very useful for making specific hybrids.

Question: T B Jones
Can the levitation /convection scheme be used on plant protoplasts which are more fragile than yeast?

Reply: I have not tried this, but there should be no problem.

Question: Hywel Morgan
Can you suggest how the cells which are concentrated in the centre of the electrode array could be removed from the system?

Reply: The cells can be removed by suction pipette. This could be done continuously if necessary. The "raft" forms at a reproducible location, therefore the removal could be done without adjustments being necessary.

Title: Single Cryptosporidium Oocyst Isolation and Capture using a Travelling Wave Dielectrophoresis Device
Speaker: A D Goater

Question:	W M Arnold

How large can you make such a device bearing in mind the field drop-off towards the centre due to tissues in the electrodes?

Reply: In the prototype device described with gold electrodes, there was little reduction in the observed oocyst velocity from the periphery to the centre of the electrode array. On increasing the electrode area for use with larger volumes or lower particle concentration any field or drop-off could be minimised by increasing the electrode thickness or by increasing the applied voltages. Alternatively using multilayer fabrication with via holes and busbars there would be no limit to the size of the device.

Title: Characterisation of the Dielectrophoretic Movement of DNA in Micro Fabricated Structures

Speaker: D J G Bakewell

Question: A G Bailey

How important is the suspending liquid conductivity bearing in mind that high fields may result in joule heating?

Reply: Thank you for your question. The conductivity of the suspending liquid is very important for at least two reasons. As you correctly point out, finite conductivity results in joule heating from high electric fields. High conductivity also compromises the DEP force by two mechanisms. (i) It limits the maximum applied voltage, for a chosen frequency (or limits the minimum allowable frequency for a chosen voltage) for hydrolysis not to occur - hydrolysis invariably damages the electrodes. (ii) Generally, a high conductivity for positive DEP, lowers the DEP force as evident from the Claussius-Mosotti factor.

Question: Allel Bouziane

The presence of DNA modifies the electric field $E(x,t)$. Do you take this into account in your model?

Reply: As a <u>first</u> step in modelling the spatial - temporal evaluation of DNA concentration, we did <u>not</u> include modifications, or changes to the electric field as the DNA accumulates on the electrode edges. This was because we wanted to keep our model simple, at the outset. Your point is very pertinent, and it is our intention to include the above mentioned changes to the electric field as we refine our model. Thank you for your question.

Question: W M Arnold

What are the units of the DNA dielectric decrement? OR what was the DNA concentration?

Reply: The volume faction of DNA in solution was estimated to be approximately 1%.

Title: Electrostatic Properties of Metered Dose Inhalers
Speaker: J Peart

Question: S Gerard Jennings
Did you measure particle size or was it assumed?

Reply: We measured particle size using laser diffraction and found it to be 2.69 μm in diameter.

Question: Martin Taylor
There is an assumption made that the fine particles passing through the aerosol electrometer do not undergo further tribocharging - any comment?

Reply: That is correct, it is assumed that the fine particles passing through the aerosol electrometer do not undergo further tribocharging.

Question: A G Bailey
Semiconducting or conducting spacer devices made of treated materials (as in the microelectronics industry) which increase conductivity can help to reduce charged droplet loss.

Reply: The significance of charge on inhaled particles with respect to deposition does not depend solely on charge level. Particle size and inhalation flow rate are just as important.

Section C

Atomisation
and EHD

Inst. Phys. Conf. Ser. No 163
Paper presented at the 10th Int. Conf., Cambridge, 28–31 March 1999

Electrostatic atomization - questions and challenges

A J Kelly

Charged Injection Corporation, 11 Deer Park Drive, Monmouth Junction, NJ 08852, USA

Abstract. Electrostatic fuel preparation represents enabling technology for steady-state, low flow rate burners, and radically improves mixing in a diesel engines. A crucial need for charged plume patternation and combustion modeling exists if this technology is to be widely accepted and applied. Additional work is required to fully describe the behavior of high charge density sprays particularly in the "transition" region at droplet radii of ~1 μm where future atomizers will operate. Analysis of quadrupole mass spectrometer data for Octoil sprays operating in this muli-modal transition region reveals that it marks the boundary between the Rayleigh and emission limited regimes. Further, these data indicate that the demarcation occurs when the relative charging level is approximately half the limiting value. As a consequence, emission electric field strength can be fundamentally directly proprtional to surface tension. The challenge is to understand the fundamental behavior of this complex transition zone in the vicinity of 1 μm for Octoil, and by implication all hydrocarbon liquids.

1. Introduction

By any measure, electrostatic fuel atomization has been, and is, of marginal technological significance. This is confirmed by a review of the electrostatic spray combustion literature. Aside from the pioneering work of Weinberg and Thong, the more recent detailed studies of Gomez and his students, and the attempts to engineer electrostatic combustion systems referenced by Kelly (1984), remarkably little attention has been devoted to this topic. Given the current emphasis on development of clean, efficient combustion systems and the advent of compact, high flow rate ($\gg$mL/s) capacity electrostatic fuel atomizers (Kelly 1984, Lehr and Hiller) this lack of meaningful effort at exploitation is curious to say the least.

The reasons for questioning this lack of acceptance are compelling. Electrostatic atomizers have been demonstrated (Lehr and Hiller, Kelly 1998) to provide an array of highly desirable attributes that are unavailable from conventional (non-charged) spray systems. Among these attributes are:

- true electronic control of droplet size
- self-dispersive droplet plumes
- droplet size insensitivity to fluid properties
- minimal electrical power input
- low pressure drop atomization
- adaptability to existing nozzle envelopes

By way of example, two electrostatic fuel systems are discussed: a relatively low flow rate (~0.02 mL/s) steady state Jet, diesel fuel burner, and an electrostatic diesel nozzle (~100 mL/s). In both instances, fundamental issues concerning the application of this enabling technology arise. These issues raise challenges that must be addressed if the transformation of this technology from being an enduring laboratory curiosity to commercialization is to be facilitated.

This is a plea to the electrostatics community to hasten the day when, to paraphrase Michael Faraday (Kendall), electrostatic fuel atomization can take its rightful position in our society as an accepted technology by being taxed.

2. Steady state electrostatic fuel burner

By virtue of their aromatic content, relatively high viscosity and subdued volatility, the so called "logistics fuels" (Jet, diesel) are notoriously difficult to burn cleanly. Even with the availability of copious quantities of precisely controlled air, or extreme pressure differentials, the challenge has proven to be difficult to meet. Outwardly the task is impossible to accomplish in quiescent air, with pressure differentials less than about 100 kPa, and with the requirement that combustion be complete and soot free.

Nevertheless, these latter conditions, which are associated with the operation of the so-called "Pocket Stove", can be handily met, c.f. the image of Figure 1. In this instance, electrostatic atomization represents "enabling technology" for this ~$^1/_2$ kW tuna-fish can sized burner. Blue flame combustion of Jet-A, in the combustor rig illustrated in Figure 2, is a direct consequence of atomization at a fuel input pressure of 70 kPa, a flow rate of 0.02 mL/s, and an input voltage of ~5 kV, corresponding to a charge density (ρ_e) of ~6 C/m^3.

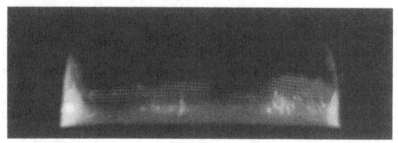

Figure 1: Side view, burner (see Figure 2) operation with Jet-A fuel in quiescent air.

Spray currents in 100 to 150 nA range, corresponding to a milliwatt level power requirement, is sufficient to generate an intensely self-dispersive spray. Since droplet sizes of the vigorously dispersing plume are sufficiently small (cf. the maximum size distribution of Figure 3) combustion proceeds in much the same manner as for a premixed gas flame. In other words, a majority of the electrostatic spray attributes noted earlier, contribute directly to achieving complete combustion of difficult fuels under exceptionally stringent conditions.

Despite its outward simplicity, the burner depicted in Figure 2 is the culmination of a long and involved empirical test effort. This work highlighted the need for a quantitative engineering model of combusting, space-charge dominated plumes. An absolute prerequisite for the development of such a meaningful combustion model is quantitative knowledge of the droplet size distribution immediately downstream of the nozzle.

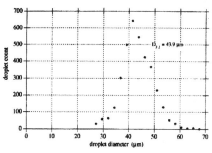

Figure 3
Aerometrics PDPA Jet-A droplet size distribution obtained 20 mm above the atomizer and 5 mm from the centerline during operation at a charge density of ~6.3 C/m³

Figure 2
Side view
76 mm diameter by 100 mm tall
test burner

Providing the droplet sizes are larger than about ten microns, corresponding to charge density levels less than approximately 56 C/m³, the monomodal droplet size distribution and the minimum energy atomizer model (Kelly 1994) will suffice. However, as discussed in Section 4, spray behavior undergoes a charging transition, accompanied by the appearance of multimodal size distributions, precisely in the droplet size regime now of interest for burner applications. Any future model has to include these complex, poorly understood and sparsely documented distributions as input conditions.

3. Electrostatic Diesel Nozzle

The diesel engine is a prime candidate for application of electrostatic atomization. Detailed imaging of conventional diesel atomization by Seiber, using a transparent diesel engine and sophisticated laser probing, demonstrates that atomization (droplet size) is not the controlling parameter; emissions and performance are firmly linked to mixing of the well atomized plume with the chamber air. This conclusion is forcibly emphasized by the theoretical work of Bellan and Harstad. Comparison of electrostatic dispersion *vis a vis* turbulence and swirl, under quite general conditions, shows that mechanical mixing is irrelevant in the presence of modest, readily achievable droplet charging. Electrostatic droplet dispersal completely overwhelms all other dispersive/mixing effects.

Graphic confirmation of this has been provided by the collaborative effort undertaken by CFDRD Corporation and Charged Injection Corporation. Using the qualified CFD-ACE advanced flow simulation code for conventional diesel atomization described by Giridharan , et al as a basis, CFDRD has added self-field droplet force terms to represent charged plume behavior. To ease the computational burden introduced by the pervasive long range force driving plume dispersion, this first simulation has been restricted to non-combustion behavior. Such factors as droplet shattering, and Rayleigh bursting have been neglected, but evaporation is included. In addition, the simulation has been limited to the maximum pressure condition at top-dead-center.

Starting with the theoretical droplet size distribution corresponding to readily achievable charge density level of 4 C/m^3 (Kelly), and using the actual parameters for a well characterized Caterpillar diesel truck engine (cf. Table 1) two sets of simulations were recently reported (Kelly, Avva). Figures 4 and 5 represent injection at a nominal 268 m/s with and without charging. An overlay of electrostatics, provided by a modified injector of the type illustrated in Figure 6 provides a modest, but meaningful increase in mixing.

Table 1

Engine Parameters		Spray Size Distribution	
		size (μm)	vol%
cylinder bore	137 mm	22.5	1.2
piston stroke	65 mm	27.5	7.9
compression ratio	15:1	32.5	7.3
fuel per injection	228 mm^3	37.5	13.0
injection duration	24.8°crank	42.5	9.7
instantaneous flow rate	100 mL/s	47.5	15.6
compressed air temperature	885 °K	52.5	28.0
cylinder pressure	4.43 MPa	57.5	17.3

It is only when operation at lower injection pressures is considered that the true value of electrostatic atomization is manifest. Computational limitations restrain simulation to an injection velocity of 100 m/s, even though there is no reason that lower injector pressure drops cannot be used. Albeit, at 100 m/s the required pressure drop is a seventh of the base-line 268 m/s case, droplet dispersion is profoundly changed, as illustrated by comparison of the simulations of Figures 7 and 8.

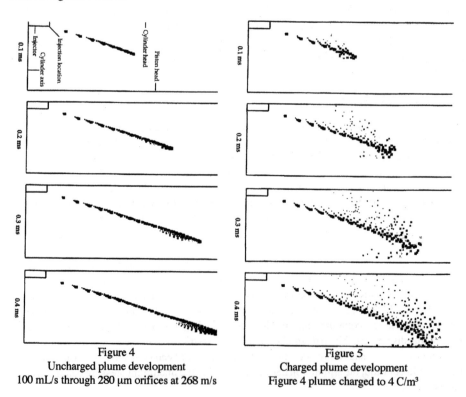

Figure 4
Uncharged plume development
100 mL/s through 280 μm orifices at 268 m/s

Figure 5
Charged plume development
Figure 4 plume charged to 4 C/m³

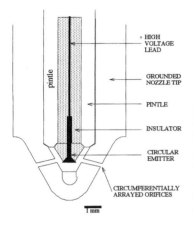

Figure 6
Electrostatic multi-orifice
diesel injector

Contrary to conventional, uncharged sprays wherein inertial forces dominate penetration and mixing, charged droplet trajectories are controlled by the mutual droplet repulsive force impeded by viscous drag. This means that charged sprays disperse essentially independent of gas density, reaching a terminal velocity (U, m/s), independent of droplet size, that is solely dependent on the local electric field (E, V/m)):$U=1.6 \cdot 10^{-10} E/v$, where v is chamber air viscosity (kg/ms). For example, assuming a relatively modest local field strength of 1 MV/m (well away from the maximum at the plume centerline), and the 610°C air of the example, the electric field driven droplet velocity is slightly in excess of 4 m/s. Close to the spray centerline the comparable velocity is well over an order of magnitude higher. This provides a general context to place the simulations in perspective. While the details of the simulation can be quibbled with, there is no argument that mixing is dramatically enhanced by modest droplet charging. If nothing else, this work indicates that by taking advantage of the ability to rapidly control droplet charging, electrostatic atomization opens the way for the direct electronic manipulation of droplet size, placement and mixing. The way is clear for the development of an adiabatic engine in which low pressure atomizers are used to precisely control both the temporal and spatial placement of fuel during injection, restraining heat release to a prescribed volume away from the cylinder walls.

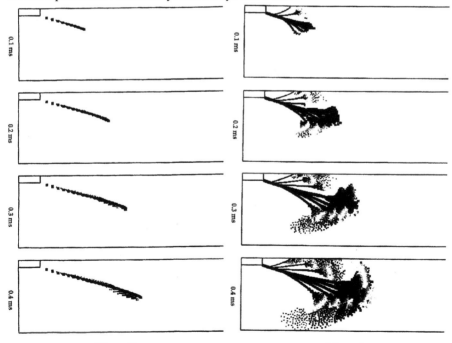

Figure 7
Uncharged 100 m/s plume development

Figure 8
Charged (4 C/m³) 100 m/s plume development

104

A further benefit of electrostatic diesel nozzles is to be found when cold start is considered. Figure 9 displays simulations of cylinder pressure histories for initial operation of the diesel engine used in the example. Here, the model has been extended to include a moving piston and intake air flow fields. Normal, (non-electrostatic, room temperature) operation is characterized by ignition somewhat prior to dead top center at 360° immediately followed by a rise to a peak pressure of almost 5 MPa. By contrast, after prolonged cold soak at -40°C, compression heating is inadequate to ignite the injected fuel. The pressure simply rises to a peak and falls without ignition.

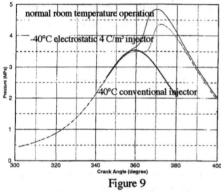

Figure 9
Cold soak diesel engine operation

Cold-soak behavior is dramatically altered when electrostatic nozzles capable of charging the fuel to 4 C/m³ are substituted for the conventional injectors. Computational limitations permitted only the electrostatic droplet size distribution to be used, and no attempt was made to optimize timing, nevertehless ignition does occur. Although initial operation is only ~40% of nominal room temperature performance, it is sufficient to sustain engine operation. Inclusion of the full electrostatic dispersal and droplet shattering terms and optimization of timing will show markedly better start behavior.

The challenge is to understand the true behavior of charged atomization in realistic diesel engine environments. Existing simulations and theoretical studies, although sparse, unambiguously demonstrate that electrostatic atomization represents a practical means for the full electronic control of diesel atomization, and will permit the development of clean and highly efficient engines. A crucial need exists for more detailed charged plume combustion simulations and for engine testing to verify this promise. To this point, the need for this has not been taken seriously. I emplore you to change this.

4. Fundamental Questions

Work on the micro-burner system, no less than diesel simulation effort has focused attention on the droplet size range below about 10 μm. This realm is of particular interest for several reasons. MEMS (Micro Electro Mechanical Systems) thermal systems require efficient, compact combustion that can only be provided by the effectively premixed gaseous behavior of ≤10 μm droplets. Since this droplet size regime is dominated by a transition in spray behavior it is of fundamental interest and import.

Relegating details to the references (Weis, Kelly 1994, Okuda and Kelly), evidence obtained using a quadrupole mass spectrometer-charge detector apparatus confirms the existence of a complex (multimodal) transition in charging behavior in the micron droplet size range. Droplet sprays having larger mean sizes exhibit a simple proportionality between droplet mean charge number (N_e) and mean spray plume diameter (d). The proportionality constant, stated in terms of the characteristic Lagrangian multipliers of the descriptive model ($-\alpha'/\beta'$) is approximately 55 eV, (Lehr and Hiller, Kelly 1994, Lehr 1993 private communication). Within ~1% accuracy, this "larger droplet" regime can also be characterized in terms of the first Bohr radius (a=5.29 10^{-11} m) as N_e=d/a, implying a quantum-mechanical basis for charged droplet development.

Figure 10, quadrupole mass-spectrometer/electrometer data for Octoil obtained in this droplet size range (Weis, Okuda and Kelly), reveals complex behavior for charge-to-mass ratios >0.01 C/kg (droplet radii below ~10 μm). Each of the data points plotted on the experiment parameter space (Q/m and droplet charge population N_e) were obtained during operation at a fixed quadrupole operating state (fixed Q/m value). Figure 11 is typical of the multimodal distributions observed in this regime. For clarity, only the peak value from each of the distributions have been plotted in Figure 10 to represent the distribution, with each peak of a multiple peak distribution plotted separately. Repeated and futile attempts to obtain data in the obvious gap regions led credence to the idea that these regions are not artifacts of the experiment, but reflect actual spray behavior. The data indicate a rich physical regime that must be explored if spray behavior in this droplet size region is to be understood.

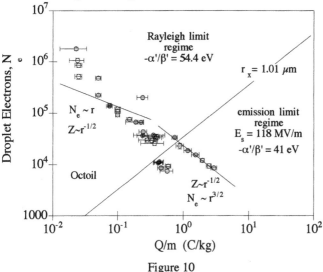

Figure 10
Quadrupole mass spectrometer/electrometer data for Octoil

Spray behavior in the vicinity of this transition cannot be quantitatively described until the available spray equations (Kelly 1984) are reformulated by relaxing the simplifying assumption. As currently constituted, the descriptive equations are valid starting a micron or so on either side of the transition where the charging level is only weakly dependent on droplet number density. This is certainly not the situation in the immediate vicinity of the transition where abrupt shifts, visible as data gaps in droplet charge level, are the rule. A quantitative description of the transition is needed.

As evidenced by the departure of the data from the $N_e = d/a$ line representative of the large droplet regime, the transition starts at charge densities of ~0.01 C/kg (~10 C/m³). This is just where microburner atomizers are now operating. An important clue concerning spray behavior in the post-transition (higher Q/m) region is provided by the data trend line. As noted on Figure 10, the data are approximately correlated by an inverse relationship

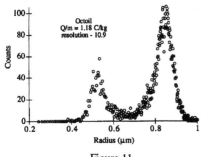

Figure 11
Multi-modal distribution

$Q/m\sim 1/N_e$, implying $N_e \propto r^{3/2}$. Superficially it is tempting to claim that this is simply what is to be expected if the Rayleigh limit were controlling ultimate droplet charging level. However, if this is true, why is the larger droplet behavior represented by $N_e \propto r$? Cast in terms of the relative charging level Z, the ratio of droplet charge level to the limiting value, a more consistent picture emerges. In Rayleigh limit regime, $N_e \propto r^2$ and $Z \propto r^{-1/2}$. Presuming $Z \propto r^{-1/2}$ also prevails below the transition, $N_{e\ emission} \propto r^2$, and $N_e \propto r^{3/2}$, precisely the observed trend.

Consistency of behavior forces the conclusion that beyond the transition, droplet surface emission field strength, and not the Rayleigh limit, is the controlling factor. This immediately explains the observed complexity of this region; each region will have three energetically favored branches. Only well away from the transition will two of the three branches become energetically improbable, resulting in a mono-modal distribution.

Analysis indicates that the relative charging level increases consistently with decreasing droplet size attaining a value of approximately half at the transistion. Under the assumption that Z is precisely $1/2$ at the transition, and that the statistical equilibrium model equations are operative despite their acknowledged approximation in this regime, permits the following to be written: $E_s = 2\gamma q / [\varepsilon_o(-\alpha'/\beta')]$. Alternatively, written in terms of the first Bohr radius, $E_s = [4\varepsilon_o h^2/(m_e q^3)]\gamma = 4144\ \gamma$ MV/m. This places the transition at $0.0262/\gamma$ or 0.85 μm for Octoil and other fluids having a surface tension of 0.031 N/n. Also, the limiting electric field is 118 MV/m, or about an order of magnitude smaller than the previously accepted value of 10^9 V/m for electron emission (Kelly 1978). By way of reference, it should be noted that the highest surface tension value, that of molten pure tungsten in vacuum is 2.5 N/n corresponding to $E_s = 1.04 \bullet 10^{10}$ V/m, the highest commonly accepted value for emission limited field strength.

This implies that as the spray fluid temperature is increased, the accompanying decrease in surface tension will shift multimodal spray behavior of the transition region to progressively larger droplet sizes. In fact, it is entirely realistic to anticipate encountering multimodal populations having as many as three distinguishable peaks.

As diagrammed in Figure 10, the data suggest that the transition occurs at a radius of just about 1 μm, corresponding to a limiting electric field of 118 MV/m, and $-\alpha'/\beta' = 41$ eV. Just why the characteristic energy in the emission limited region is just 3/4 of the Rayleigh limit regime value (a decrease of ~13.5 eV-the ionization level of hydrogen) is a mystery that should be addressed.

5. Conclusions/Recommendations, Questions/Challenges

The vast majority of liquid fuels are atomized preparatory for combustion. Performance and emissions generated by these combustion systems are intimately associated with the manner in which the fuel is presented to the combustion zone. By any quantitative measure of excellence, be it droplet dispersion, electronic control of droplet size, insensitivity to fuel properties, efficiency or pressure drop, electrostatic atomization is far and away superior to all conventional, non-charged atomizer systems.

Two "real-world" examples have been presented to illustrate the importance of electrostatic fuel preparation for combustion. These examples indicate the need for the development of more astute simulation models of combusting charged sprays, and the need to understand the physical processes prevalent at the high charge densities and small droplet size levels provided by current atomizers. While it is unquestionably important to fill these gaps in our understanding and computational capability, these unknowns in no way can account for the pervasive lack of acceptability of the technology. In the last analysis, the challenge is to overcome "cultural" impedimenta that have for too long prevented the adaptation of this fundamentally important technology. I look to the electrostatics community for support in this goal.

I also ask that the community take seriously the implications of a quantum mechanical basis for the electrostatic spray process and that the physics of two dimensional droplet surface charge underlies spray development. From a technological perspective, unlimited quantities of free charge can now be simply generated. Virtually nothing is known concerning the electrical, chemical, pharmacological, physiological, psychoactive or entertainment properties and uses of this new form of matter. The challenge is to address these matters.

References

Bellan J and Harstad K 1998 *Atomization and Sprays* **8** 601-624

Chen G and Gomez A 1997 *Combustion and Flame* **110** 392-404

Giridharan M G, Pindera M Z and Keenan J 1997 *AIAA paper 97-0693*

Gomez A and Tang K 1994 *Phys. Fluids,* **6** (1) 404-414

Gomez A. and Chen G 1994 *Combust. Sci and Tech.* **96** 47-59

Gomez A 1996 *Proc. Eastern States Section: The Combustion Institute, 1996 Fall Technical Meeting*

Kelly A J 1984 *Proc. Second International Conference on Liquid Atomization and Spray Systems*

Kelly A J 1994 *J. Aerosol Sci.* **25** (6) 1159-1177

Kelly A J and Avva R K 1998 *AIAA Aerospace America, Engineering Notebook*

Kelly A J 1998 *Proc. DoE sponsored Diesel Engine Emissions Reduction Conference*

Kendall J 1955 *Michael Faraday, Man of Simplicity* p 14

Lehr W and Hiller W 1993 *J. of Electrostatics* **30** 433-440

Lehr W 7/3/93 private communication

Okuda H and Kelly A J 1996 *Phys. Plasmas* **3** (5) 2191-2196

Seibers D L 1998 *SAE technical paper 980809*

Tang K and Gomez A 1994 *Phys. Fluids* **6** (7) 2317-2332.

Thong K C and Weinberg F J 1971 *Proc. R. Soc. London Ser. A* **324** 201

Weis D M 1994 *MS Thesis* MAE Department Princeton University

Inst. Phys. Conf. Ser. No 163
Paper presented at the 10th Int. Conf., Cambridge, 28–31 March 1999

Viscosity effect on Electrohydrodynamic (EHD) spraying of liquids

A. Jaworek[1], W. Machowski[2], A. Krupa[1], W. Balachandran[2]

[1]*Institute of Fluid Flow Machinery, Polish Academy of Sciences, Fiszera 14, 80-952 Gdansk, Poland. Telephone: (048)58 3460881 ext.151, Fax: (048)58 3416144, ajaw@imppan.imp.pg.gda.pl*

[2]*Department of Manufacturing & Engineering Systems, Brunel University, Uxbridge UB8 3PH, UK. Telephone: (044) 1895 274000, Fax: (044) 1895 81255 , e-mail: emsrwwm@brunel.ac.uk*

Abstract
The electrohydrodynamic (EHD) atomization of liquids is an effective method for generating electrically charged aerosol. The liquid is atomized only due to electrical forces without additional mechanical energy applied. The effect of viscosity on the spraying modes and the droplet size distribution has been investigated experimentally for distilled water and ethylene glycol mixture, with viscosity varied from 1mPa*s to about 22mPa*s. When the viscosity increased droplet size decreased in the spindle mode of atomization. The Sauter mean diameter $D_{3,2}$ is inversely proportional to the flowrate, but in the cone jet mode $D_{3,2}$ is directly proportional to the flow rate.

1. Introduction

A growing interest in EHD spraying is spurred mainly by new applications such as electrostatic painting, fuel atomization in combustion systems, drug delivery, emulsification, microincapsulation, crop spraying, scrubbing etc., as charging of droplets can enhance the efficiency of these processes. The EHD spraying also extends the droplet size available from conventional mechanical atomizers to a lower size range. The droplets generated by the EHD method possess electrical charge, usually equal to a fraction of the Rayleigh limit, i.e., a few orders of magnitude greater than the elementary charge. The charge on the droplets significantly influences the electrodynamic properties of an aerosol.

In recent years many papers have been published, summarising the modes of spraying, among them by Hayati et al. [1], Cloupeau and Prunet-Foch [2], Grace and Marijnissen [3], Shiryaeva and Grigorev [4]. The role of viscosity of the liquid was theoretically considered by Ganan-Calvo [5]. He determined only the effect of liquid viscosity on the electric current in the cone-jet mode. Some experimental results were given by Ogata et al. [6], and Rosell-Llompart and Fernandez de la Mora [7]. Ogata et al. reported that the specific charge of the droplets (q/m) depends on viscosity to the power -1/4 to -3/4 for semi-conducting liquids and Reynolds number lower than 1000. They also noticed that the volume surface mean diameter of the droplets is proportional to viscosity, $\eta^{2/9}$. Rosell-Llompart and Fernandez de la Mora investigated the effect of viscosity on the normalized current.

The atomization process can be categorized into two main groups. In the first group the dripping, microdripping, spindle, multispindle and ramified meniscus modes were observed. The second group comprises of cone-jet, precession, oscillating-jet, multi-jet and ramified jet modes. In the first group only fragments of liquid are ejected from the capillary by deforming and detaching the liquid meniscus. In the second group, the liquid is elongated into a long fine jet, free end of which disintegrates into droplets.

It is well known that the atomization process is governed by the physical properties of the liquid such as surface tension, density, viscosity, conductivity and permittivity as well as the parameters of the spraying system. i.e., the capillary diameter, excitation voltage and liquid flow rate. However, it was hitherto not recognized the combination of these parameters to generate the droplets of required size and charge.

The spraying process is governed by the mass and momentum conservation equations into which electrical and mechanical forces are included. The body forces and stress tensor on the meniscus and the jet are given by the following equations:

$$\frac{\partial \rho_l \bar{v}_l}{\partial t} = \rho_l \bar{g} - \bar{\bar{\Phi}}_D - \nabla \bullet \left(\bar{\bar{1}} \left(p_l - p_g \right) + \eta_l \nabla \bar{v}_l + \rho_l \bar{v}_l \otimes \bar{v}_l \right) + \bar{L}_l \tag{1}$$

$$\sigma_l \nabla \times \bar{\bar{\Xi}}^1 = \bar{\bar{1}} \left(p_l - p_g \right) + \eta_l \nabla \bar{v}_l + \rho_l \bar{v}_l \otimes \bar{v}_l + \bar{\bar{\Lambda}}_l \tag{2}$$

with subscript l and g refers to the liquid and gas phases, respectively. p_l and p_g are the pressures on both sides of the interfacial surface, ρ_l, η_l, σ_l are the liquid density, viscosity and surface tension, respectively. The symbol $\otimes$ denotes the dyadic product of the vectors yielding a tensor. $\bar{\bar{\Xi}}^1$ is dimensionless surface tension tensor normalized to σ_l.

The first term on the right hand side of equation (1) is the gravitational component of the force on the liquid jet, the second is the drag force on the moving jet, resulting from the surrounding gas viscosity, opposing jet acceleration. The first term in brackets is the hydrostatic pressure difference on the liquid surface, the second refers to the pressure caused by the viscosity of liquid resulting in the drag force to the jet, and the third term is the pressure caused by inertial forces of flowing liquid. The last term in (1) is the electromagnetic body force on the charged jet.

The first term on the right hand side of equation (2) is the stress tensor due to pressure difference on the interfacial surface. The second one is the stress due to liquid viscosity. The third term is the stress caused by inertial forces acting on the flowing liquid. The last term is the electrodynamic stress tensor on the interfacial surface.

The body forces on the bulk jet, resulting from the electric field cause the liquid jet to move as a whole. The tangential component of the electrodynamic stress tensor causes jet elongation from the body of liquid issuing from the capillary, while the normal component of the electric field balances the hydrostatic pressure. Similar to pressure difference and the gas drag force. The viscosity of the liquid opposes the jet formation and its movement.

The role of liquid viscosity can be predicted from equations (1) and (2). Converting these equations to dimensionless form, a time constant of the jet formation can be derived:

$$\tau_\eta = \frac{\eta_l D_0}{\sigma_l} \tag{3}$$

Since the viscose forces opposes the jet formation, it can be interpreted as determining a limiting value of velocity with which the jet can be stretched out from the bulk liquid, without its disruption.

The effect of viscosity on the jet elongation and its disruption into droplets, and the droplet size distribution has been investigated experimentally, and is reported in this paper.

2. Experiment

The experiments were carried out using the arrangement shown schematically in Fig.1. This system consists of a stainless steel capillary (hypodermic needle) of inner diameter of 0.25mm, and outer diameter of 0.4mm, mounted in a vertical plane, and an earthed ring of inner diameter of 9mm made of brass wire of diameter of 1mm in a horizontal plane. The distance between the tip of the capillary and the plane of the ring electrode was kept constant and was equal to 6mm. The liquid was fed to the capillary needle by a syringe pump system providing a constant volume flow rate. High voltage power supply was connected directly to the capillary needle.

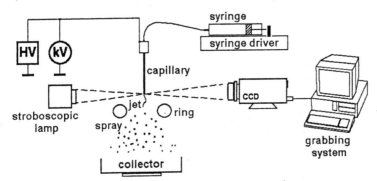

Fig.1. Schematic diagram of the experimental arrangement.

The spraying process was observed with the aids of a CCD camera equipped with a long distance microscopic lens and stroboscopic illumination of 30μs of flash illumination. All recordings were processed using a frame grabber and stored in a computer. Distilled water, ethylene glycol, and their mixtures were tested during the course of the experimental programme. The viscosity of the investigated liquids varied from 1mPa*s to about 22.7mPa*s. The physical properties of the liquids used are presented in Table 1. All observations were performed at normal pressure and ambient air temperature of about $16^{\circ}C$, in steady gas conditions. The droplet size distribution was measured using the Malvern - Mastersizer S.

Table 1. Physical properties of the liquids used in the experiments (T=16±1°C).

Liquid name	density	viscosity	surface tension	conductivity	relative permittivity
(volume fraction)	kg m^{-3}	MPa*s	N m^{-1}	S m^{-1}	-
distilled water	1000	1.02	0.0740	1.4×10^{-4}	80.4
water 75% + glycol 25%	1040	2.2	0.0711	4.5×10^{-4}	-
water 50% + glycol 50%	1080	4.2	0.0614	4.2×10^{-4}	-
water 25%+glycol 75%	1100	9.6	0.0555	2.4×10^{-4}	-
ethylene glycol	1110	22.7	0.0510	2.7×10^{-5}	38.8

3. Results

The photographs of the jet formation for liquids of different viscosity are shown in Fig.2. It can be seen that an increase in liquid viscosity causes the jet to become longer and thinner. The cone-jet mode develops from the spindle mode for lower voltages as the liquid viscosity increases.

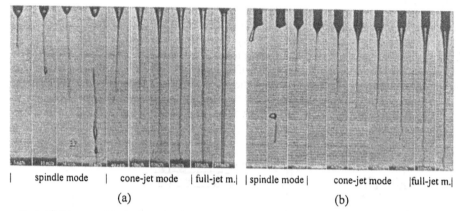

| spindle mode | cone-jet mode | full-jet m.| | spindle mode | cone-jet mode | |full-jet m.|

(a)   (b)

Fig.2. CCD images of the jet formation for distilled water (a) and ethylene glycol (b). Potential - U=10kV.

Two spraying modes were observed for the tested liquids: the spindle mode and the cone-jet mode.

In the spindle mode the meniscus elongates in the direction of the electric field, taking the shape of a thick jet, which detaches as a vast spindle-like fragment of liquid. In this mode the applied electric field detaches a fragment of liquid before a continuous jet is formed. After its detachment, the spindle can disrupt into several smaller droplets of different sizes because of the electrostatic forces, while the meniscus contracts to its initial shape, and a new jet starts to be formed. The spindle mode operates at higher voltages than the dripping mode, and it differs from the dripping mode in that no regular droplet is ejected from the meniscus but only elongated fragments of liquid of different sizes. Droplets of two distinct sizes (see Fig.3) are generated by the spindle mode: large droplets which are different in size, with diameters varying in the range from 200μm to 400μm, and small droplets, usually smaller than 70μm. The fraction of smaller droplets increases with the increase in volume flow rate and also with the increase of liquid viscosity.

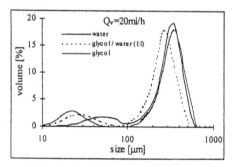
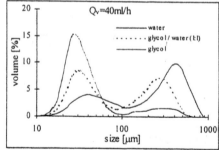

Fig.3 Typical size distribution of droplets generated in spindle-mode of atomization (Q_v<50ml/h).

In the cone-jet mode the liquid issues from the capillary in the form of a regular, axisymmetric cone which gradually changes into a thin jet. The diameter of the jet estimated from the photographs is of about 140μm (for distilled water), and decreases with an increase in liquid viscosity to about 90μm (for ethylene glycol), as measured at the same distance of 4mm from the capillary outlet. The end part of the jet undergoes kink instabilities moving irregularly off the system axis, and breaks up into fine droplets due to electrical and inertial forces. It was observed that when the applied voltage was increased, the jet diameter and length decreased proportionally. In the cone-jet mode only fine droplets of most probable diameter smaller than about 50μm for liquid of low viscosity (distilled water) and 30μm for high viscosity liquids (ethylene glycol) were observed. The effect of liquid viscosity on mean droplet size distribution for different volume flow rates is shown in Fig.4. It can be seen that the increase in viscosity causes the droplet size to decrease.

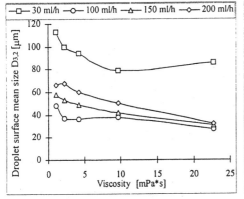

Fig.4. Mean droplet size distribution versus liquid viscosity.

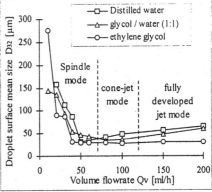

Fig.5. Droplet mean size $D_{3,2}$ of the generated liquid droplets versus applied volume flowrate. (DC potential applied to the capillary - 10kV)

The effect of the volume flow rate of the liquid on the droplet mean size is presented in Fig.5. Under a defined electric field condition, it was observed that, initially, when the flowrate increased, the droplet mean size ($D_{3,2}$) decreased due to the fact that the liquid jet elongates and gradually changes to the cone-jet mode. When the flowrate increased above a threshold value, fully developed jet was formed. The Sauter mean diameter $D_{3,2}$ was observed to be inversely proportional to the flowrate in the spindle mode of atomization. In the cone jet mode and the fully developed jet mode, $D_{3,2}$ was found to be directly proportional to the flow rate. From Fig.5 it can also be seen that as the viscosity increases droplet size decreases.

4. Conclusions

The fundamental equations governing the spraying process are presented in the paper. Two types of forces are distinguished; bulk forces such as electrostatic, drag, hydrostatic and viscosity forces which govern jet motion, and surface tension forces that are responsible for the jet deformation.

It was experimentally shown that an increase in the liquid viscosity causes the jet diameter to decrease and the jet length to increase. Also the diameter of the generated droplets decreases as the liquid viscosity increases.

Acknowledgements

This paper has been supported in part by the Brunel University and in part by the Institute of Fluid Flow Machinery, Polish Academy of Sciences. The scientific cooperation was supported in part by the State Committee for Scientific Research of Poland (KBN) and in part by the British Council within the frame of the British-Polish Joint Research Collaboration Programme WAR/992/152.

114

References

[1]   Hayati I., Bailey A.I., Tadros Th.F., Investigations into the Mechanisms of Electrohydrodynamic Spraying of Liquids. J. Coll. Interface Sci. 117 (1987) No 1, 205-30

[2]   Cloupeau M., Prunet-Foch B., Electrostatic Spraying of Liquids. Main Functioning Modes. J. Electrostatics 25 (1990), 165-84

[3]   Grace J.M., Marijnissen J.C.M., A Review of Liquid Atomization by Electrical Means. J. Aerosol Sci. 25 (1994), No.6, 1005-19

[4]   Grigoriev A.I., Shiryaeva S.O., The Theoretical Consideration of Physical Regularities of Electrostatic Dispersion of Liquids as Aerosols. J. Aerosol Sci. 25 (1994) No 6, 1079-92

[5]   Ganan-Calvo A.M., The Role of the Viscosity in the EHD Spraying of Liquids in Cone-Jet Mode. Inst. Phys. Conf. Ser. No 143, (1995), 61-8

[6].   Ogata S., Hatae T., Shoguchi K., Shinohara H., The dimensionless correlation of mean particle diameter in electrostatic atomization. Int. Chem. Eng. 18 (1978) No.3, 488-93

[7]   Rosell-Llompart J., Fernandez de la Mora, Generation of monodisperse droplets 0.3 to 4 μm in diameter from electrified cone-jets of highly conducting and viscous liquids. J. Aerosol Sci. 25 (1994) No.6, 1093-1119

Inst. Phys. Conf. Ser. No 163
Paper presented at the 10th Int. Conf., Cambridge, 28–31 March 1999

Scaling laws for droplet size and current produced in the cone-jet mode.

R.P.A. Hartman, D.J. Brunner, K.B. Geerse, J.C.M. Marijnissen, B. Scarlett.

Delft University of Technology, Faculty of Applied Sciences, Waterman Institute for Precision Chemical Engineering, Particle Technology Group, Julianalaan 136, 2628 BL, Delft, The Netherlands.

Abstract. Electrohydrodynamic Atomization (EHDA) is a method to produce very fine droplets from a liquid by using an electric field. Depending on the properties of the liquid, the flow rate of the liquid and the voltage applied to create the electric field, different modes of EHDA can occur.

The mode of most interest is the cone-jet mode. In this mode droplets are produced with a narrow size distribution. From a total physical model describing this mode, scaling laws for the current and the droplet size are derived. These scaling laws are presented in this paper.

In Electrohydrodynamic Atomization (EHDA) in the cone-jet mode a liquid is supplied to a nozzle. When a strong electric field is applied, then the electric stresses transfrom the shape of the droplet into a conical shape. At the cone apex, a highly charged jet occurs. This jet breaks up into a number of main droplets and a number of smaller secondary droplets.

The jet break-up has been investigated by means of a High Speed Spray Imaging System. This system consists of a digital camera connected to a computer, which is equipped with a frame grabber. A long distance microscope lens is fixed to the camera. The illumination is provided by short laser pulses. The optical system allows to see droplets of a few micrometers.

Figure 1. Photograph of EHDA in the cone jet mode.

The measurement results showed that the produced droplet size in the cone-jet mode depends on the jet break-up mechanism. Which was found to depend on the ratio of the electric stress over the surface tension stress. At low stress ratios, the jet breaks up due to varicose perturbations. At higher stress ratios, the jet showed a whipping motion.

Fernández de la Mora (1994) and Gañán-Calvo (1997), derived current scaling laws for highly conducting liquids. The difference between these two relations is the influence of the relative permittivity of the liquid.

Simulations with the cone-shape model, Hartman (1997), showed that the influence of permittivity was less than 5% in a range for the relative permittivity from 1.25 to 50. That is why, the influence of the permittivity is removed from the current scaling law. The following equations show the derivation of this new scaling law

$$\pi r_{j*}^2 E_z K = 2\pi r_{j*} u_z \sigma \quad \sigma \sim \left(\frac{\gamma \varepsilon_0}{r_{j*}}\right)^{\frac{1}{2}} \quad E_z \sim \frac{\sigma}{\varepsilon_0} \quad u_z = \frac{Q}{\pi r_{j*}^2} \quad r_{j*}^3 \sim \frac{Q \varepsilon_0}{K}$$

$$I = \frac{4\sigma Q}{r_{j*}} \sim \left(\frac{\gamma \varepsilon_0 Q^2}{r_{j*}^3}\right)^{\frac{1}{2}} \sim (\gamma K Q)^{\frac{1}{2}} \quad I_0 = \left(\frac{\varepsilon_0 \gamma^2}{\rho}\right)^{\frac{1}{2}} \quad Q_0 = \frac{\varepsilon_0 \gamma}{K \rho} \quad \frac{I}{I_0} = b\left(\frac{Q}{Q_0}\right)^a \quad (1)$$

where r_{j*} is the jet radius where charge convection is equal to the charge conduction [m], E_z is the electric field strenght in the axial direction [Vm^{-1}], σ is the surface charge [Cm^{-2}], u_z is the liquid velocity [ms^{-1}], K is the conductivity [Sm^{-1}] ,ε_0 is the permittivity of vacuum [C^{2}N^{-1}m^{-2}], Q is the liquid flow rate [m^{3}s^{-1}], γ is the surface tension [Nm^{-1}], I is the current [A] and ρ is the density [kgm^{-3}].

Table 1. Parameters a and b for the current scaling law.

Q/Q_0	a	b
<50	0.493	2.215
50-250	0.518	1.931
>250	0.427	3.203

For a nozzle with a diameter of 8 mm and for n-butanol and ethylene glycol with various conductivities, the scaling law was fitted to the current measurements. The result is shown in table 1 and in figure 2. It has to be noted that the current also depends on the electrode configuration and the type of ions used. So, measurements might yield deviations from the scaling laws up to 30%.

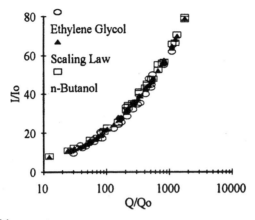

Figure 2. Comparison of the current.

The results of the measurement showed that the droplet size in the varicose jet break-up regime scales with the liquid flow rate as $d_d \sim Q^{0.48}$. In the whipping jet break-up regime the droplet size scales with $d_d \sim Q^{0.33}$. The scaling laws of Fernández de la Mora (1994) and Gañán-Calvo (1997) only scaled with $Q^{0.33}$. In the varicose jet break-up regime, Rayleigh's jet break-up theory can be used to derive the jet break-up time constant τ_d [s]. If the kinetic energy term is balanced with the tangential electric stress term of the Navier-Stokes equation, then a relation for the droplet diameter d_d [m] can be derived.

$$\tau_d \sim \left(\frac{r_j{}^3 \rho}{\gamma}\right)^{\frac{1}{2}} \quad \frac{E_z I}{Q} = \rho \frac{Q}{\pi r_j{}^2} \frac{\partial v}{\partial z} \sim \rho \frac{Q}{\pi r_j{}^2} \frac{1}{\tau_d} \quad d_d \sim r_j \sim \left(\frac{\rho \varepsilon_0 Q^4}{I^2}\right)^{\frac{1}{6}} \sim \left(\frac{\rho \varepsilon_0 Q^3}{\gamma K}\right)^{\frac{1}{6}} \qquad (2)$$

where r_j is the jet radius [m]. This equation shows that the droplet diameter scales with $d_d \sim Q^{0.5}$.

When the jet breaks up in the whipping jet break-up regime, then the Rayleigh Charge Limit was found to be the determining factor for the droplet size. The stress ratio of the electric stress over the surface tension stress on the droplets produced in the whipping jet break-up regime was about equal to 0.8. Gañán-Calvo (1997) reported a value of 0.64. The droplet size in the varicose and in the whipping jet break-up regime can then be calculated from preferably the measured or otherwise from the calculated current.

$$d_{d,\text{varicose}} = c\left(\frac{\rho \varepsilon_0 Q^4}{I^2}\right)^{\frac{1}{6}} \qquad d_{d,\text{whipping}} = \left(0.8\frac{288\varepsilon_0 \gamma Q^2}{I^2}\right)^{\frac{1}{3}} \qquad (3)$$

118

The constant c was fitted to the experiments and was found to be equal to 2.05 for the measurements with the camera, and was found to be equal to 1.76 when compared to a result of Gañán-Calvo (1997). Figure 3 shows this last result. The difference can be attributed to the measurement accuracy of the camera system. The droplet size equation (3) that yields the smallest droplet is the one that has to be used.

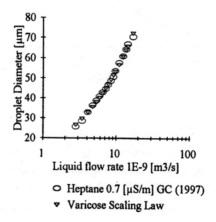

Figure 3. Comparison of the droplet size.

Acknowledgements.

We like to thank NOW/SON/STW for funding this project.

References.

Fernández de la Mora J., Loscertales I.G. (1994) *J. Fluid Mech.* **260**, 155-184

Gañán-Calvo A.M., Davila J., Barrero A. (1997) *J. Aerosol Sci.* **28**, 249-275

Hartman R.P.A., Marijnissen J.C.M., Scarlett B. (1997) *J. Aerosol Sci.* **28 suppl. 1**, 527-528

Inst. Phys. Conf. Ser. No 163
Paper presented at the 10th Int. Conf., Cambridge, 28–31 March 1999

Electrostatic Atomization of Ultra Fine
Spray of Ceramic Solution.

P. Miao, Z.A. Huneiti, W. Machowski and W. Balachandran
Department of Manufacturing & Engineering Systems
Brunel University, Uxbridge, Middlesex UB8 3PH, UK

P. Xiao
Department of Materials Engineering
Brunel University, Uxbridge, Middlesex UB 8 3PH, UK

J.R.G. Evans
Department of Materials, Queen Mary and Westfield College, London E1 4NS

Abstract. An electrostatic atomization technique has been developed to generate ultra-fine spray droplets of ZrO_2 ceramic suspension in a range of a few micrometers. Preliminary results have shown that for low through-put atomization, cone-jet mode is the most suitable method to produce a monodispersed fine charged aerosol. It was found that the applied electric fields have significant influence on the breakup process of the fine jet, hence the size distribution. Initial deposition experiments showed that the diameter of the ceramic powder particles deposited on silicon substrates is between 0.1 and 1.0 μm which is the original size of the powder. The density of the powder particles on the substrates was found to be related to droplet size which can be controlled by applied electric fields.

1. Introduction

Over the past two decades, there has been a significant growth of interest in research and development in the advanced ceramic industry. In order to obtain high quality ceramic thin films and coatings, mouldless fabrication technology is desirable. The quality of ceramic thin films and coatings is heavily dependent on the size and size distribution of the ceramic powder particles during deposition process [1]. This is because fine particles with a narrow size distribution have the effect of limiting the size of voids between particles, which enables these ceramic thin films and coatings to be deposited with high homogeneity and without flaws. This, in turn, determines the mechanical and electrical properties of the ceramic materials. In searching for a new micro-fabrication technology of ceramic thin films and coatings, electrostatic atomization in cone-jet mode exhibits a very promising potential with its inherent capability of producing fine droplets in micro and sub-micrometer range with a narrow size distribution [2-5]. Compared with other thin film deposition technologies like chemical vapour deposition (CVD), physical vapour deposition (PVD) and plasma spraying (PS), electrostatic atomization is relatively inexpensive and also has the added advantage of having the potential to control and deposit multi-layer coatings. Recently, Teng and co-workers [6] have demonstrated the feasibility of elecrostatically atomizing a suspension containing a ceramic filler to produce fine droplet less than 5 μm. The ceramic powder particles deposited on silicone release paper were between 2 and 15 μm which is still considered not to be narrow enough in size distribution. The droplet size in their work was roughly estimated from the relics of the particles on the substrate. The real size and size distribution of the spray during transportation to the substrate was not measured.

In this paper, the behaviour of ceramic (ZrO_2) particles deposited on silicon substrates by electrostatic atomization of ceramic suspension was investigated. The influence of flow rate and applied electric fields on cone-jet, droplet size and distribution of the ceramic particles on the substrates will be reported. A high shutter CCD camera with a fast image grabbing system, a laser diffraction technique and scanning electron microscope were employed for monitoring the cone-jet mode, measuring droplet size distribution during transportation and imaging the morphologies of the deposited powder particles on substrates respectively.

2. Experimental details

The experimental configuration is schematically shown in Figure 1. The electrospray system is composed of a stainless steel capillary with an inner diameter of 0.23mm and an outer diameter of 0.51mm. The high voltage (3.5-8kV) was applied to the capillary. A concentric ring electrode with an inner diameter of 15mm was arranged 8mm away from the capillary tip. This electrode was connected to ground via a 20MΩ resistor. A metallic container which also served as an electrode was grounded. Ceramic suspension was fed to the capillary at a flow rate of 0.3 to 2ml/hr, using a Harvard 11 syringe pump. The suspension spray process was monitored using a high shutter speed TM-765 CCD camera together with a fast image grabbing software (Image 1.42) installed in a Macintosh computer. The size distribution of ceramic suspension (ZrO$_2$) during transportation was measured using a HELOS particle size analyser. Silicon substrates were positioned 10mm below the ring electrode for 2 minutes for deposition. The morphologies of the powder particles deposited were examined by scanning electron microscope.

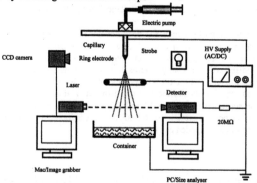

Figure 1 The schematic diagram of the experimental arrangement.

Formulation of ceramic suspension is the first and key stage for electrospray deposition, since the properties of the suspension play an important role in electrostatic atomization. For example, conductivity of the suspension must fall in a suitable range in order to obtain a stable cone-jet [7]. In this work, fine zirconia powder HYY-3 with an average particle size of 0.47μm was used. The suspension was prepared by dispersing 5 % ZrO$_2$ powder in weight in a solution of distilled water and ethanol mixture of a ratio of 1:1 by volume. This suspension was subjected to ultrasonic agitation and electrical stirring with a magnetic bar for 30 minutes each. The suspension was found to be very stable.

3. Results and discussions

Figure 2 shows the influence of flow rate of the ceramic suspension delivered by the syringe pump on the formation of liquid jet at a constant DC potential of 5.6kV. At the lowest flow rate, 0.3ml/hr (Figure 2(a)), the suspension was sprayed in an unstable cone-jet mode with wider conical angle (170°). When the flow rate was increased to 0.6ml/hr, (see Figure 2(b), a very stable cone-jet was formed at a conical angle of about 70°. Further increase in flow rate (1.0ml/hr, Figure 2(c)) causes a decrease in cone angle which was observed to be about 60° and a significant increase in jet diameter. Further increase in flow rate generated a fully developed jet. It was found that an increase in flow rate helped to stabilise the jet, but at the expense of an increase in jet diameter, which results in larger diameter of droplets. When flow rate was kept at 0.6ml/hr, and when the applied potential was raised from 3.5 to 8.0kV, the jet changed from a dripping mode to a cone-jet mode, and finally to a multi-jet mode. The number of the jets along the capillary rim increased with increase in applied potential. Similar behaviour was also observed by Hayati *et al* [9], but the liquid they employed was a mixture of isopar M and butanol. This implies that the presence of the ZrO$_2$ ceramic powder particles in the suspension does not change the basic process of electrostatic atomization, provided the suspension is stable and its physical and electrical properties fall within the desired range.

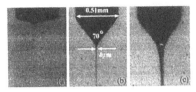

Figure 2 CCD images of jet formation at a constant applied potential of 5.6kV. (a) 0.3ml/hr (b) 0.6ml/hr (c) 1.0ml/hr.

From the CCD image (Figure 2(b)), the diameter of the jet close to the tip of the cone was estimated to be about 4μm and the length of the jet was measured to be about 200μm. The volume mean diameter (VMD) of the generated spray was measured to be 0.467μm with a narrow size distribution (0.1-1.5μm). The calculated ratio of the diameter of the droplets to the diameter of the jet (ϕ_D/ϕ_J) is 0.125. The conventional Rayleigh theory predicts that the ratio of (ϕ_D/ϕ_J) is about 1.89 for uncharged liquid jets of low viscosity. For charged jets, theoretical study by Neukermans [8], Bailey and Balachandran [9] and experimental study by Cloupeau *et al* [5] show that the ratio of (ϕ_D/ϕ_J) does not change significantly provided the surface charge density is not too high. However, in our case, the ratio of (ϕ_D/ϕ_J) was measured to be much less than 1.89. This could be because of the evaporation effect of the organic solvent (ethanol) in the ceramic suspension during transportation, resulting in higher surface charge density. Furthermore the presence of the ceramic particles in the liquid jet may influence the dynamic behaviour of the jet.

Figure 3 shows influence of applied potential on droplet size distribution, when the flow rate was kept constant at 0.6ml/hr. The VMD was measured to be about 0.5 and 2.5μm respectively at 5.6 and 5.0kV. Increase in applied voltage shortens the height of the cone and decreases the diameter of the jet, which in turn leads to smaller droplet size. This observation is in agreement with previous work by others [4, 5 and 7].

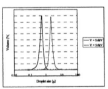

Figure 3 Comparison of size distributions atomized at 5.0 and 5.6kV, the flow rate was kept constant at 0.6ml/hr.

Balachandran and co-workers [10] investigated the influence of AC field superimposed on DC field on the jet breakup process for conducting liquids (e.g. water) in fully developed jet. They observed that the presence of AC field helped to improve the stability of the jet breakup process significantly and it is possible to generate monodispersed spray. They explained that the AC field can neutralise excessive surface charge which leads to the Rayleigh type of breakup process. In the present work, the influence of the AC field on the spray of ceramic suspension in cone-jet mode is also presented. Experimentally, it was found that the AC field with a frequency higher than 4kHz and a peak-to-peak amplitude lower than 2kV helped to improve the breakup process. It appears that the size distribution obtained using AC superimposed on DC field is narrower than that using DC field only.

SEM characterisation of deposited ZrO_2 ceramic thin films and coatings on silicon substrates is shown in Figure 4. From Figure 4(a) and (b), it can be seen that the ceramic

122

Figure 4 SEM images of ceramic particles deposited on silicon substrates at different applied potentials.

particles are reasonably evenly distributed on the surfaces. The particle size is estimated to be in the range of about 0.1 – 1.0 μm which, in fact, is the original size of the ceramic powder particles. The size of the deposited ceramic particles equal to the size of the original powder particles could be the evidence of one particle in one droplet. If this is not the case, the particles would not be evenly distributed on the surfaces, since the particles from the same droplet would be grouped together, even form bigger particles by agglomeration. This was observed by Teng and co-workers [6]. The ceramic particles in Figure 6(a) and (b) were deposited at the same flow rate of 0.6ml/hr for the same duration of 2 minutes, but at different applied voltages of 6.0 and 5.0kV respectively. From Figure 4(a) and (b), one can find that the density of those particles on the surfaces is different. A higher applied voltage leads to a less dense deposited layer. The reason for this is probably that when the droplet size is smaller, these droplets are likely to have a higher charge-to-mass ratio [2], hence larger space charge effect. Furthermore smaller droplets have smaller inertia. All these factors are favourable for smaller droplets to spray at a larger spray angle after breakup of the jet. This clearly demonstrates that the dispersion and deposition patterns on substrates can be controlled by controlling the applied potential to the capillary.

4. Conclusions

An initial experimental investigation on the deposition of ceramic thin films and coatings on silicon substrates using electrostatic atomization technique has been carried out. The results demonstrate that the fine ceramic droplets in the range of a few micrometers, even sub-micrometer can be achieved. The presence of the ceramic particles in the liquid does not affect basic electrospray process significantly. The flow rate and applied voltage were observed to have significant effects on the formation of the spray mode, hence droplet size. AC field superimposed on DC field was found to be helpful for stabilising the jet breakup process, which narrows the size distribution. Initial attempt of depositing ceramic thin films and coatings indicates that the morphologies of deposited particles on substrates can be controlled by applied potential.

Acknowledgements

We are grateful to Dr Z.C Wang in Department of Materials Engineering for his kind help in preparation of ceramic suspension. We would like to thank Sympatec GmBh for providing us HELOS size analyser for size measurements.

References
[1] Balachandran W, Machowski W and Ahmad C N 1994 *IEEE Trans. Ind. Appl.* **30** 850-855
[2] Tang K and Gomez A 1994 *Phys. Fluids* **6** 2317-2332
[3] Tang K and Gomez A 1995 *J. Colloid Interface Sci.* **175** 326-332
[4] Tnag K and Gomez A 1996 *J. Colloid Interface Sci.* **184** 500-511
[5] Cloupeau M and Prunet-Foch B 1989 *J. Electrostatics* **22** 135-159
[6] Teng W D, Huneiti Z A, Machowski W, Evans J R G, Edirisinghe M J and Balachdran W 1997 *J. Mat. Sci. Lett.* **16** 1017-1019
[7] Hayati I, Bailey A I and Tadros Th F 1987 *J. Colloid Interface Sci.* **117** 205-221
[8] Neukermans A 1973 *J. Appl. Phys.* **44** 4769-4770
[9] Bailey A G and Balachandran W 1981 *J. Electrostatics* **10** 99-105
[10] Balachandran W, Huneiti Z, Machowski W and Hu D 1995 *Electrostatics* **143** 55-60

Inst. Phys. Conf. Ser. No 163
Paper presented at the 10th Int. Conf., Cambridge, 28–31 March 1999
© 1999 IOP Publishing Ltd

Analysis of charged plumes with Coulomb repulsion

P.A. Vázquez††, A.T. Pérez‡ and A.Castellanos‡

† Dpto de Física Aplicada, E. S. I., Avda. de los Descubrimientos, s/n, 41092 Sevilla, Spain.

‡ Dpto. de Electrónica y Electromagnetismo, Facultad de Física, Avda. Reina Mercedes s/n, 41012 Sevilla, Spain.

Abstract. The flow of EHD plumes is described including the effect of Coulomb repulsion in the spread of the charge region. Using integral methods, a set of three equations is found, describing the evolution of the velocity at the axis of the plume and the width of the hydrodynamic and charged regions.

1. Introduction

EHD plumes arise when sharp edges or points immersed in dielectric liquids are subjected to high electric fields. Injection of charge occurs, and the electric field pushes the charge away from the injector surface, putting the liquid into motion. The structure of the motion is similar to that of thermal plumes observed above a linear or punctual source of heat.

Several cases can be envisaged: if charge diffusion dominates over Coulomb repulsion, self-similar solutions can be found[1]. If Coulomb repulsion dominates over charge diffusion, but is negligible in front of charge convection, self-similar solutions can still be found[2, 3]. In this case, the charged region is modeled as a layer of zero width. Finally, if the effect of Coulomb repulsion is of the same importance as charge convection, self-similar solutions do not exist. Here we consider this case. A set of three integral equations is deduced describing the evolution of the width of the charged zone, the width of the hydrodynamic boundary layer and the fluid velocity along the axis of the plume, including the effect of Coulomb repulsion. We focus our analysis in the axisymmetric case, when a point electrode faces a plate.

2. Formulation of the problem

2.1. Hydrodynamic equations

The flow structure in EHD plumes is sketched in figure 1. It takes the appearance of a jet, originated at the point or the blade, that impinges upon the plate. We examine the problem in the asymptotic region, that is, far from both the injector and the plate. In this asymptotic region the flow has a double boundary-layer structure, with an inner charged core of width δ_q, and an outer hydrodynamic layer, of width δ_l. It is $\delta_q, \delta_l \ll d$, where d is the distance between the electrodes. Thus, the usual boundary-layer approximations are applicable. The continuity and the Navier–Stokes equations in stationary conditions take the form,

$$\frac{\partial yu}{\partial x} + \frac{\partial yv}{\partial y} = 0, \tag{1}$$

$$u\frac{\partial u}{\partial x} + v\frac{\partial u}{\partial y} = \frac{\nu}{y}\frac{\partial}{\partial y}\left(y\frac{\partial u}{\partial y}\right) + \frac{qE_x}{\rho}. \tag{2}$$

Here, u and v are the axial and transverse velocity of the fluid, q is the density of charge, E_x is the axial electric field, and ρ and ν are the density and the kinematic viscosity of the fluid, respectively.

124

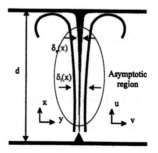

Figure 1. Flow structure of the EHD plumes. The motion is jet-like in the asymptotic region, far from both electrodes, with a double boundary layer structure.

2.2. Electric equations

The electric field is described by means of the Poisson equation and the charge conservation equation, in stationary conditions,

$$\nabla \cdot E = \frac{q}{\epsilon}, \tag{3}$$

$$\nabla \cdot j = 0. \tag{4}$$

Here ϵ is the dielectric permittivity of the fluid and j is the current density. In dielectric liquids j has three contributions, the ion drift by the electric field, KqE (K the ionic mobility), the charge convection by the fluid, qu and the charge diffusion $-D\nabla q$ (D the diffusion coefficient). As it has been showed in [2] and [3], diffusion of charge is negligible in EHD plumes. In this case, the charge conservation equation can be integrated to give[4]

$$q(t) = \frac{q_0}{1 + t/\tau_r}, \tag{5}$$

with q_0 the initial value of the charge density and $\tau_r = \epsilon/Kq_0$ the relaxation time, t is the elapsed time from injection. Essentially, it is the convective time, $t \approx \int dx/u$, because in most EHD flows $KE_x \ll u$, as it is our case.

2.3. Boundary conditions

The boundary conditions are imposed by the geometry of the flow

$$\frac{\partial u}{\partial y} = \frac{\partial q}{\partial y} = v = 0 \quad y = 0, \qquad u = 0 \quad y \to \infty. \tag{6}$$

Besides of this, the electric current must be the same at any plane transversal to the plume. Using $KE_x \ll u$ we get

$$I = 2\pi \int_0^{+\infty} quy \, dy. \tag{7}$$

3. Integral solutions for axisymmetric EHD plumes

We have applied and integral method to find the evolution of the three more relevant magnitudes of the flow: the fluid velocity at the center of the plume, the width of the hydrodynamic zone and the width of the charged core, following the previous works[5, 6]. The radius of the charged core, δ_q, is supposed to be smaller than the radius of the hydrodynamic boundary layer, δ_l.

Firstly, we consider the equations of conservation of the momentum (2) and the kinetic energy. This one can be obtained multiplying (2) by the radial coordinate, y, and the axial velocity, u,

$$\frac{\partial}{\partial x}\left(\frac{1}{2}u^3 y\right) + \frac{\partial}{\partial y}\left(\frac{1}{2}u^2 vy\right) = \nu\left[\frac{\partial}{\partial y}\left(y\frac{\partial}{\partial y}\left(\frac{1}{2}u^2\right)\right) - y\left(\frac{\partial u}{\partial y}\right)^2\right] + \frac{yuqE_x}{\rho}. \tag{8}$$

Here, equation (1) has been used. The left-hand side of (8) represents the convective transport of kinetic energy, the first viscous term is the lateral transport of kinetic energy and the second viscous term represents the viscous dissipation. Finally the last term is the power injected by the electric field into the system.

The conservation of electric charge is considered by means of the boundary condition (7). If δ_q is small enough it is

$$I = 2\pi \int_0^\infty quy\,dy \approx u_0(x)2\pi \int_0^{\delta_q} qy\,dy \approx u_0(x)Q(x), \tag{9}$$

with $Q(x) = 2\pi \int_0^\infty qy\,dy$, the density of charge per unit length, and $u_0(x)$ the velocity at the axis of the plume.

Equations (2) and (8) are integrated over a plane transversal to the axis of the plume. Taking into account that $\delta_q \ll \delta_l$ we assume a velocity profile $u(x,y) = u_0(x)f(\eta)$, with $\eta = y/\delta_l(x)$, $f(0) = 1$ and $f'(\eta) \le 0 \ \forall \eta$. From (9) we obtain another equation to describe the evolution of $\delta_q(x)$. In this way, we find a set of three differential equations that describe the evolution of $u_0(x)$, $\delta_l(x)$ and $\delta_q(x)$

$$u_0\frac{d}{dx}\left(u_0^2\delta_l^2\right) = \frac{1}{A_1}\frac{IE_0(x)}{2\pi\rho}, \tag{10}$$

$$A_2\frac{d}{dx}\left(\delta_l^2 u_0^3\right) = -A_3\nu u_0^2 + \frac{IE_0(x)}{2\pi\rho}, \tag{11}$$

$$u_0\frac{d}{dx}\left(\delta_q^2 u_0\right) = \frac{KI}{\pi\epsilon}. \tag{12}$$

The constants A_1, A_2 and A_3 are

$$A_1 = \int_0^\infty \eta f^2(\eta)d\eta, \qquad A_2 = \int_0^\infty \frac{1}{2}\eta f^3(\eta)d\eta, \qquad A_3 = \int_0^\infty \eta(f')^2(\eta)d\eta. \tag{13}$$

If $f(\eta) = \exp(-\eta)$ their numerical values are $A_1 = 1/4$, $A_2 = 1/18$, $A_3 = 1/4$. For a given dependence of the axial field on x, the equations (10), (11) and (12) along with their corresponding initial conditions describe the evolution of the axisymmetric EHD plumes.

3.1. Asymptotic solutions for uniform electric field

In the case of an uniform electric field $E = E_0 i_x$, i_x the unit vector in x direction, asymptotic solutions of these equations for large values of x can be found. These solutions are

$$u_0^\infty = \left(\frac{A_1 - A_2}{A_1 A_3}\frac{IE_0}{2\pi\rho\nu}\right)^{1/2}; \qquad \delta_l^\infty = \left(\frac{A_1 a_3^3}{(A_1 - A_2)^2}\frac{2\pi\rho\nu^3}{IE_0}\right)^{1/4}x^{1/2};$$

$$\delta_q^\infty = \left(\frac{2A_1 A_3}{A_1 - A_2}\frac{\rho\nu K}{\epsilon E_0}\right)^{1/2}x^{1/2}; \qquad \left(\frac{\delta_q}{\delta_l}\right)^\infty = \left(\frac{2A_1(A_1 - A_2)}{\pi A_3}\frac{\rho K^2 I}{\epsilon^2\nu E_0}\right)^{1/4} \tag{14}$$

Comparing these expressions with the self-similar solutions given in [3], we observe that the evolution of the velocity at the axis of the plume and the width of the hydrodynamic layer are similar to those obtained from the self-similar expressions. The integral model gives the evolution of the charged zone, that it is not provided by a self-similar one. For typical values of the involved magnitudes ($E_0 = 10^6$ V/m, $I = 10^{-7}$ A/m, $\rho = 850\,\text{gr/cm}^3$, $\nu = 2 \times 10^{-5}\,\text{m}^2/\text{s}$, $\epsilon_r = 2$ and $x = 1$ cm) it is

$$u_0^\infty = 1.8, \text{ m/s} \qquad \delta_l^\infty = 380\,\mu\text{m}, \qquad \delta_q^\infty = 80\,\mu\text{m}, \qquad (\delta_q/\delta_l)^\infty = 0.2. \tag{15}$$

These values are similar to those obtained from the self-similar model, for the velocity and the hydrodynamic thickness. They are in accord, in order of magnitude, with the experimental values for u_0 and δ_l[7]. To our knowledge, there are no experimental measures of the charged layer thickness.

126

4. Integral equations for two-dimensional EHD plumes

A similar analysis can be done for two-dimensional EHD plumes. In this case this three equations similar to (10), (11) and (12) are obtained.

When the electric field depends on x as $E_0(x) = \bar{E}x^m$, solutions as powers on x can be sought. We obtain

$$u_0 = C_u \left(\frac{J^2 E^2}{\rho^2 \nu} \right)^{1/5} x^{(1+2m)/5}, \qquad \delta_l = C_l \left(\frac{\rho \nu^3}{JE} \right)^{1/5} x^{(2-m)/5},$$

$$\delta_q = C_q \frac{K}{\epsilon} \left(\frac{J\rho^4 \nu^2}{E^4} \right)^{1/5} x^{(3-4m)/5}, \qquad \frac{\delta_q}{\delta_l} = C_{ql} \frac{K}{\epsilon} \left(\frac{J^2 \rho^3}{\nu E^3} \right)^{1/5} x^{(1-3m)/5}. \tag{16}$$

For a uniform electric field it is $m = 0$. In this case we reobtain the evolution of the center velocity and the hydrodynamic width given by the self-similar model in [2]. Using the same values than in the axisymmetric case for the physical quantities involved, with $J = 10^{-7}\,\text{A/m}$, we have

$$u_0 = 13\,\text{cm/s}, \quad \delta_l = 1.8\,\text{mm}, \quad \delta_q = 5.3\,\mu\text{m}, \quad \delta_q/\delta_l = 3 \times 10^{-3}. \tag{17}$$

The values of the velocity and the hydrodynamic thickness are similar to those experimentally obtained [2, 5]. Similarly to the axisymmetric case, there are no experimental measures of the thickness of the charged region.

5. Conclusion

The flow of axisymmetric and two-dimensional EHD plumes has been analyzed, including the effect of Coulomb repulsion in the spread of electric charge. Integral methods have been applied to obtain three differential equations that describe the behavior of the center velocity of the plume and the thicknesses of the hydrodinamic and charged regions. It is found that the velocity at the center of the plume and the hydrodynamic width are well described by the self-similar models, but it is not the case for the charged region. The integral method gives the quotient between these two thicknesses. For an uniform electric field, in the axisymmetric case, the thickness of the charged core turns out to be one fifth of the thickness of the hydrodynamic boundary layer. The quotient between both thicknesses is 3×10^{-3} in the two-dimensional case.

Acknowledgments

This research has been supported by the Spanish Government Agency DGES under contract N0. PB96-1375.

References

[1] Zhakin A 1984 Electroconvective jets in liquid dielectrics. *Mekhanica Zhigkosty i Gaza. Isvestia Academy Science USSR* 6 13–19

[2] Pérez A T, Vázquez P A and Castellanos A 1995 Dynamics and linear stability of charged jets in dielectric liquids. *IEEE Transactions on Industry Applications* 31 761–8

[3] Vázquez P A, Pérez A T and Castellanos A 1996 Thermal and electrohydrodynamical plumes. a comparative study *Physic of Fluids,* 8(8) 2091–6

[4] Castellanos A (Editor) 1998 *Electrohydrodinamics* (New York: Springer-Verlag)

[5] McCluskey F M J and Pérez A T 1992 The electrohydrodynamic plume between a line source of ions and a flat plate *IEEE Trans. on electrical insulation,* 27 334–41

[6] Atten P, Malraison B and Zahn M 1993 Electrohydrodynamic plumes in point-plane geometry *Proc. IEEE 11th ICDL* (Baden, Switzerland)

[7] Takashima T, Hanaoka R, Ishibashi R and Ohtsubo A 1988 I-v characteristics and liquid motion in needle-to-plane and razor blade-to-plane configurations in transformer oil and liquid nitrogen. *IEEE Transactions on Electrical Insulation,* 23 645–57

Inst. Phys. Conf. Ser. No 163
Paper presented at the 10th Int. Conf., Cambridge, 28–31 March 1999

Enhancement of Heat Transfer and Mass Transport in Thermal Equipment with Electrohydrodynamics

J. Seyed-Yagoobi and J.E. Bryan

Department of Mechanical Engineering
Texas A&M University
College Station, Texas 77843-3123

Abstract. Significant work on EHD enhanced heat transfer and mass transport has been performed by researchers around the world for the last twenty years with a focus on industrial applications. There is still a significant amount of applied and fundamental research required before the EHD phenomena can be utilized in an industrial setting. However, it is expected that within the next decade some EHD enhanced industrial applications will become a reality.

1. Introduction

Electrohydrodynamics (EHD) deal with the interaction between electric fields and fluid flow. This interaction, for example, can result in electrically induced pumping, mixing, or enhancement of heat transfer. The application of EHD to heat transfer and mass transport problems continues to show promise in industrial applications where special requirements and restrictions are imposed, but enhanced heat transfer and mass transport are required (Seyed-Yagoobi and Bryan (1999)). One example is enhanced performance and control of two-phase heat transfer in refrigeration and process industries. This includes evaporation and condensation heat transfer. Another example is mass transport in microgravity environments where single-phase liquid or two-phase liquid-vapor transport is required with high reliability (few if any moving parts), and low or no vibration. The advantages of EHD enhancement are: 1) rapid and smart control of enhancement by varying the applied electric field; 2) non-mechanical and simple in design; 3) suitable for special environments (space); 4) applicable to single and multi-phase flows; and 5) minimal power consumption. However, the implementation of high voltage in these promising industrial applications still poses design and economic challenges. Furthermore, the implementation of the EHD phenomena to heat transfer and fluid dynamics introduces a complex interaction of many interdependent variables.

The field of EHD, especially its application to heat transfer and mass transport has been studied for some time. Jones (1978) provided an excellent review of EHD enhanced heat transfer in liquids. Most of the research at that time dealt with the fundamental demonstrations supported by some theoretical models. Since that time, researchers have demonstrated EHD enhanced heat transfer and mass transport which are more closely related to industrial applications. However, the fundamental understanding and theoretical model development have become much more complicated and difficult. Because of this, a complete fundamental understanding does not exist and only limited theoretical models are available. A comprehensive review of past research on EHD enhanced single and two-phase heat transfer and EHD induced flow, has been presented by Allen and Karayiannis (1995).

Recently, Seyed-Yagoobi and Bryan (1999) provided a review on enhancement of heat transfer and mass transport in single-phase and two-phase flows with EHD. The purpose of this review was to provide perspective researchers with an initial background on EHD enhanced heat and mass transfer and information and approaches to studying electric field, fluid flow field, and heat transferred coupled problems.

2. EHD Enhancement of Boiling Heat Transfer

From the 1980's to the present, EHD enhanced nucleate boiling has been studied and applied on boiling surfaces more closely related to industrial applications. In the nucleate boiling, the charge relaxation time, τ_e, plays an important role during bubble growth and departure in an electric field. Enhancements in nucleate boiling heat transfer are expected if τ_e is smaller than the bubble departure time. The EHD forces acting on the liquid in the nucleate boiling regime can be responsible for a significant portion of the heat transfer enhancement. More fundamental studies on the influences of the EHD forces on the bubble growth dynamics, the effects of varying applied electric fields, long term effects, and a better understanding of the influence of fluid properties and impurities are needed. The issues of technology transfer also need to be addressed, such as tube bundles, cost, and manufacturability. Additionally, design methodology, models, and/or correlations are needed so EHD enhanced evaporators can be studied at the integrated system level.

EHD enhancement of convective boiling has only recently been investigated. There have been only a few researchers in the last seven years to investigate EHD enhanced convective boiling. The limited amount of research is, in part, due to the lack of fundamental knowledge on EHD enhanced heat transfer and the complicated nature of convective boiling. The heat transfer which occurs in convective boiling is a function of not only the fluid properties but also the mass flux, quality, flow regime, heat flux, operating temperature, and geometrical characteristics of the heat transfer surface. By introducing the EHD force the heat transfer is now a function of the above variables as well as the fluid electrical properties, electrode geometry, and the applied electric field. In convective boiling, the EHD force in many situations will be perpendicular to the bulk fluid motion and the influence of this force will be strongly dependent on the variables mentioned above. As the quality changes, the flow regime will change and the heat transfer becomes either or both nucleate boiling dominated and forced-convection dominated. This same EHD force which will be acting primarily perpendicular to the fluid flow can also result in increased flow resistance, there by increasing the pressure drop. Therefore, it is important to understand the interdependence of the EHD force and the convective boiling process if an EHD force is to be produced which will maximize the heat transfer and minimize the pressure drop penalty for a given flow condition.

3. EHD Enhancement of Condensation Heat Transfer

The internal convective condensation and external condensation processes are both ideal for application of EHD. Assuming that no free charges are generated or introduced, the dominant electric body force density acting on the condensate will be due to the permittivity gradient. This force can be utilized to remove the liquid condensate, which has a higher permittivity than the surrounding vapor phase, away from the condensation surface (electrically grounded and low electric field) toward the electrodes (high electric field). This will produce a so-called extraction mechanism resulting in enhancement of condensation heat transfer by removing the condensate and exposing the condensation surface to the vapor. The research in this area has been primarily based on the work by Yabe et al. (1987). Compared to the large number of studies conducted on EHD enhancement of pool boiling, few investigations have been published that deal with EHD augmentation of condensation heat transfer.

More work, experimentally and theoretically, is needed in this area to design optimum electrodes under various operating conditions and working fluids. An accurate theoretical model will reduce the need for extensive experimental work. Future work in this field should especially focus on tube bundles. Furthermore, better understanding of EHD enhanced condensation heat transfer will allow for proper design of enhanced tubes that will be ideal from EHD application standpoint.

4. EHD Pumping

Pumping is achieved when electric fields are generated that drag the charges along in a given direction. Therefore, EHD pumping is a phenomenon that has two basic requirements. First, the dielectric fluid being pumped must contain free charges. Second, an electric field must be present to interact with the free charges in the fluid. In EHD pumping, the Coulomb force is the main mechanism of this interaction. Within the last ten years, many unique EHD pumps have been built with various applications in mind. It is very likely that the commercialization will become a reality within the next decade for some special applications. Primarily for the purpose of outer space applications, Bryan and Seyed-Yagoobi (1997) investigated heat transport enhancement of a monogroove heat pipe with EHD pumping. The EHD pump was located on the liquid channel in the adiabatic section of the heat pipe. The heat pipe fluid used in all experiments was R-123. Over 100% enhancement in the transport capacity was achieved using the EHD pump operating at 20 kV. This enhancement was maintained with less than 0.98 W of electric power to the EHD pump. The EHD pump was also able to provide immediate recovery from dryout when the heat pipe had been experiencing progressive evaporator dryout for over 70 min at 400 W.

5. Enhancement of Heat Transfer with Corona Wind

Heat and mass transfer between a surface and the surrounding gas can be significantly enhanced through the application of a high electric field. When a high voltage is applied to a small diameter electrode, such as a fine wire or sharp needle, gas in the vicinity of the electrode becomes ionized. The free ions are forced away from the electrode along the electric field lines, and are drawn towards the nearest electrical ground. Collisions between the ions and the gas transfer momentum to the fluid, and create a plume-like gas flow. The resulting transport enhancement requires little power, causes virtually no sound or vibration, and can be controlled by simply changing the applied voltage. It is important to recognize that the heat transfer enhancements with corona wind in fluid flow situations can be significant primarily in laminar conditions. The additional fan power required due to presence of electrode(s) should be carefully considered. Future studies in this area should focus on identifying problems relevant to industrial applications. For example, drying, cooling, or heating of food products and biological substances with corona wind may have certain advantages.

6. Conclusions

The areas showing the most promise for EHD augmentation in the next decade are pool boiling, microchannel convective boiling, internal and external condensation, frost control, and pumping in microgravity applications. It is also expected that other novel applications using the EHD phenomena will be identified in the years to come.

7. References

Allen PHG and Karayiannis TG 1995 *Heat Recovery Systems & CHP*. **15** 5 pp 389-423
Bryan JE and Seyed-Yagoobi J 1997 *AIAA Journal of Thermophysics and Heat Transfer*, **11** 3 pp 454-460
Jones TB 1978 *Advances in Heat Transfer* (Academic Press) **14** pp 107-148
Seyed-Yagoobi J and Bryan JE 1999 *Advances in Heat Transfer* (Academic Press) **33**
Yabe A Taketani T Kikuchi K Mori Y and Hijikata K 1987 *Heat Transfer Science and Technology*, Bu-Xuan Wang, ed. (Hemisphere Publishing Corporation) pp 812-819

Inst. Phys. Conf. Ser. No 163
Paper presented at the 10th Int. Conf., Cambridge, 28–31 March 1999
131

Flow electrification in high power transformers
(Effect of temperature gradient and charge accumulation)

G Touchard[1], G. Artana[2], O. Moreau[3]

[1]LEA UMR 6609 CNRS, 40 avenue du Recteur Pineau, Poitiers, France
[2]CONICET, UBA., Fac Ing. Paseo Colon 850-1063 Buenos Aires Argentina
[3]EDF DER, Service ERMEL 1 Av. du Général de Gaulle, Clamart France

Abstract. This paper deals with flow electrification in transformers. An experimental set-up has been built to analyse the influence of the transverse thermal gradient, and the surface pressboard resistance. It is found that the transverse thermal gradient has a great influence on the phenomenon as the wall current is greatly increasing in terms of this parameter. In an other hand, the charge build-up at the oil/pressboard interface reduces the surface resistance.

1. Introduction

It is now well recognise that flow electrification could be suspected to be responsible of major failures in power transformers [1].

Important programs have been entered upon as well in USA [2], managed by E.P.R.I., as in France, managed by the French Electricity Corporation (E.D.F.). Taking part in this program the L.E.A. is working on the subject for several years [3&4]. Thus, a set of large scale equipment has been perfected in order to analyse the influence of different parameters on the phenomenon.

Nevertheless, the phenomenon of flow electrification is not yet well understood, especially the magnitude of the charge convected cannot be predicted. This is due to the interfacial physicochemical process itself which is not well understood. For instance the magnitude of the current generated by a given liquid flowing through a given tube cannot be predicted at the present time. The reasons of these difficulties are due to the fact that only a few amount of impurities (few P.P.M.) in the liquid or in the solid composition, changes radically the value of the charge convected.

The case of oil flowing past impregnated pressboard is even worth as the oil contains unknown impurities and the pressboard is an insulating material, more it is porous and contains unknown impurities also.

Thus, investigations made all over the world in this difficult domain enhance the general knowledge of the process and help to solve, but very often empirically, the safety problem induced by this phenomenon. In this context, the L.E.A. is working for more than five years under contract with the French electricity corporation (E.D.F.) in conjunction with a E.P.R.I. program, and of course, if this contractual research is still going on it is because , still now, transformers in different countries including, U.S. Japan and France have major failures due to this phenomenon.

The equipment perfected (figure 1) allows to measure the charge convected in cylindrical pressboard ducts inside which oil is flowing. These ducts are supposed to be

132

representative of the flow path inside high power transformers like washers separated by spacers. Successive equipment configurations have been made up. First, to analyse the effect of duct diameter and entry length. Then, we have compared different flow geometry : parallel flow and impinging immersed jet on a pressboard target. We analysed also the influence of the pressboard insulation. For each configuration two kinds of flows was tested : a flow loop and a single path. In this paper two different parameters are observed, the effect of the transverse thermal gradient through the pressboard on the charge generation and the evolution of the surface resistance due to the charge accumulation.

2. The thermal transverse gradient influence.

To measure the effect of a transverse thermal gradient, in conjunction with the mean flow velocity, on the charge convected, a special part of the mean equipment has been perfected (figures 2&3), it is a cylindrical pressboard channel of rectangular cross section divided in three different sections which could be heated and investigated separately. Each section contains a set of heating resistors disposed above and below the pressboard duct.

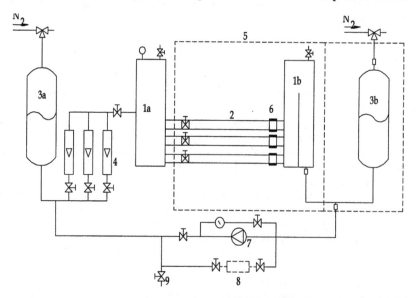

Figure 1 : Experimental Equipment
1 Stainless steel vessel, 2 Channels, 3 Pressure vessels, 4 Flowmeter, 5 Faraday cage, 6 Insulators, 7 Pump, 8 Conductivity measurement system, 9 Oil sampler.

The temperatures are measured with the help of thermocouples at the outer pressboard side and at the inner one (oil-pressboard interface). Oil sampling is also available in each section to control the moisture evolution. Such configuration was perfected in order to observe the mean temperature, the longitudinal thermal gradient and the transverse thermal gradient. In fact the effect of the longitudinal thermal gradient is very difficult to observe as the temperature variation at the oil/pressboard interface varies very little from one section to the other even if only one section has its heating resistors working.

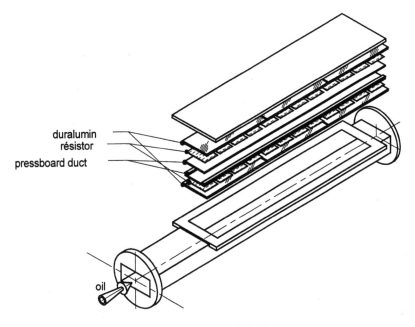

duralumin
résistor
pressboard duct
oil

Figure 2 : Scheme of the duct

Here we examine the evolution of the pressboard surface resistance in terms of the mean temperature and the effect of the transverse thermal gradient on the electrification process. Indeed, impregnated pressboard is a good thermal insulator, and heating with the resistors leads rapidly to a stable temperature at the contact between the pressboard and the resistors (outer face of the pressboard duct), while the oil/pressboard interface, as the oil is flowing and because of the important volume of oil in the installation, remains to an other stable temperature. So a thermal gradient takes place through the pressboard.

The wall current generated by flow electrification is measured by the help of electrodes placed at the oil pressboard interface and compared to the thermal gradient evolution. We can see figure 4 , that it is greatly increasing with the thermal gradient and that a kind of hysteresis phenomenon appears. Also, even if the evolution is more important for the lower flow, the influence of flow velocity on the phenomenon does not seem to be very important. The hysteresis effect is not really surprising as the temperature gradient induces a very slow diffusion process which does not follow so rapidly the thermal gradient changes, then the wall current is of course in correlation with this diffusion process.

3. The surface resistance evolution.

The influence of flow electrification on the surface resistance of a piece of immersed pressboard is made in a metallic vessel in which a pressboard sheet is placed parallel to the oil flow (figure 5). This vessel is a part of the whole experimental equipment which is placed between the flowmeters (4) and the vessel (1a). Thus it is a online measurement. Resistance at the oil pressboard interface can be measured with the help of four electrodes inserted at the pressboard surface.

134

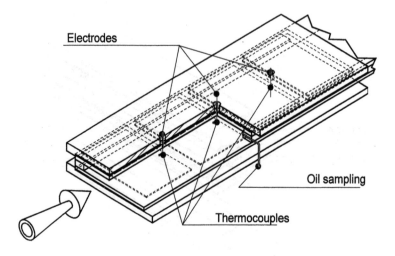

Figure 3 : One section of the duct

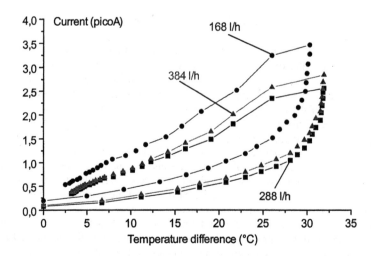

Figure 4 : Evolution of the wall current in terms of the thermal gradient for different flow.

The resistance between two electrodes of the same side give the surface resistance, while the resistance between vis-a-vis electrodes gives the impregnated pressboard resistivity. In this paper we are interested in the surface resistance evolution in terms of time, mean temperature and flow velocity. The reason of this investigation concerns the charge build up on a part of pressboard very well insulated, as it has been very often related that failures are initiated in a region of the transformer where the pressboard is well insulated. Thus in this region flow electrification leads to a charge accumulation on the pressboard then successive crawling discharges could occur like brush discharges but these discharges change themselves the interfacial process increasing electrification, finally there is a kind of avalanche phenomenon.

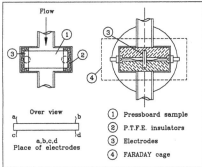

Figure 5 : Experimental set-up

The mean to analyse the effect of the charge accumulation is to observe the evolution of the pressboard surface resistance.

Thus experiments show that the surface resistance is not greatly dependent on the flow velocity (figure 6), but is decreasing to an asymptotic value with flow duration and temperature (figures 7 and 8), then when the flow is stopped it increases to recover the initial value (figure 9).

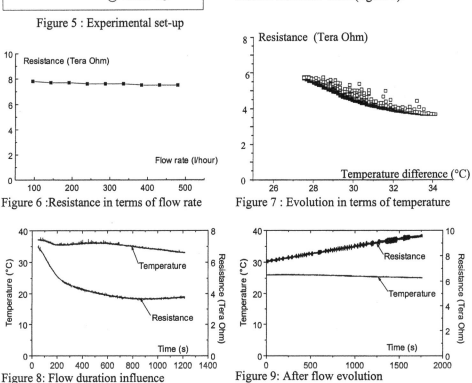

Figure 6 :Resistance in terms of flow rate

Figure 7 : Evolution in terms of temperature

Figure 8: Flow duration influence

Figure 9: After flow evolution

Thus it seems that the charge accumulates in terms of flow duration on the pressboard surface to an asymptotic value which probably corresponds to the equality between the wall current and the discharge current, more the level of this equilibrium depends on the temperature.

References

[1] EPRI TR-105019 1995 Project 3334-78 Proceedings May.
[2] Howells E., Zahn M., Lindgren S.R. 1990 *IEEE-Transactions on Power Delivery* **5** 1000-1006.
[3] Moreau O., Artana G., Touchard G. 1996 *Proc CEIDP 96* 186-189.
[4] Artana G., Touchard G., Moreau O. 1997 *Journal of Electrostatics* **40&41** 129-134.

Inst. Phys. Conf. Ser. No 163
Paper presented at the 10th Int. Conf., Cambridge, 28–31 March 1999

137

Fluid flow driven by a.c. electric fields in microelectrodes

A. Ramos†[1], A. González‡†, A. Castellanos†, H. Morgan§ and N.G. Green§†

† Dpto. de Electrónica y Electromagnetismo (Fac. Física, Universidad de Sevilla) Reina Mercedes s/n, 41012 Seville (Spain)

‡ Dpto. de Física Aplicada (E.S.I., Universidad de Sevilla) Camino de los Descubrimientos s/n, 41092 Seville (Spain)

§ Dept. Electronics and Electrical Engineering, University of Glasgow, Rankine Building, Oakfield Avenue, Glasgow G12 8LT, Scotland, UK

Abstract. During recent experimental observations of dielectrophoretic behaviour of sub-micron particles under a.c. electric fields, a new type of fluid flow has been observed. This fluid flow occurs close to, and across, the electrode surface. The magnitude of the velocity is frequency dependent, showing a maximum and going to zero for high and low frequencies. Evidence is presented to suggest that this fluid flow has its origin in the electrical stresses acting on the electrolyte/electrode double layer. A simple model based on the treatment of the double layer as a distributed impedance is used to predict the basic aspects of the behaviour of the system.

1. Introduction

The manipulation and separation of particles in suspensions using a.c. electric fields has developed rapidly in the last few years. This has been achieved through the application of micro-fabrication methods in the manufacture of microelectrode structures which generate high electric fields from low potentials. During the a.c. electrokinetic manipulation of particles in suspension on microelectrode structures, fluid flow was observed. We arrived at this conclusion because particles of different sizes and properties were observed moving at the same velocities [1].

In order to characterize this fluid flow experiments were performed using microelectrodes consisting of two parallel coplanar plates on glass slides. The plates were 2 mm long, 100 μm wide and separated by a gap of 25 μm (see fig. 1a). The electric field was generated using a variable frequency/amplitude signal generator. Fluorescent latex beads 282 nm in diameter were suspended in KCl solutions with conductivities of 2.1 mS/m, 8.4 mS/m and 84 mS/m. The particles were used as markers to observe the fluid flow arising when an a.c. voltage is applied to the electrodes. A sketch of the observed trajectory of the particles is also shown in figure 1a. In order to quantify the

[1] E-mail: ramos@cica.es

138

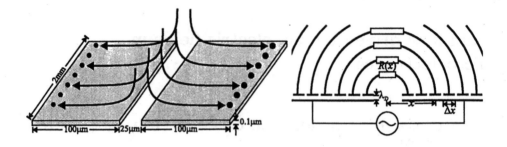

Figure 1. Schemes of electrodes and fluid flow (a), and of the distributed circuit (b).

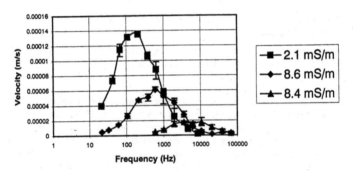

Figure 2. Velocity as a function of frequency for three conductivities at 2 V pp.

particle velocity, an arbitrary point 10 μm in from the edge of the electrode was chosen and only particles moving in the focal plane of the microscope objective, placed at the level of the electrodes, were recorded.

Figure 2 shows the magnitude of the velocity as a function of frequency for three conductivities, with an applied signal of 2 volts peak to peak. As can be seen from the figure, the graph of the velocity as a function of the frequency has a bell shape: it goes to zero for high and low frequencies and has a maximum. It can also be seen that the frequency of the maximum velocity increases with increasing conductivity of the liquid and that the magnitude of the velocity decreases with increasing conductivity.

We postulate that the driving force for this fluid arises from the interaction of the non-uniform electric field with the charge in the diffuse double layer. We refer to the resulting continuous fluid flow as a.c. electroosmosis.

2. Theoretical model

A bipolar electrolyte between two electrodes is electroneutral except for a layer (of width the *Debye Length*, λ_D), near the electrodes, called the electrical double layer [2]. We are going to use the linear approximation of the double layer to get the qualitative aspects of the fluid flow.

We model the double layer as an array of parallel plates capacitors (fig. 1b). In the simplest version, the capacitances per unit area of these capacitors are given by ε/λ_D, where ε is the permittivity of the liquid (if needed the model can include a capacitance due to the Stern layer). To model the liquid bulk, we use resistors of different lengths, following the field lines, that we assume circular arcs. The liquid bulk can be modelled as a resistor if the frequency is low enough to neglect the displacement current, i.e. $\omega\varepsilon/\sigma \ll 1$ (ω = driving frequency, σ = conductivity). We then have an array of elements in parallel, with each element composed of two capacitors and a resistor. Taking an elemental width Δx we have the following capacitances, resistances and impedances per unit length

$$C = \frac{\varepsilon\Delta x}{\lambda_D}, \qquad R = \frac{\pi x}{\sigma\Delta x}, \qquad Z = \frac{2}{i\omega C} + R = \frac{2\lambda_D}{i\omega\varepsilon\Delta x}.$$

The voltage drop across the double layer is then

$$V_D(x) = \frac{V_0}{2 + \frac{i\varepsilon\pi\omega x}{\sigma\lambda_D}},$$

where V_0 is the amplitude of the a.c. potential difference between electrodes. In this model, the volume charge is substituted by a surface charge placed at an average distance λ_D from the electrodes. The magnitude of this surface charge at a position x is given by the capacitance times the voltage drop across the double layer at x, and the tangential field is given by the changes in the potential from one capacitor to the next. The results are

$$\sigma_s = \frac{\Delta Q}{\Delta x} = \frac{\varepsilon}{\lambda_D}V_D(x) \qquad E_x = -\frac{\partial V_D}{\partial x} = \frac{i\varepsilon\pi\omega}{\sigma\lambda_D}\frac{V_0}{\left(2 + \frac{i\varepsilon\pi\omega x}{\sigma\lambda_D}\right)^2}$$

If $\omega \to 0$ (d.c. voltage) the tangential field vanishes, whereas if $\omega \to \infty$ both the charge and the field go to zero. In the former case, the main voltage drop occurs in the double layer, while in the latter it occurs in the bulk.

The interaction between the surface charge and the electric field results in electrical stresses, which, in this model, are concentrated on a surface at a distance λ_D from the plate. Both the charge and the field are oscillating magnitudes, and their product has a non zero average, which will produce a continuous liquid motion. The mean stresses are $\langle M \rangle = (1/2)\mathrm{Re}\,(q_s E_x^*)$, that gives a net force in the direction of increasing x.

To these stresses we must add the electroosmotic pressure gradients [3]. Very close to the plate, the liquid is almost at rest, and the field is normal to the plate. In this case, we can approximate the pressure by $p = p_0 + (1/2)\varepsilon E^2$. The pressure is higher near the center and decreases with the horizontal distance, and the pressure gradient acts in the same direction as the electrical stresses. To a first approximation the stresses due to the pressure are

$$-\frac{\partial p}{\partial x} = -\frac{1}{2}\frac{\partial}{\partial x}\left(\frac{\varepsilon|V_D(x)|^2}{\lambda_D^2}\right).$$

This gradient acts on the whole diffuse layer, but, in the same approximation, we can integrate these stresses and consider them acting on the surface $y = \lambda_D$. It results in a term equal to the electrical stresses, thus the total stresses are twice this quantity. These stresses must be balanced by the jump in the viscous stresses. If we assume a linear profile in the double layer, and a uniform one outside it, the outer velocity results

$$U = \frac{\lambda_D}{\eta}\langle M \rangle = \frac{\lambda_D}{\eta}\mathrm{Re}\,(q_s E_x^*) = \frac{V_0^2 \varepsilon x}{4\eta\left(1+\left(\frac{\varepsilon\omega\pi x}{2\sigma\lambda_D}\right)^2\right)^2}\left(\frac{\omega\varepsilon\pi}{2\sigma\lambda_D}\right)^2. \tag{1}$$

3. Discussion and conclusions

The velocity predicted by the model gives the qualitative aspects of the observed motion. The direction of the flow and the bell shape of the velocity as a function of frequency are in accordance with the experiments. The expression shows a maximum for the velocity when $x\omega\varepsilon/\sigma\lambda_D = 2/\pi$, giving frequencies for the maximum of 400 Hz, 750 Hz, and 2.5 kHz for the conductivities of the experiment. These values compare reasonably well with the experimental ones that can be seen in figure 2.

However, the figure also shows a decrement in the velocity with the conductivity, that is not described by (1). The model is oversimplified. It produces good results for the expected trends but fails to give accurate estimates for the velocity magnitudes (they are higher than in the experiments). This means that the model probably captures the physical mechanism that move the liquid, but a more detailed analysis is required.

It is possible to refine the model introducing new circuit elements. An improvement would be to include the effects of the Stern layer as another capacitor in series with the diffuse double layer capacitor, and the nonlinear effects should also be considered. When the voltage is high, the interaction between the charge and the field in the diffuse layer produces an accumulation of charge next to the electrode. In addition, in the nonlinear limit, the electric stresses act on a narrower layer, near the electrode. Both effects result in a reduction of the expected liquid velocity, which would depart from the quadratic behaviour for high voltages.

Acknowledgements

This work has been supported by the Spanish DGES under contract PB96–1375 and Acciones Hispano-Británicas HB1996-0230, the European Union, through a Marie Curie fellowship (contract BIO4-CT98-5010 (DG12-SSMI)).

References

[1] A. Ramos, H. Morgan, N.G. Green and A. Castellanos, Ac electrokinetics: a review of forces in microelectrode structures, *J. Phys. D: Appl. Phys.* 31, 2338-2353 (1998).

[2] J.O'M Bockris and A.K.N. Reddy. *Modern Electrochemistry*. Plenum Press, New York (1973).

[3] V.G. Levich, *Physicochemical hydrodynamics*, Prentice-Hall (1962).

Inst. Phys. Conf. Ser. No 163
Paper presented at the 10th Int. Conf., Cambridge, 28–31 March 1999

Global Instabilities in an electrified jet

Guillermo Artana† and Bruno Seguin †‡

† CONICET, Dept. Ingenieria Mecanica, Fac. Ingenieria, Universidad de Buenos Aires, Argentina, gartana@fi.uba.ar

‡ CEFIS-INTI-P.T. Miguelete, San Martin, Argentina

Abstract. This article describes the influence of an electric field on the stability of an axially accelerated liquid jet. We present the fundamental data to determine the global stability for any case and particularly study the case when the jet is accelerated mainly by gravitational forces. This research indicates that dangerous situations of global stability loss occur for the cases of strong electric field close to the jet exit and low velocity jets.

1. Introduction

In the stability analysis of electrified jets as a function of the type of perturbation imposed to the flow, two different problems, the temporal and spatial one, have usually been considered. The temporal problem considers that the perturbation is applied to the flow in a certain region of the space at time equal zero, while the spatial problem considers that the perturbation is applied in a certain region and follows a given function in time. These kind of analyses have usually been restricted to the local stability of the velocity profile in a typical streamwise station that was invariant in the direction of the wave propagation. However, when the basic flow differs significantly from one streamwise station to the other, the stability analysis requires to consider the stability not only of the local velocity profile in a typical station (local stability analysis), but of the entire flow field (global stability analysis). In the particular case of electrified jets results obtained from global analysis are of great interest. The streamwise stations may differ from each other as a result, among other phenomena, of gravity acceleration, of initial velocity profile relaxation, of viscous action of the surrounding atmosphere and of the limited time of charge relaxation that leads to a non equipotential jet surface.

Previous research dealing with electrified jets has tried to gain some insight of this kind of problem, and the effect of the acceleration of the jet due to gravitational forces on the growth rate of perturbations imposed to the flow has received special attention [1] . However, at that time the theory of global instability was not fully developed and concepts of locally and globally stable flows were not applied. Recently the global response of a gravitational jet flow acting either as an amplifier that selectively amplifies extrinsic noise (global stability) or as an oscillator of self-excited time-growing oscillations at any fixed location (global instability) has been established [2-3]. It is the objective of this article to identify in electrified jets the flow and electric field configuration that leads to either behaviour.

2. Description

We analyze the case of a downwards jet issuing from a circular nozzle coaxial with a cylindrical electrode. Hypothesis considered are: the surface of the liquid jet is an electrical equipotential; the magnetic effects can be disregarded; the jet is inviscid, incompressible and isothermal and there is no mass transfer between the jet and the surrounding atmosphere.

2.1. Local Analysis : Absolute and Convective flows

The link between the local stability properties and the global behaviour of the flow is still controversial. Nevertheless, most observations support the idea that the formation of local absolute instability pockets is a necessary condition for the appearance of a global instability. The absolute or convective nature of an instability is related to the asymptotic impulse response for time t→ ∞ . Convectively unstable flows give rise to wave packets that move away from the source and ultimately leave the medium in its undisturbed state. Absolutely unstable flows, by contrast, are gradually contaminated everywhere by a point-source input. In what follows we undertake this analysis by using the dispersion equation obtained in [4] considering negligible the effect of the sourrounding atmosphere and also that the electrode radius is much larger than the jet radius. This equation $D(k, \omega, n, We, Eue) = 0$ establishes a relationship between the complex wavenumber k, the complex frequency ω, the mode number n, the Weber number We (the ratio between inertial and surface tension forces) and the Electric Euler number Eue (the ratio between the electric forces and inertial forces). Using this equation, and analysing the spatial branches in the k-plane [5] it is possible to obtain the absolute growth rate that characterizes the temporal evolution of the wave number of zero group velocity at a fixed station in the limit $t \to \infty$. A positive absolute growth rate is therefore associated with an absolute instability, and a negative or zero value with a convective one. Figure 1 represent this absolute growth rate (non dimensionalized with jet radius and velocity) as a function of mode number for different $Eue = \frac{\epsilon_0 E_n^2}{\rho U^2}$ (where ϵ_0 is the dielectric constant in vaccum, E_n the electric field at the jet surface, ρ the liquid density and U the mean jet velocity). It can be observed in this figure that, as the first mode is the one which has the highest absolute growth rate, in general we can limit the analysis to this mode to observe the absolute/convective unstable behaviour of the electrified jet. Figure 2 shows the absolute growth rate as a function of the ratio of electric forces with surface tension forces ($Rae = Eue * We$) for different Weber number ($We = \frac{\rho U^2 a}{\gamma}$, with a the jet radius and γ the surface tension). This figure shows that for low Electric Euler number the absolute growth rate is null and also how by increasing the electric field it is possible to obtain positive values of the absolute growth rates. In Figure 3 we represent the critical Electric Euler Number $cr.Eue$ as a function of Weber number (critical Euler number is the one that separates absolute from convective behaviour). This last figure indicates that higher Weber number are associated with lower $cr.Eue$ ratios.

2.2. Global modes in weakly inhomogeneous flow

Under the umbrella of the hypotheses cited above the shape of the jet for a given nozzle diameter and flow rate will depend on the relative magnitude of gravitational, surface tension and electrical forces. Analytical expressions are difficult to obtain as one of the boundary conditions is part of the solution, and so numerical methods seem the most adequate alternative to obtain refined solutions. However in some limit scenarios simplified analytical solutions can be easily obtained. Results that are given below correspond to the extreme case occurring when the acceleration of the jet is determined mainly by gravitational forces. To obtain the jet shape we use the expressions deduced in [5] where it is determined that the ratio $\tau = a/a_0$ can be obtained from:

$$\left(\frac{1}{\tau^4} - 1\right) + \frac{2}{We_0}\left(\frac{1}{\tau} - 1\right) = 2gz\left(\frac{\pi a_0}{q}\right)^2$$

with We_0 and a_0 the Weber number and the jet radius at $z = 0$, g gravity acceleration, z axial coordinate, and q the flow rate. Links between local and global analysis are obtainable when on the scale of the instability wavelength the base flows develops slowly in the streamwise direction [2]. This condition is satisfied in this case if $ga_0^3/q^2 \ll 1$. Using the above expression and considering an electric field corresponding to corona inception we show in figure 4 the non-dimensional growth rate as a function of the axial position. This figure indicates that pockets of absolute instability may appear and so global stability be lost if the electric field strongly stresses the jet surface close to the jet exit.

In some cases where gravitational forces are not predominant, it may be of interest to have some rough estimations of the limits of global stability of electrified jets without undertaking numerical analysis to obtain the velocity and electric field variation with the axial coordinate. Looking at figure 3 we may see in this figure that it seems reasonable for many cases of interest (We close to or lower than 100) to accept as a limiting value for global stability the ratio $Eue \cong 0.35$. By doing that and considering at the jet surface the maximum ionization field in air, it can be stated that no global instability should appear for values of $Uo \leq \sqrt{227.7/\rho}$ with the liquid density ρ in SI units. In contrast, if it is desired to design an experiment to observe global instabilities by stressing a water jet with an electric field, by using the above criteria the jet velocity U_0 should at least be lesser than a value of 0.47m/s but it should be pointed out that the nozzle radius should assure an Weber number large enough to avoid dripping regime ($We \succeq 3.1$).

3. Conclusions

This research work establishes the set of Weber and electric Euler number that assures dampening of the global mode for the case of highly conducting inviscid liquids. We have considered a simple case of predominant gravity acceleration of the jet. For this case pockets of absolute instability that could lead to a global instability may appear only for strong electric field close to the jet exit. This research also indicates that in general, flows with low velocity jets stressed with strong electric fields are situations of special attention where global stability must be analysed carefully.

This research has been undertaken with CONICET Grant PEI 175/97

144

References

[1] J. Crowley 1968 *Physics of Fluids* **8** 2172-2178

[2] P. Monkewitz 1990 *Eur. J. Mech. B/ Fluids* **5** 395-413

[3] S. Le Dizes 1997 *Eur. J. Mech. B/Fluids* **16(6)** 761-778

[4] G. Artana et al 1997 *J. of Electrostatics* **40&41** 33-38

[5] P. Huerre and P. Monkewitz 1990 *Annual Rev Fluid Mechs* **22** 473-537

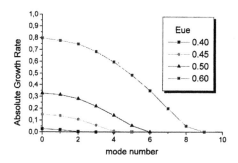

Figure 1. Non dimensional absolute growth rate as a fuction of mode number n for different Electric Euler Number

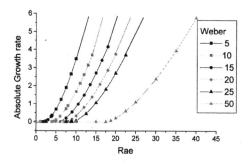

Figure 2. Non dimensional absolute growth rate as a fuction of Rae for different Weber number

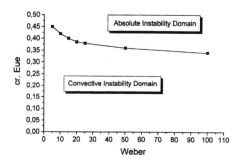

Figure 3. Critical Electrical Euler Number as a function of Weber number

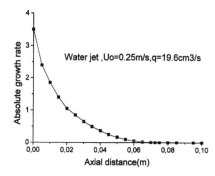

Figure 4. Non dimensional absolute growth rate vs axial distance for a gravitational jet

Inst. Phys. Conf. Ser. No 163
Paper presented at the 10th Int. Conf., Cambridge, 28–31 March 1999

·

Study of the changes in the flow around a cylinder caused by electroconvection

Guillermo Artana† Gaston Desimone† and Gerard Touchard ‡

† CONICET, Dept. Ingenieria Mecanica, Fac. Ingenieria, Universidad de Buenos Aires, Argentina, gartana@fi.uba.ar

‡ L.E.A., U.M.R. 6609 du C.N.R.S., Universite de Poitiers, Poitiers, France

Abstract. In this work we analyze the d.c. corona discharge between a wire-plate electrode configuration placed on the surface of an insulating cylinder. The influence of different parameters are analysed in order to characterize the discharge and to determine the best configuration to study the possibility of drag reduction of the flow acting in immersed bodies. A generalized glow regime seems the most suitable to undertake studies with this goal.

1. Introduction

The research on corona discharge when the electrodes that enabled the discharge are far away from any extraneous body has been analysed extensively. However, the knowledge of the physics of corona discharge occurring in a fluid close to an insulating surface is of interest in different engineering applications. In the mechanical engineering domain, the effect of the corona discharge on the fluid mechanics close to a limiting solid surface appears as a promising technique for flow and instabilities control. The ions injected on a fluid flow and subjected to coulombian forces will migrate from one electrode to the other exchanging momentum with the fluid particles modifying the fluid velocity field in the vecinity of the surface. Prior research work has analysed the effects of this forced electroconvection on the main flow like the one occuring on an electrostatic precipitator and other flow types like the wake downstream a cylinder with a point-to-plane electrode arrangement [1], boundary layers on flat plates with razor blade electrodes [2] and Poiseuille flows [3]. These researches show that the secondary flow caused by the exchange of momentum between injected ions and fluid particles can modify the main flow and thus it can be used as a control parameter. In view of applications leading to a drag reduction of immersed bodies some of the authors cited above have indicated a lack of knowledge of the physics of a corona discharge near an insulating surface, a problem that becomes essential to achieve optimal devices. It seems possible that in this special configuration the new surface will modify some aspects of the discharge close to the electrodes or cause the appearance of new phenomena close to the insulating surface.

Though no general theory is proposed here, it is one of the objectives of this article to report a characterization of a corona discharge occurying close to the surface of an electrical insulating cylinder. From this we will also try to establish the conditions when forced electroconvection around a cylinder should lead to more dramatic effects.

2. Experimental study

In our study the injection of ions is obtained by a d.c. corona discharge between a wire type electrode (275 mm length) and a plane electrode of aluminum foil (25x275mm), both located on the surface of a Plexiglas cylinder and parallel to the cylinder axis. Our experimental device has been placed in a wind tunnel (0-20 m/s, 0.45 x 0.45 m rectangular cross section) with the cylinder axis normal to the main flow and horizontally placed (see figure 1). Two different H.V. sources (0-30kV,-8kV-0V) enabled to impose voltage differences between both electrodes. A current-voltage amplifier circuit using an electrometer operational amplifier with very low input bias current was used to measure the discharge passing through the electrodes. With this device we have studied the influence on the corona discharge of different parameters (voltage, diameter of the cylinder (30-40-50 mm), diameter of the wire, air velocity (0-20 m/s)). In our experiments one of the electrodes was kept at constant potential -8kV and we increased the voltage of the other between a range of 0-30kV. By doing that, the influence of any surrounding grounded object was avoided and different discharge regimes could be observed. Typical voltage-current curves are shown in figures 2 and 3 for different cylinder and wire diameters. We have tested two different arrangements: wire positive-plate negative opposed at 180 degrees (positive corona) and wire negative-plate positive opposed at 180 degrees (negative corona). These figures show that the current depends not only on electrode spacement but on electrode radius and on polarity. For large voltage difference current is higher in positive coronas and for larger radii. As electrode voltage is increased different regimes can be observed :

Spot type: (figure 4) The discharge is concentrated in some visible spots of the wire and by increasing the voltage difference they can increase in number. Some of them may ionize in a plume like type or may lead to a channel quite attached to the cylinder surface. In positive corona the border of the plate can also show a luminescence, and the discharge is noisy, while in negative coronas the luminescence is not observed and the discharge is more quiet.

Generalized glow: (figure 5) At higher voltage differences, for wires with the larger diameters the spots occupy almost all the wire. Then, a regime characterized by a luminescence similar to the one appearing in a glow discharge occurs, occupying the whole arc distance between both electrodes . This discharge is quite homogeneous and it extends almost all along the electrode length, and when it occurs the cylinder surface looks like supporting a thin film of ionized air. For highly intense discharge this film is simmetric from the wire and the glow does not move along it. The phenomena is noisy and the current quite stable with time. In the voltage current curves this regime appears as a large increase in the slope of the curve (see in figure 2 results close to 21 kV for configuration cylinder diam. 30 mm and wire diam. 0.9mm). The glow is more difficult to be achieved with negative coronas where the discharge is similar to the so called pulseless corona of the point-to-plane arrangement. For the larger cylinder diameters (40-50mm) it was not possible to obtain this discharge when the electrodes were opposed at 180 degrees, however when placing them on these cylinders at arc distances similar to the one of the smaller cylinder (30 mm diam.) this regime appeared once again.

Streamer type: (figure 6) This regime occurs preceded of the generalized glow regime or directly from spot type regime. It can be characterized by some points of the wire with a concentrated discharge in an arborescent shape or in filament type. In them the discharge seems to be concentrated to a small channel. This filaments could achieve the whole arc distance in the positive corona case, while in the negative corona case they can interrupt between both electrodes. By further increasing the voltage some localized sparks with sharp noises appeared and then tests were stopped.

The influence of the surrounding air velocity on the discharge corona depends on the regime considered. As air velocity was increased (0-20m/s) in the spot or streamer regime it could be observed that the number of points increases, while in the glow discharge regime it extended the lenght of the wire supporting a visible discharge. No degradation of the polymer's surface could be observed after our series of experiments.

3. Analysis of results

The generalized glow discharge regime cited above can only be explained by an electric field configuration that enables the ionization all along the arc distance. In order to have a first approach of this we have solved Laplace's equation for the electrical potential in air in a two dimensional domain neglecting the effect of the space charge. The boundary conditions considered on the electrical potential function are: a linear dependence with angular coordinate of the potential function on the cylinder surface having extreme values at the electrodes (similar to the one used in [2]), and null value at a very large radius R_∞ for any value of angular coordinate. To simplify the analysis both electrodes have been considered of the wire type. The solution of the problem can be considered as a superposition of two solutions: one corresponds to an electrode configuration $\Phi = \pm V_1$ and the other to a constant potential cylinder at $\Phi = V_2$. The potential function Φ obtained is

$$\Phi = \frac{4V_1}{\pi^2} \sum_{n=1}^{\infty} \left(\frac{a}{r}\right)^n \frac{(-1)^n - 1}{n^2} \cos(n\theta) + V_2 \frac{\ln(r/R_\infty)}{\ln(a/R_\infty)}$$

with a the cylinder radius, r and θ the radial and angular co-ordinate. From this equation the electric field E can be easily obtained as $E = -\nabla\Phi$ and figure 7 gives an image of the electric field prior to ion injection or when the space charge is very low. In this figure we can see that the electric field is very large in regions close to the surface in coincidence with the expressed above.

3.1. Effects of the discharge on the Flow Field configuration

We have undertaken a visualization of the flow field configuration with smoke injection when the wire was placed on the frontal stagnation point. Though this technique may induce to some errors due to the possibility of charging of the smoke particles it can be considered as a first approach to compare the spot or streamer regime with the generalized glow regime. From our results, like from those in [1], it is observed that

150

the wake is largely modified and that vortex streets can be modified and controlled by electroconvection. In the spot or streamer regime, and under certain conditions, new instabilities can be observed, of the Kelvin-Helmholtz type, which do not occur in a cylinder without electroconvection or in the configuration of [1]. An analysis of the characteristics of this instability due to the shear flow could enable an indirect estimation of the velocity of the secondary flow. In the generalized glow regime the effects are much more dramatic and the filaments of the smoke tracers are perturbed in almost the whole section of the wind tunnel downstream the cylinder. The effect of the dc corona discharge in this regime are important not only because of its intensity but also because the homogeneity of the discharge occurring all along the wire that leads to a two-dimensional flow field configuration. Thus the main flow presents very distinctive characteristics given that, as it attacks the solid body it finds a slip condition.

4. Conclusions

This work shows that it is possible to obtain an homogeneous discharge ionizing the air in the whole distance between electrodes of the wire-plate type placed on the surface of an insulating cylinder. This kind of discharge occurs in a very thin region and the effect of an air flow on the range of velocities tested was to homogeneize the discharge. Flow visualization by smoke injection indicates that the generalized glow leads to the more dramatic effects in the wake downstream the cylinder. As a result, this regime seems to be the more convenient to study the control by electric forces of the drag of an air flow on immersed bodies.

References

[1] C. Noger et al 1997 *Proc. 2nd Intnl Symp Plasma Techn Polution Cont, Bahia* 136-141

[2] S. El-Khabiry, G. Colver 1997 *Physics of fluids* **9(3)** 587-599

[3] M. Malik et al 1983 *AIAA 21st Aerospace Sci. Meeting, Reno Nevada* AIAA-83-0231

This reasearch has been undertaken with grants of the UBACYT IN-003 and ANPCYT 12-02177 of the Argentine government

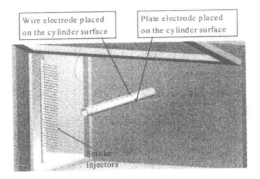

Figure 1. Experimental device: Cylinder placed on a wind tunnel (section 450×450mm)

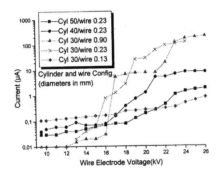

Figure 2. Positive corona-I(V)

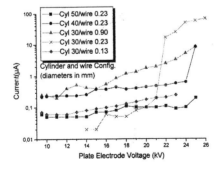

Figure 3. Negative corona-I(V)

Figure 4. Spot type discharge

Figure 5. Generalized glow discharge

Figure 6. Streamer type discharge

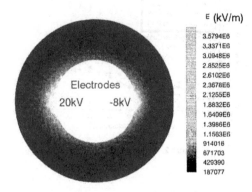

Figure 7. Electric field configuration for a cylinder with a 30mm diameter

Inst. Phys. Conf. Ser. No 163
Paper presented at the 10th Int. Conf., Cambridge, 28–31 March 1999

A New Method of Analysing Electroosmosis

S.Watanabe*, Y.Hobo*, N.Hayashi*, T.Sumi*, D.Dykes**, G.Touchard ***

* *Aichi Institute of Technology, Yachikusa Yakusa-cho Toyota 470-0392 Japan.*
** *Yokkaichi University, Kayo-cho, Yokkaichi, 512-8045 Japan.*
****Université de Poitiers, 40 av. du Recteur Pineau, Poitiers 86022 France.*

Abstract. *Electroosmosis*, first discovered by Reus in 1808, is the phenomenon that appears when the liquid in a capillary tube is displaced by an electric charge. The constant properties of the phenomenon were investigated by Wiedeman and Quincke. The quantity Q of liquid displaced relates to the charge according to the equation [1]
$Q = \varepsilon \zeta I / \eta \kappa$
Hence, in order to increase the liquid displacement, one may either increase one of the three factors ε (relative permittivity), ζ (potential) or I (current between the electrodes), or else reduce one of the two factors η (viscosity) or κ (conductivity). In practice, the best way to control the flow is by adjustments in the current, as it is extremely difficult to vary the other factors over a very wide range. The current between the electrodes has been observed to reach a maximum at some point after the application of a voltage, and thereafter to decrease. The cause for this has been found to lie in the movement of impurity particles, so-called *electrophoresis*. If this displacement of particles could be controlled, it ought to be possible to increase the amount of flow still further. As one means of achieving this, the authors have experimented with placing a grid electrode between the upper and lower electrodes. If the displaced particles are regarded in a similar way to electrons, the same method of analysis can be used as for a triode. In tests conducted after the fitting of a grid, the flow was found to increase to 1.5 times the amount obtained under conventional conditions. The results of the experiment are reported in this paper.

1. Introduction

In recent years, the disposal of industrial and household refuse has become a major social problem. In order to permit the limited capacity of disposal plants to be exploited to the maximum, it is necessary to reduce the liquid content of the waste, and to recycle as many materials as possible. Methods of reducing the liquid content include natural filtration by means of gravity, and compression filtration by mechanical means. Methods like these, however, while effective enough for water in the free state, are inadequate for the extraction of water from waste products in colloid or gelled states. In the case of gelled conditions, one effective procedure is to apply an electric voltage so as to set up a field. The authors have successfully conducted experiments on this principle [2].

In the present research, they have attempted to measure the electrical properties obtained by varying the applied voltage, wave-form and frequency of the field. Further, they have tried inserting a grid electrode and studied its effectiveness in extracting water.

2. Experimental use of electroosmosis for solid-liquid separation

For methods of solid-liquid separation using electroosmosis, the amount of water released is usually expressed by the Helmholtz-Smoluchowski equation, relating factors such as the amount of liquid released, the permittivity of the mixture, the ζ potential, the electric current between the electrodes, the viscosity and the conductivity of the mixture. Therefore, in order to regulate the amount of liquid released, it is necessary to vary either the permittivity, the ζ potential and current factors, or else the viscosity and conductivity. By far the easiest factor to vary is the current. In previous experiments, an increase in current was registered immediately after the application of voltage between the electrodes, but this was followed by a diminution tendency after a certain lapse of time[3]. For a reliable control of the amount of liquid released, we need to investigate the factors influencing this increase and decrease of the current.

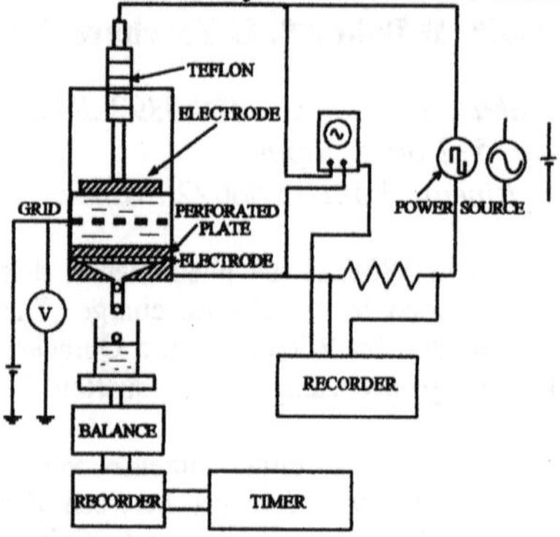

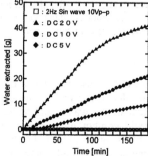

Fig.1 Apparatus for measurement

3. Experimental apparatus and results

The experimental apparatus consisted of an upper electrode secured by a teflon block, a filter paper, a grid electrode, a perforated disc and a lower electrode. A diagram of this apparatus is shown in Fig. 1. The experimental liquid used was of solid-liquid separation 10 grams of bentonite powder (a ceramic material with an average Stokes diameter of 3-5 μ m) suspended in 90 grams of distilled water and left to settle for 3 days.

4. AC, DC and water extraction

After this, DC and AC voltages were applied, and the quantity of water extracted was measured. In the case of the AC voltage, no extraction could be observed at all, a result that can be understood in light of the Helmholtz-Smoluchowski equation. In the case of the DC voltage, the extraction performance was found to increase with an increase in the voltage applied. The results are shown in Fig. 2. In these tests, the grid electrode was not yet in place.

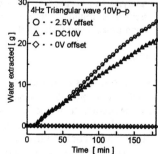

Fig.2 Time dependence of leakage on DC voltage

Fig.3 Time dependence of leakage on offset voltage

5. Offset voltage and water extraction

In a second series of tests, a voltage of triangular wave-form was introduced between the electrodes, and a varying offset voltage of $0 \sim 2.5$ V was also applied. In these tests, too, the grid electrode was not used. The results are given in Fig. 3. A comparison of the results shows that the extraction performance obtained from the offset triangular-wave voltage is superior to that obtained from the simple direct voltage. Where no offset voltage was added, no water extraction effect could be observed.

6. Applied voltage wave-form and water extraction

Voltages of sine, triangular and rectangular wave-form were applied between the electrodes and measurements were made of the water extracted, using a balance. The voltages were all 5 V, offset. The grid electrode was not installed. The results are shown in Fig. 4. Up to a certain point of time, all three curves show the same slope, but then the triangular-wave voltage gives a better performance, followed by the sine-wave and rectangular-wave voltages, in that order. However, the current reaches its peak at exactly the same time in all three cases, irrespective of the wave-form. The size of the current at this peak moment is greatest in the case of the rectangular wave-form, followed by the triangular-wave and then the sine-wave.

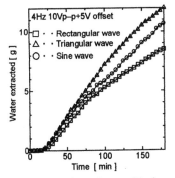

Fig.4   Time dependence of leakage
for three different wave-forms

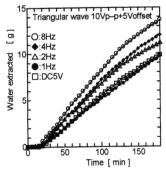

Fig.5   Time dependence of leakage
for different frequencies

7. Applied voltage frequency and water extraction

10 Vp-p voltages of rectangular, triangular and sine wave-form were applied between the electrodes in four frequencies of 1, 2, 4 and 8 Hz. As in the preceding test, the voltage was 5 V, offset, and no grid was used. One example set of results (for the triangular wave-form) can be seen in Fig. 5. It appears that the water extraction increases with each rise in frequency.

8. Duty ratio of the rectangular-wave voltage and water extraction

Two different voltages were applied between the electrodes: first a 5 V DC voltage, and secondly a 2 Hz, 10 Vp-p rectangular-wave voltage. In the second case, measurements were taken while varying the duty ratio (t/T). The results are shown in Fig. 6. With a duty ratio of 5/10, no change was found in the current between the electrodes. But when the duty ratio was varied, a peak phenomenon was observed, as in the case of the applied DC voltage. The extraction rate was found to increase as the duty ratio approached 10/10 (i.e., the DC condition).

9. Grid electrode and water extraction

Upon measuring the current between the electrodes, it was found in all of the above experiments that a peak appeared. Generally, with electroosmosis phenomena, no variation in current is found. Therefore, it seems likely that some kind of electrophoresis process may also be at work. Electrophoresis is a transfer of particles resulting from the action of an electric field. In the present case, it would seem that bentonite particles are being transferred. Bentonite particles have a ζ potential of -1 mV in pH 7 liquid[4], and the observed current variation can be explained by assuming that it is the transfer of these particles that causes the peak to appear in the current. This also means that if it is possible to control the flow of particles, it is also possible to regulate the current. In order to control

156

both the particle and current flow, the method adopted was to insert a grid between the upper and lower electrode plates. This grid electrode represents a type of triode. By varying the grid voltage, it is possible to control the anode current. Fig. 7 gives experimental results obtained from applying a 10 V DC voltage to the upper electrode and +5 V and -5 V DC voltages to the grid. In the figure, the ●, ○ and ◇ symbols show the respective effects of applying DC grid voltages of +5 V, -5 V and 0 V to an anode voltage of 10 V. The largest observed values in this test were obtained from the +5 V grid voltage.

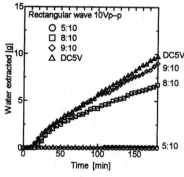

Fig.6 Time dependence of leakage
for different Duty ratios

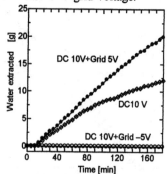

Fig.7 Time dependence of leakage
on grid electrode polarity

10. Conclusion

In this series of tests concerning water extraction, it was established that two phenomena, electroosmosis and electrophoresis, simultaneously play a part. In order to maximise the extraction rate, it is necessary to control the transfer of particles at the same time as increasing the current between the electrodes. One extremely effective means of controlling the movement of particles is to insert a grid. Visualising the experimental apparatus as a triode system, the upper electrode corresponds to the anode, the central grid to the triode grid, and the lower electrode to the cathode.

If variations in the grid voltage had been used as a means of varying the anode current within the triode system, it is probable that a similar phenomenon of electroosmosis would have been observed. An optimal water extraction can be obtained by seeking the most effective anode and grid voltages for mutual conductance. In order to analyse the characteristics of a given triode, it is necessary to consider three constant values: mutual conductance, anode resistance and amplification[5]. Taken together, the measurement of these three factors and the simultaneous control of the anode current should provide us with a new indicator for electroosmosis. In future experiments the authors intend to pursure further investigations into the above three factors, in the light of the findings reported in this paper.

References
[1] Kitahara A., Watanabe M. 1978 : *Electrokinetic Phenomena at the Liquid-Solution Interface*, (Kyoritsu Shuppan) 109
[2] Yamamoto T., Hayashi N., Watanabe S., Dykes D., Touchard G. 1995 : *Electrostatics 1995*, (Institute of Physics Conference Series **143**) 243-246
[3] Okazaki K., Maeda Y., Hayashi N., Watanabe S. 1989 : *National Convention Record I.E.E. Japan*, **13**, 204
[4] Watanabe S., Yamamoto T., Hayashi N., Sumi T., Dykes D., Touchard G. 1996 : *7th World Filtration Congress*, 619-623
[5] Yano M., Suga H., Kawabata K., Tanaka T., Kotera M., Tanaka M. 1997 : *Electronic Devices*, (Sangyotosho) 83

Inst. Phys. Conf. Ser. No 163
Paper presented at the 10th Int. Conf., Cambridge, 28–31 March 1999
© *1999 IOP Publishing Ltd*

Transient Electrohydrodynamic Stability of Dielectric Liquids Subjected to Unipolar Injection

R Chicón†, A Castellanos‡ and A T Pérez‡

† Departamento de Física Aplicada, Universidad de Murcia. Apartado 4021, Murcia 30071, Spain

‡ Departamento de Electrónica y Electromagnetismo, Universidad de Sevilla. Avenida de Reina Mercedes, Sevilla 41012, Spain

Abstract.
This paper presents numerical simulations on the problem of transient Electrohydrodynamic stability of dielectric liquids subjected to unipolar injection. A stability criterion for the transient does not exist. The velocity perturbation always increases during the transient, the time it takes the injected charge to reach the collector. When the voltage is above the linear stability limit for steady conditions, the motion will continue after the transient. Conversely, if the voltage is below the linear stability limit the velocity increases during the transient and, when the front of charge reaches the collector, dies out.

1. Introduction

Fluid motion driven by Coulomb forces may appear in an insulating liquid when ions are injected by various mechanisms. This is a common occurrence in many natural and industrial situations. The problem of the stability of the stationary state of conduction under unipolar injection of ions has been extensively studied in the last three decades, both in the weak injection and space charge limited current (SCLC) regimes [1, 2, 3]. There are two critical voltages corresponding to the linear and nonlinear criteria, together with a subcritical bifurcation diagram formed by one unstable and two stable branches, with the corresponding associated hysteresis loop.

There is a discrepancy between the theoretical and experimental values for the stability criteria. Some attempts to understand this discrepancy have been made by considering that, while the stability problem was formulated for the stationary conduction state, some of the experiments were performed with an initial transient regime [4]. However, the results obtained through a quasi-stationary approximation for the transient did not solved the discrepancy in these kind of experiments [4, 5].

In this work we study the transient regime corresponding to the application of a voltage step to an initially uncharged liquid layer, which is subjected to unipolar injection. The charge conservation equation is solved by the superparticles method, and the Navier-Stokes equation is solved by a finite difference method. Since the driving term in the vorticity equation is proportional to $\nabla q \times \mathbf{E}$ (where q and $\mathbf{E}$ represent charge density and electric field, respectively), the existence of a front of charge during the transient must substantially increase the importance of that term with respect to the stationary state. Therefore, a faster growth would be expected for perturbations during the transient than for perturbations of the stationary state.

The results confirm that expectation. Growing rates are an order of magnitude higher during the transient than in the following period when the liquid is already completely charged. However, these growing rates are not high enough to trigger measurable velocities during the transient.

2. Basic Equations and Numerical Method

An insulating, incompressible and isothermal liquid is considered, with density ρ, dynamic viscosity η and permittivity ϵ. It is confined between two parallel electrodes of infinite extent separated a distance d. A voltage step of value ϕ_0 is applied at an initial time, and unipolar injection of ions takes place at the anode (injector). It is also supposed that the ions discharge instantaneously when they reach the opposite electrode (collector). The equations governing the problem in non dimensional form are[6]:

$$\nabla^2 \phi = -q \tag{1}$$

$$\frac{\partial q}{\partial t} + \nabla \cdot \mathbf{j} = 0 \tag{2}$$

$$\nabla^2 \psi = \omega \tag{3}$$

$$\frac{T}{M^2} \left(\frac{\partial \omega}{\partial t} + \mathbf{v} \cdot \nabla \omega \right) = \nabla^2 \omega + T (\nabla q \times \mathbf{E})_y \tag{4}$$

taking d, ϕ_0, $\epsilon \phi_0 / d^2$ and $K \phi_0 / d$, where K is the ionic mobility, as the units for distance, potential, charge density and velocity, respectively. In these equations q denotes the charge density, ϕ the electric potential, $\mathbf{E}$ the electric field, $\mathbf{v}$ the velocity of the liquid and $\tilde{p}$ the sum of the mechanical and electrostriction pressures. The current density is given by $\mathbf{j} = q(\mathbf{E} + \mathbf{v})$, where the first term accounts for the migration of the charge carriers under the action of the electric field and the second term for the convection by the liquid. A diffusion term has not been included, since it is negligible compared to the migration term [8]. Three non-dimensional parameters appear in the equations: $T = \epsilon \phi_0 / \eta K$ is a measure of the relative importance of electric and viscous forces. The parameter $C = q_0 d^2 / \epsilon \phi_0$, where q_0 is the charge density at the injector, measures the injection strength and $M = (\epsilon / \rho)^{1/2} / K$ is the ratio of the hydrodynamic to the ionic mobility. The motion of the fluid is assumed to be two-dimensional, and a stream function $\psi(x, z)$ is introduced such that $v_x = \partial \psi / \partial z$, $v_z = -\partial \psi / \partial x$, and thus the continuity equation is automatically satisfied. Only the y-component of the vorticity $\omega = \nabla \times \mathbf{v}$ is different from zero.

The set of Eqs.(1),(2),(3) and (4), together with the conditions of fixed potential and rigid boundary conditions for the fluid velocity at the electrodes ($z = 0$ and $z = 1$), are to be solved numerically, starting from a suitably specified initial state. The two Poisson equations and the evolution equation for the vorticity can be solved by a usual finite differences scheme, which has been proven to be stable and convergent in the SCLC regime, i.e. $C >> 1$ considered in this work. The quantities q, ϕ, $\mathbf{E}$, ω, ψ and $\mathbf{v}$ are defined on a rectangular grid of mesh points, non-uniform in the z-direction to tackle with the strong variation of the charge density with z near the injector.

Steep gradients of charge are present in the convective cell. During the transient after the application of a voltage step the sharp gradient is inherent to the advancing front of charge. When finite-amplitude electroconvection has established, the sharp charge gradients appear due to the existence of internal boundaries surrounding regions empty of charge. Although finite difference schemes have been successfully employed to study the transient regime [7], they are not adequate to study the steady convective state due to numerical diffusion, even when specially designed antidiffusive schemes are used [8]. For this reason the charge conservation equation is solved by a superparticle-type method developed in a previous work [3]. The discretization of that equation leads to the concept of computational superparticles, which can be thought of as clouds of ions [9]. Superparticles are characterized by a conserved attribute, its electric charge, and variable position and velocity. Thus, conservation of charge is automatically satisfied, and the partial differential equation (2) is substituted by a large set of ordinary differential equations, namely the equations of motion of the superparticles:

$$\frac{d\mathbf{r}_i}{dt} = \mathbf{v}(\mathbf{r}_i, t) + \mathbf{E}(\mathbf{r}_i, t) \tag{5}$$

where $\mathbf{r}_i$ denotes the position of particle i. Fluid velocity and electric field at $\mathbf{r}_i$ are interpolated from their values at neighbouring mesh points. Likewise, the charge of superparticles is assigned to neighbouring mesh points to obtain the charge density q that takes part in Eqs.(1) and (4).

Superparticles are injected at $z = 0$ at each time step, and special care is taken to comply with the injection boundary condition. Although the particle method can be implemented for an arbitrary

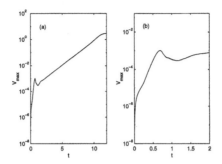

Figure 1. Maximum velocity vs. time during the transient, for $T = 200$, $M = 10$, $C = 10$. Two different rates of growth are visible: the first correspond to transient amplification, the second to the linear instability of volume charge. Plot b) is a zoom of plot a) that shows the evolution during the time the front of charge needs to reach the collector.

injection law, in this work it is assumed that the charge density remains constant at the injector, independently of the value of the electric field. This is referred to as autonomous injection, and is known to be a good approximation to the stability problem in simple geometries [1].

3. Results and Discussion

Numerical results are presented for a case with $C = 10$, well within the SCLC regime. Periodic boundary conditions are imposed on the x-axis, such that $L = 0.61$ is the half-wavelength of the convective cell [1]. The method has been tested by its application to two situations where analytical results are available. First, the classical linear stability analysis of the steady conduction state has been reproduced. A linear criterion $T_c = 165$, independent of M, is obtained in excellent agreement with the theoretical result [1], above which small perturbations of the steady state grow up until finite-amplitude electroconvection is reached.

The convective state for $T > T_c$ is characterized by a velocity field of a roll-type, though it lacks the perfect symmetry assumed in previous works where a self-similar roll was considered instead of solving the Navier-Stokes equation. The fluid velocity is higher in the (narrower) upward than in the (wider) backward part of the flow, thus resulting in a lateral displacement of the center of the roll. The fluid velocity shows small fluctuations around a mean value, with spectral properties analogous to those obtained under the assumption of a self-similar roll [3].

Secondly, the transient conduction solution is reproduced. Now the calculation starts from an initial state where the liquid is discharged and a voltage step is applied at $t = 0$. The results obtained for the advance of the front of charge and the charge density $q(z, t)$ and electric field $E(z, t)$ distributions show a very good agreement with the analytical results [10].

The transient stability is studied by giving a small perturbation in the fluid velocity at $t = 0$ and following its evolution as the front of charge travels from injector to collector. The effect of such a perturbation, taking into account that $\mathbf{v} = \mathbf{0}$ at the electrodes, is just a tiny variation in the first movement of the superparticles injected at $t = 0$, compared to their movement under the effect of the electric field alone. This notwithstanding, the perturbation evolves as shown in Figure 1 for $T > T_c$. Starting from a very small value, there is a very fast initial increase of the perturbation, followed by a period of exponential growth. An inflexion in the evolution occurs when the front of charge approaches the collector (the non-dimensional crossing time is $t_c = 0.82$), and the perturbation decreases for a period around t_c. After that, with the liquid already fully charged and starting from a still very small value of the velocity, the evolution of the system is perfectly linear, with an exponential growth of the perturbation until the velocity takes appreciable values and the finite-amplitude electroconvection regime is established.

160

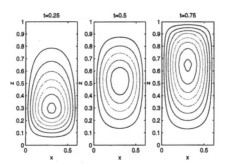

Figure 2. Streamlines at three different times: t=0.25, t=0.5 and t=0.75. The center of the streamlines moves with the charge front. The relevant parameters take the same values as in figure 1.

The growth rate during the latter linear regime is exactly the same obtained from the linear stability analysis of the steady state. On the other hand, the growth rate associated to the transient exponential regime is an order of magnitude larger, as it may be appreciated in the plot. This can be understood as a direct consequence of the existence of the front of charge. A sharp gradient of charge at the front implies large values for the driving term $T\nabla q \times \mathbf{E}$ in the vorticity equation, substantially larger than in the case of the perturbed steady state. The driving term takes maximum values around the front of charge.

The velocity field reaches maximum values around the front of charge too, and it accompanies the traveling front, as shown in Figure 2. While the amplitude of the velocity perturbation increases as the front is advancing, its overall aspect is not drastically altered. This has been checked by considering initial perturbations corresponding to one or several rolls. All rolls subsisted at the end of the transient, and the amplitude of the perturbation during the transient always evolves as in Figure 1. When the front of charge approaches the collector the velocity perturbation bounces off the collector, and ends up centered equidistant from the two electrodes.

It has been found that the initial perturbation always increases during the transient for *any* value of T. The behaviour is always similar to that shown in Figure 1, with the amplitude of the perturbation monotonically increasing with time until the front of charge approaches the collector. Strictly speaking, this means that there is no transient stability threshold. This result differs from previous works on transient stability, based on the use of a quasistationary approximation to study the transient regime [4, 5].

This research is supported by DGES (Spanish Government Agency) under Contract No. PB96-1375.

References

[1] Atten P and Moreau R 1972 *Journal de Mécanique* **11** 471–520

[2] Atten P and Lacroix J C 1979 *Journal de Mécanique* **18** 469–510

[3] Chicón R, Castellanos A and Martín E 1997 *Journal of Fluids Mechanics* **344** 43–66

[4] Atten P 1974 *The Physics of Fluids* **17** 1822–1827

[5] Watson P K , Schneider J M and Till H R 1970 *The Physics of Fluids* **13** 1955–1961

[6] Castellanos A 1991 *IEEE Transactions on Electrical Insulation* **26** 1201–1215

[7] Theodossiou G , Nelson J K and Odell G M 1986 *Journal of Physics D: Applied Physics* **19** 1643–1656

[8] Pérez A T and Castellanos A 1989 *Physical Review A* **40** 5844–5855

[9] Hockney R W and Eastwood J W 1981 *Computer Simulation using Particles* (McGraw-Hill)

[10] Zahn M, Tsang Ch F and Pao S 1974 *Journal of Applied Physics* **45** 2432–2440

Inst. Phys. Conf. Ser. No 163
Paper presented at the 10th Int. Conf., Cambridge, 28–31 March 1999

Effect of an electric field on the response of a liquid jet to a pulse type perturbation

Guillermo Artana† and Bruno Seguin †‡

† CONICET, Dept. Ingenieria Mecanica, Fac. Ingenieria, Universidad de Buenos Aires, Argentina, gartana@fi.uba.ar

‡ CEFIS-INTI-P.T. Miguelete, San Martin, Argentina

Abstract. In this work we analyze the influence of an electric field on the response of a circular liquid jet excited by a pulsed signal externally imposed to the mean flow. The influence on the limits of the wavepacket and on the growth rate are established showing that the electric field spreads and destabilizes a pulse type perturbation.

1. Introduction

In an unstable flow the response to a pulsed disturbance can be characterized by two important parameters, the growth rate of the perturbation and the velocity of propagation of the disturbance in the flow. Concerning this last parameter prior research work has established [1] that an electric field reduces the velocity of propagation, making possible in theory to control with an electric field the absolute or convective nature of the instability in a jet. However with this analysis neither the spreading of the wave packet nor the growth rate at the different streamwise stations as a function of time can be specified.

The objective of this work is to establish the influence of an electric field on the boundaries of the wavepacket and on the growth rate of the disturbances as a function of time for the different axial coordinate positions.

2. Description

The geometry of the problem we have analysed is a circular jet coaxial with a cylindrical electrode at a different electric potential. We have considered the following simplifying hypothesis: -The surface of the liquid jet is an electrical equipotential. -The magnetic and gravitational effects can be disregarded -The jet is inviscid, incompressible and isothermal -There is no mass transfer between the jet and the surrounding atmosphere.

-All streamwise stations have the same plug velocity profile independent of the axial coordinate in the basic state.

We will analyze the motion resulting in this basic flow from a pulse input following a linear analysis. Considering a pulse perturbation giving rise to two-dimensional travelling wave modes of the type

$$\Psi(r, z, t) = \Phi(r)e^{i(kz-\omega t)}$$

with both k and ω complex values, Gaster [2] deduced that in any ray z/t within the wave packet (z being the axial coordinate of the jet, t: interval of time elapsed after the application of the pulsed signal) the response is dominated by a specific complex wave number k^* and the wave motion appears to grow in space and time with an amplification rate

$$-\left[k_i^*\frac{z}{t} - \omega_i(k^*)\right]t$$

where k^* is obtained from the conditions on the group velocity

$$\frac{\partial \omega_r}{\partial k_r}(k^*) = \frac{z}{t}$$

$$\frac{\partial \omega_i}{\partial k_i}(k^*) = 0$$

and where the dispersion equation defines the behavior of ω in terms of k. The set of four equation constituted by these and the corresponding equations resulting from the splitting of the dispersion equation in its complex and real part enables to obtain for any z/t ray the values of k^* and of $\omega(k^*)$ and consequently the growth rate for any z/t ray. By representing the growth rate as a function of the ratio z/t it is possible to obtain the boundaries of the wave packet that can be defined by the rays z/t along which the amplification is zero.

In the following paragraph we show results obtained from the dispersion equation obtained in [1]. Considering only the axisymmetric case and disregarding the effect of the sourrounding atmosphere the equation $D(k, \omega, We, Eue) = 0$ establishes the relationship between the complex wavenumber k, the complex frequency ω, the Weber number $We = \frac{\rho U_0^2 a}{\gamma}$ (the ratio between inertial and surface tension forces) and the Electric Euler number $Eue = \frac{\epsilon_0 E_n^2}{\rho U_0^2}$ (the ratio between the electric forces and inertial forces). In these expressions $\rho, U_0, a, \gamma, \epsilon_0, E_n$ are respectively the liquid density, the mean jet velocity, the jet radius, the interfacial tension, the dielectric constant in vaccum and the electric field at the jet surface.

2.1. Results and Discussion

The set of equations cited above has been solved numerically for different convective unstable flows (the ray $z/t = 0$ is not involved by the perturbation propagation).

Figure 1 shows the non dimensional growth rate (non dimensionalized with the jet radius and jet velocity) as a function of z/t (non dimensionalized with jet velocity) for different Weber Numbers. We can see that as the Weber number increases the boundaries of the wavepacket get closer and that the maximum growth rate corresponds to the ray $z/t = 1$. A propagation upstream of the pulse source occurs when the ray $z/t = 0$ has a positive growth rate (absolute unstable flows) and this occurs for low velocity jets and at Weber number of 3.1 in agreement with [3]. Analyzing these figures it can be concluded that inertial forces stabilizes a pulse type perturbation and leads to a more compact pulse propagation. The effect of surface tension forces is the opposite. Figure 2 shows the non dimensional growth rate as a function of the non dimensional ratio z/t for different Eue. The symmetry of the spreading along the non dimensional ray $z/t = 1$ is kept even when an electric field is applied but the pulse type perturbation may propagate upstream if the electric field is intense enough . It can be observed that larger electrical Eue (larger electric fields) lead to a more spread pulse, as the boundaries of the wavepacket are more separated. The effect of the electric forces differs from that of inertial forces: they destabilize a pulse type perturbation but lead to a more spread pulse propagation.

Regarding experiments to confirm the theoretical predictions one must consider a convective flow and it is possible either to have an image of a jet region including the pulse or to study the radius evolution as a function of time as the pulse passes through a fixed station with an experimental setup as described in [4]. By these experiments a three dimensional graph of the radius as a function of the axial position and time $r(z,t)$ can be obtained for either t constant or z constant, and from this graph a set of points of growth rate as a function of z/t easily deduced. Figure 3 shows a diagram of the propagation of the perturbation. The figure represents the perturbation at two different times in the z-t domain with the lines identified by $(z/t)_I$ and $(z/t)_{II}$ indicating the head and tail ray limits of the wavepacket boundary. Outside the region delimited by this two rays the perturbation is dampened and it is amplified inside it. The pulsed perturbation can be generated with a square signal of low frequency applied to a piezoelectric that excites the jet but it should be mentioned that measurements should be undertaken in positions where there is no interaction between two successive pulses. Looking at figure 3 it can be easily deduced that to satisfy this for a jet of a lenght L (or for the largest distance pulse source-streamwise station observed) the time elapsed between the excitation of two successive pulses should be larger than the interval $\Delta t_{\min}$ given by

$$\Delta t_{\min} = L \left(1 - \frac{(z/t)_I^{-1}}{(z/t)_{II}^{-1}}\right)(z/t)_I^{-1}$$

As example for a water jet of a radius of $100\mu m$, $We = 5$ and $Eue = 0.5$ at a distance L of 1 cm this time should be larger than about 11 ms.

3. Conclusions

In a jet excited by a pulse type perturbation the effect of electric forces is different from the action of inertial and surface tension forces. An increase in inertial forces stabilizes a pulse type perturbation in z/t rays but leads to a more compact pulse propagation, while an increase of surface tension forces leads to an opposite effect. Our results indicate that the effect of an electric field stressing the jet surface on the response to a pulse type perturbation is to increase the separation of the boundaries of the wavepacket (even upstream for large enough electric fields) and also to increase the growth rate at which the perturbation is amplified in each ray z/t. This action is reduced for high Weber number jets (or large jet velocities).

This research has been undertaken with CONICET Grant PEI 175/97.

164

References

[1] G. Artana et al 1997 *J. of Electrostatics* **40&41** 33-38

[2] M. Gaster 1968 *Physics of Fluids* **11(4)** 723-727

[3] S. Leib and M. Goldstein 1986 *J. Fluid Mech.* **168** 479-500

[4] H. Taub 1976 *Physics of Fluids* **19(8)** 1124-1129

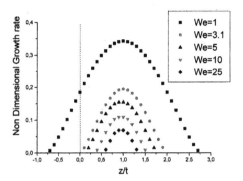

Figure 1. Non dimensional absolute growth rate as a fuction of z/t for different Weber Numbers

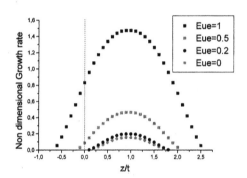

Figure 2. Non dimensional absolute growth rate as a fuction of z/t for different Eue Numbers. We=5

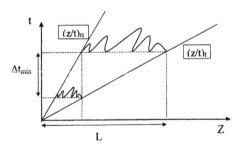

Figure 3. Diagram of the pulse proagation

Inst. Phys. Conf. Ser. No 163
Paper presented at the 10th Int. Conf., Cambridge, 28–31 March 1999

167

Drop charging during liquid dispersion

Yanyang Wang and Adrian G Bailey

Department of Electrical Engineering, University of Southampton,
Southampton, SO17 1BJ, UK

Abstract. The electric charge carried by water drops breaking away from the tips of various vertical cannulae at low potential differences (-1 V to +1 V) between the cannula and a surrounding metal tube is measured: charge level can be explained by the ordinary laws of electrostatic induction, with allowance for contact potential. In the absence of an external field between the cannula and the surrounding tube, by varying ionic concentration and types of ion, charge profiles are generated to determine the influence of these parameters on the mechanism of the single drop charge generation processes. The effects can be explained by taking into account the contact potential between solution and the cannula setting up a small local electric field at the drop surface. Induction charging then occurs.

1. Introduction

When nebulisation and other liquid dispersion processes produce aerosols the droplets within the aerosol clouds carry electrostatic charge. This is of interest in many fields, including the pharmaceutical industry. The natural charging processes of aqueous liquid are exceedingly complex involving hydrodynamic and hydrostatic effects.

To date no completely satisfactory quantitative spray charging model has been formulated although some qualitative understanding of the charging process of bubbling [1], drop break-up [2] and liquid ligaments break-up [3] into drops has been developed.

To help elucidate the complex charging processes that occur during liquid dispersion, we have broken the processes down into a sequence of related, simpler sub-processes. Our initial studies are of single drops forming at and breaking away from the tips of cannulae. The drop forming processes are so slow that any hydrodynamic effect can be neglected. Every drop assumes its maximum volume by the balance between the electrostatic, gravitational and surface tension forces. With an applied electric field between the metal cannula and the surrounding metal tube, which in our case both are stainless steel, drop charging processes can be explained by electrostatic induction.

In the absence of an applied electric field between the cannula and the surrounding tube, i.e. when they are at the same potential the drops still acquire a small charge. This is due to contact potential between solution and the cannula setting up a small local electric field at the exposed surface of the drops, thereby inducing charge.

In order to investigate the influence of contact potential between solution and the cannula on the single drop charging process, various electrolyte solutions have been used.

2. Experimental arrangement

As shown in Fig.1, the apparatus for producing drops is vertically mounted. A micrometer screw device connected to the piston of the syringe allows liquid to be delivered, in a controlled manner. Cannulae of various sizes may be fitted to the syringe; in this work cannulae of outside diameters 0.33mm, 0.41mm, 0.46mm, 0.56mm, 0.71mm, 0.89mm, 1.24mm, 1.47mm, 1.65mm and 3.40mm are used.

During all experiments, the drop drip rate is set to 0.5 drop/s so that not only every single drop reaches its maximum size very slowly before it is broken away from the tip of the cannula, but also, at this low flow rate, electrokinetic effects on drop charging processes can be ignored. DI water (deionised water, ~ 0.182 MΩm) and aqueous solutions are used. The charged drop detaching from the tip of the cannula falls into the Faraday cup. The chart recorder indicates the arrival of a drop in the Faraday cup as a step on the chart recorder and a train of steps is produced as more drops arrive. Drop charge is averaged over the individual charge steps. In our case nine steps are taken to average the drop charge. The drop size is calculated from the liquid flow rate and the drop frequency. The amount of liquid delivered during a certain time is obtained from the displacement of the piston.

Two kinds of experiment are carried out. One is with an external electric field applied between the cannula and the surrounding tube, single drop charge is measured using DI water and cannulae of various outer diameters are tested. In this case, the cannula and the surrounding tube are connected to a power supply and voltage from -1V to +1V is applied to the cannula with respect to the grounded tube. The other is that in the absence of an applied electric field between the cannula and the surrounding tube, drop charge is measured using various electrolyte solutions at various concentrations. The field-free condition is obtained by connecting the cannula and the surrounding tube with a stainless steel wire which is grounded. The liquid dispersing and collection set-up is put in a Faraday cage. The electrolyte solutions of potassium chloride, sodium chloride, ammonium chloride and lanthanum nitrate (uni-valent and multi-valent cation electrolytes) are used and the concentration is changed from 10^{-5}M to 1M.

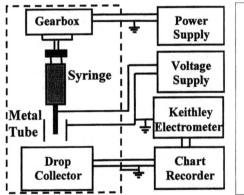

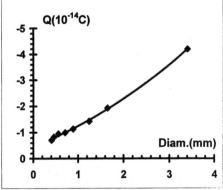

Fig.1 Diagram of Apparatus

Fig.2 DI water drop charge vs cannula outer diameter in a field-free region

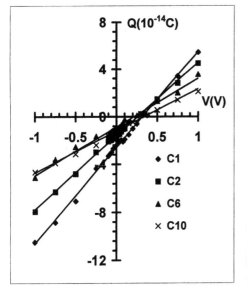

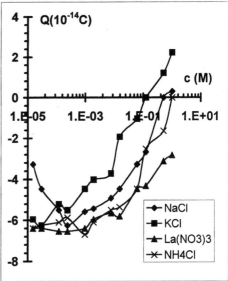

Fig.3 Charge of drop from cannulae1, 2, 6 and 10(c1, c2, c6 and c10 represent outer diameters 3.40mm, 1.65mm, 0.71mm and 0.33mm respectively) with DI water as a function of applied voltage between cannula and grounded tube

Fig. 4 Drop charge with cannula 6 (outer diameter 0.71mm) as a function of electrolyte concentration c (molarity)

3.Results

Fig. 2 shows single DI water drop charge as a function of cannula outer diameter. It is found that charge carried by drops breaking away from a cannula is always negative for DI water and increases in magnitude as the outer diameter of the cannula increases

Fig. 3 shows the drop charge as a function of applied potential difference between the cannula and the surrounding grounded tube. Deionised water is used with cannulae of outer diameters of 3.40mm, 1.65mm, 0.71mm and 0.33mm. Relationships are observed with lines biased slightly from the origin.

Fig. 4 represents the results of single drop charge obtained with solutions of sodium chloride, potassium chloride, ammonium chloride (uni-uni valent electrolytes) and lanthanum nitrate (multi-valent) at concentrations from 10^{-5}M to 1M in field-free and room temperature conditions. For all of these solutions negatively charged drops are produced at low ionic concentrations and the value of the charge is close to that of pure water ($\sim 10^{-14}$ C). As ion concentration increases drop charge reduces and then sign reversal occurs except for La(NO$_3$)$_3$. For the potassium and sodium chloride solutions drop charge reverses sign i.e. positive drops are produced, at ionic concentration of about 1 M. For ammonium chloride and lanthanum nitrate solutions similar results are obtained except that charge reversal occurs or would occur at much higher ionic concentrations.

170

4. Discussion

When an external electric field is applied between a cannula and a surrounding tube, the charging processes can be explained by the ordinary law of electrostatic induction with allowance for a contact potential between the solution and the cannula. As potential difference is applied between the cannula and the surrounding tube, an axis-symmetric electric field is established inside the tube which makes the exposed surface of the pendent drop at the tip of the cannula carry the same sign charge as the cannula. The net charge density over the exposed drop is directly proportional to the average electric field strength over the drop surface.

If the cannula and the surrounding tube are kept at the same potential, it is observed that the drop still carries charge. This is probably caused by the contact potential between the cannula and the solution, which establishes a small local electric field at the pendent drop surface. The sign and magnitude of the charge is related to the materials of the cannula and the solution. Different materials have different contact potentials.

If the contact potential hypothesis explains the field-free drop charging, then the charge sign reversal due to electrolyte concentration change can be interpreted as a polarity change of the contact potential between solution and cannula.

At this stage of the research the observed polarity change with electrolyte concentration change cannot be fully explained. Further experiments at higher capillary flow rates, when double layer shearing at the capillary tube walls will occur should help to elucidate the charging processes.

5. Conclusions

The charge of single drops produced from a vertical cannula in a field-free region may be interpreted in terms of contact potential between the cannula and the solution. The charge in an external field can be explained by induction with allowance for contact potential between the cannula and the solution.

6. Acknowledgement

The financial support from Astra Charnwood, through a studentship, is gratefully acknowledged

7. References

[1] Iribarne J V and Mason B.J. 1967 *Trans. Faraday Soc.* **63** 2234-2245,

[2] Iribarne J.V. and Klemes M. 1970, *J. Atmos.Sci.* **27** 927-936

[3] Jonas P.R. and Mason B.J. 1971 *Trans. Farad. Soc.* **64** 1971-1982

Inst. Phys. Conf. Ser. No 163
Paper presented at the 10th Int. Conf., Cambridge, 28–31 March 1999
© *1999 IOP Publishing Ltd*

Nonlinear stability of an electrified liquid jet

K Baudry, H Romat, A Sélénou Ngomsi

Laboratoire d'Etudes Aérodynamiques UMR 6609 du CNRS
Boulevard 3 - Téléport 2 - BP 179 - 86960 FUTUROSCOPE Cedex - France

Abstract. A theory for the nonlinear stability of a liquid jet in the presence of an applied electric field is presented. We impose an infinitesimal and axisymmetrical perturbation η on the surface of the jet. We use a straightforward expansion of the perturbation and we investigate its evolution with time.

1. Introduction

Rayleigh [1] gave a theory governing the linear stability of the capillary jet and developed a mathematical model for the break-up of liquid jets into drops. However, his model did not explain the experimental observations of Donnely and Glaberson [2], and Rutland and Jameson [3] which indicated that the jet broke up into a body of large main drops interspersed with smaller satellite drops. These observations motivated several researchers to formulate a nonlinear theory for the break-up of liquid jets which would show the existence of the satellites. For example, Yuen [4], Nayfeh [5], Lafrance [6], Ibrahim [7], Malik [8] worked on the nonlinear theory to explain the break-up of liquid jets.

The purpose of this paper is to investigate the theoretical nonlinear analysis of the stability of a liquid jet submitted to an electric field. We use a straightforward expansion for the surface disturbance and the velocity potential like Lafrance did and we also use a straightforward expansion for the electric potential. Our analysis differs from that of Lafrance because we don't neglect the effect of the surrounding fluid and because our jet is electrified.

2. Theory
2.1. Formulation of the problem

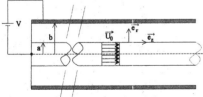

We consider an infinite liquid cylinder whose radius is a. The liquid flows away with an uniform velocity $\bar{U}_0$ into a cylindrical coaxial electrode whose radius is b. V is the value of the electric potential at b. The liquid is assumed to be incompressible and inviscid. We assume that the gaseous atmosphere situated between the liquid cylinder and the electrode is incompressible and inviscid. Moreover, we suppose that there is no mass transfer between the two phases and that the effects caused by the gravitational and magnetic fields are negligible. We will work in a cylindrical coaxial frame of reference $(\vec{e}_r, \vec{e}_\theta, \vec{e}_z)$ which moves with the jet velocity $\bar{U}_0$.

2.2. Form of the perturbation

We impose an infinitesimal and axisymmetrical disturbance η on the surface of the jet and we study its evolution with time. We assume that the surface disturbance η, the velocity potential φ and the electric potential ϕ can be expanded as follows :

$$\eta(z,t) = \sum_{m=1}^{M} \eta_0^m \eta_m(z,t), \qquad \varphi(r,z,t) = \sum_{m=0}^{M} \eta_0^m \varphi_m(r,z,t), \quad \phi(r,z,t) = \sum_{m=0}^{M} \eta_0^m \phi_m(r,z,t)$$

Given that we work with a nonlinear analysis at the second order, M=2. As Lafrance did, we use a straightforward expansion for the surface disturbance. The perturbation is expanded into a Fourier series whose second order terms are kept. This series is completed by a purely time dependent term. The perturbation is written as follows (because we work on the nonlinear stability at the second order):

$$\eta(z,t) = \eta_0\left[B_{r11}(t)\cos(kz) + B_{i11}(t)\sin(kz)\right] + \eta_0 D_1(t)$$

$$+ \eta_0^2\left[B_{r21}(t)\cos(kz) + B_{i21}(t)\sin(kz) + B_{r22}(t)\cos(2kz) + B_{i22}(t)\sin(2kz)\right] + \eta_0^2 D_2(t)$$

Our work consists in determining $B_{r11}(t)$, $B_{i11}(t)$, $D_1(t)$, $B_{r21}(t)$, $B_{i21}(t)$, $B_{r22}(t)$, $B_{i22}(t)$ and $D_2(t)$ in order to determine completely $\eta(z,t)$ and in order to study the stability of our system. The perturbation that we impose at the interface causes a modification of the velocities and of the electric fields in the gaseous and liquid phases (and we consider that the surface is equipotential). With appropriate boundary conditions we determine the velocities and electric fields after perturbation at the second order in the gaseous and liquid phases.

2.3. Equation of dispersion at the interface

The substitution of the expressions of the electric fields and of the velocities obtained in the perturbed system inside the equation of the impulsion conservation at the interface leads to an expression which has terms independent on η_0 and others dependent on η_0 and on η_0^2. The equation of dispersion gives two systems of differential equations :

at η_0 order

$$\begin{cases} C_1\ddot{B}_{r11}(t) - C_2\dot{B}_{i11}(t) + C_3 B_{r11}(t) = 0 \\ C_1\ddot{B}_{i11}(t) + C_2\dot{B}_{r11}(t) + C_3 B_{i11}(t) = 0 \end{cases} \quad (1)$$

at η_0^2 order

$$\begin{cases} D_1\ddot{B}_{r22}(t) - D_2\dot{B}_{i22}(t) + D_3 B_{r22}(t) = -D_4 \\ D_1\ddot{B}_{i22}(t) + D_2\dot{B}_{r22}(t) + D_3 B_{i22}(t) = -D_5 \end{cases} \quad (2)$$

where C_1, C_2, C_3, D_1, D_2, D_3 are constants and D_4, D_5 are functions of $B_{r11}(t)$ and $B_{i11}(t)$. With some following initial conditions, the system of differential equations (1) allows to find $B_{r11}(t)$ and $B_{i11}(t)$ and the system of differential equations (2) allows to find $B_{r22}(t)$ and $B_{i22}(t)$.

From the determination of the electric field (in the gaseous phase) and of the velocity (in the liquid phase) we find that $D_1(t)=0$, $B_{r21}(t)=B_{i21}(t)=0$ and $D_2(t) = -(1/4a)(B_{r11}^2(t) + B_{i11}^2(t))$. Because $B_{r11}(t)$, $B_{i11}(t)$, $B_{r22}(t)$, $B_{i22}(t)$ and $D_2(t)$ are known, the perturbation at second order is totally defined.

3. Numerical results

We give the numerical results of this theoretical work, that is to say the temporal evolution of the free surface displacement of the jet. In our numerical simulations, we take $\eta_0=10^{-6}$ m, $U_0=100$ m/s and $E=10^7$ V/m (electric field).

The study of the stability of our system consists in investigating the evolution of the perturbation with time. We study the evolution with time of the perturbation for a great number of wavelengths. If the amplitude of the perturbation increases with time then the system is unstable, so we can determine wavelengths for which the jet is unstable.

3.1. Numerical results of the linear analysis

We first give the numerical results of the linear analysis of the stability of a jet of diesel oil. We present the evolution of the deformation of the free surface $\eta(z,t)$ at η_0 order called η_1 for several values of time and for different wavelengths.

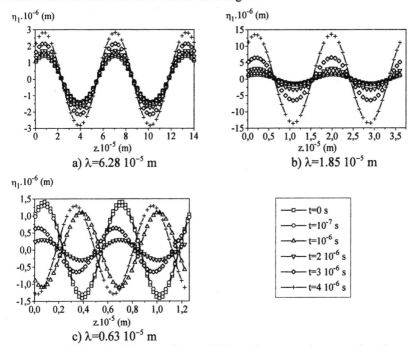

a) $\lambda=6.28\ 10^{-5}$ m

b) $\lambda=1.85\ 10^{-5}$ m

c) $\lambda=0.63\ 10^{-5}$ m

Figure 1 : Wave profiles at different times at various wavelengths

In figure 1, we represent the evolution of the perturbation with time for different wavelengths λ. The amplitude of the perturbation for $\lambda=6.28\ 10^{-5}$ m increases with time and reaches $3\ 10^{-6}$ m after $4\ 10^{-6}$ s. For $\lambda=1.85\ 10^{-5}$ m, the amplitude of the perturbation increases with time and reaches $15\ 10^{-6}$ m after the same time and the last one oscillates and never passes $1.5\ 10^{-6}$ m. In other words, the first perturbation is unstable and so is the second one and it grows more rapidly than the first one. For $\lambda=0.63\ 10^{-5}$ m, the amplitude of the perturbation oscillates and does not increase or decrease regularly.

In the case of a linear analysis, we find that for $\lambda>1.25\ 10^{-5}$ m, the perturbation is unstable with a maximum of instability for $\lambda=1.85\ 10^{-5}$ m and that for $\lambda<1.21\ 10^{-5}$ m the perturbation is stable.

3.2. Numerical results of the nonlinear analysis

Then we present the numerical results for the nonlinear analysis of the stability of a diesel oil jet. We represent the evolution of $\eta(z,t)$ at second order called η_2 (we take the terms of η_0 order and the terms of η_0^2 order) for several values of time and for different wavelengths.

For $\lambda=6.28\ 10^{-5}$ m and for $\lambda=2.17\ 10^{-5}$ m, the amplitude of the perturbation increases with time, so the perturbation is unstable. In addition to that, we notice a swelling between the crests at $t=4\ 10^{-6}$ s on the figure 2a and at $t=2\ 10^{-6}$ s on the figure 2b. For $\lambda=0.63\ 10^{-5}$ m, the amplitude of the perturbation oscillates and does not increase or decrease regularly.

174

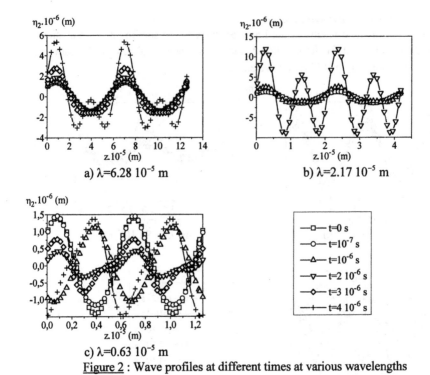

a) $\lambda=6.28\ 10^{-5}$ m

b) $\lambda=2.17\ 10^{-5}$ m

c) $\lambda=0.63\ 10^{-5}$ m

Figure 2 : Wave profiles at different times at various wavelengths

The nonlinear analysis of the stability of a diesel oil jet gives results a bit different from those of the linear stability. Indeed for $\lambda>1.43\ 10^{-5}$ m, the perturbation is unstable with a maximum of instability for $\lambda=2.17\ 10^{-5}$ m and for $\lambda<1.39\ 10^{-5}$ m the perturbation is stable.

So working with a nonlinear analysis reduces the interval of λ where the perturbation is unstable and increases consequently the interval of λ where the perturbation is stable.

4. Conclusion

We made a nonlinear analysis of the stability of a jet submitted to an electric field. We imposed an infinitesimal perturbation at the surface of the jet and we determined the evolution with time of the perturbation profiles numerically for various wavelengths. We found wavelengths for which the jet is unstable with a linear analysis but also with a nonlinear analysis. Moreover from this analysis, we can learn the influence of the velocity of the jet and of the electric field on the stability of the jet. Futhermore, we can make a rough estimate of the diameter of the first droplets produced near the nozzle [9].

References

[1] Rayleigh 1896 "Theory of sound" vol. 2
[2] Donelly R J and Glaberson W 1966 *Proc. R. Soc.London* A290 547
[3] Rutland D F and Jameson G J 1971 *J. Fluid Mech.* **46** 267
[4] Yuen M C 1968 *J. Fluid Mech.* **33** 151-163
[5] Nayfeh A H 1970 *Phys. Fluids* **13** 841-847
[6] Lafrance P 1975 *Phys. Fluids* **18** 428-432
[7] Ibrahim E A 1992 *J. of Applied Mechanics* **59** S291-S296
[8] Malik S K and Singh M 1983 *Quaterly of applied mathematics* 273-287
[9] Baudry K 1999 Thèse Université de Poitiers

Inst. Phys. Conf. Ser. No 163
Paper presented at the 10th Int. Conf., Cambridge, 28–31 March 1999
© *1999 IOP Publishing Ltd*

Asymptotic variation of radius of a viscous polymer jet in the electrospinning process

A F Spivak and Y A Dzenis

Department of Engineering Mechanics, Center for Materials Research and Analysis, University of Nebraska-Lincoln, Lincoln, NE 68588-0526, USA

Abstract. This paper presents results of analytical modeling of a steady state jet flow in the electrospinning process. A model of a weakly conductive viscous polymer jet driven by an external electrostatic field is developed. A class of polymer fluids described by nonlinear power-law rheologic constitutive equation is considered. Asymptotic decay of radius at large distances from the beginning of the jet is studied. The model can be used for the process optimization and design of better electrospinning apparatus.

1. Introduction

Motion of a liquid mass deformed and accelerated by an external electric field constitute a problem of electrohydrodynamics that has attracted considerable attention of investigators [1-4]. When liquid mass is attached to a solid object, the mass shape is called a meniscus. The meniscus deforms in the electric field. In weak fields, the meniscus is held on the surface of a solid object by surface tension. When the field increases, the meniscus elongates until, at some critical value of the field, surface tension can no longer balance electric forces and the liquid mass emits a jet [5-6]. Depending on process parameters, an intermittent or continuous jet flow can be observed. The continuous flow mode can be used to disperse liquids (atomization) [6] or to fabricate polymer fibers (electrospinning) [7-8].

In electrospinning, a jet of a polymer solution or melt is ejected from the fluid surface. The jet is elongated and accelerated by the external field, deposited on a substrate, dried and/or chemically treated, and converted into a thin fiber. Recently, electrospinning was extensively studied experimentally by Reneker and co-authors [7-8]. Over thirty synthetic and natural polymers were spun into fibers with diameters from a few nanometers to microns.

Polymer solutions or melts used in electrospinning are highly viscous, weakly conductive fluids which often exhibit nonlinear rheologic behavior. Motion of these fluids in a jet mode in presence of the electric field has not yet been sufficiently studied. In addition, the behavior of electrostatically driven jets at large distances from their origin has not been investigated. It is, perhaps, due to the fact that low-viscosity, low-molecular weight fluids, utilized in most electrostatic jet applications, break-up into droplets long before the jet reaches its asymptotic length [5-6]. This break-up is caused by surface capillary waves described by Rayleigh [1]. The Rayleigh instability is not typically observed in electrospinning where thin polymer jets can exist at large distances from the origin. Studies of peculiarities of the jet flow at large distances are important for better understanding of the electrospinning process. An asymptotic model of the process can be used to evaluate and control the diameters of resulting polymer fibers.

2. Model of steady state spinning

Consider a thin jet of a viscous weakly conductive fluid pulled by a constant external electric field (Fig.1). Governing electrohydrodynamic equations can be written as [9-10]

$$\rho(\vec{v} \bullet \nabla)\vec{v} + \nabla p = \nabla \tau_c + \nabla \tau_e$$

$$\nabla \vec{v} = 0 \qquad \nabla \vec{j} = 0 \tag{1}$$

where ρ is mass density, $\vec{v}$ is velocity, p is pressure, τ_c is viscous stress tensor, $\vec{j}$ is electric current density, and

$$\tau_e = \varepsilon(\nabla\phi \otimes \nabla\phi) - \frac{\varepsilon(\nabla\phi \cdot \nabla\phi)}{2}I$$

is Maxwell's stress tensor, ϕ is electrostatic field potential, ε is permittivity of vacuum, I is unit tensor.

$2R_o$

E

1mm

z

Figure 1. A schematic of the jet: an experimental micrograph of the 3% aqueous PEO solution electrospun at 15kV.

Consider elongational and axisymmetric jet flow in the direction of the external electric field, z. Due to the weak conductivity of the fluid and small jet diameter the influence of the internal electric field can be neglected. By analogy with [11], assume that the total electric current is due primarily to the convective flow of the surface charges with the jet particles. These charges interact with the external electric field creating the pulling force responsible for jet acceleration. In addition, they cause transverse electric repulsion. The boundary condition for the mechanical part of the problem is:

$$(pI + \tau_c)\big|_L = \sigma_s(k_n + k_t)I - \tau_e\big|_L ,$$

where σ_s is the surface tension coefficient, k_n, k_t are the principal curvatures of the jet surface, L is the perimeter of the jet cross section.

The overall electrostatic field potential, ϕ, can be written as a sum of the potentials of the surface charges, ϕ_s and the external field, $\phi_{ext} = -Ez$, where E is the constant external electric field. Electrostatic boundary condition on the jet surface [9-10] completes the problem formulation. Rheologic behavior of many polymer fluids can be described by the power-law constitutive equation, known as Oswald-deWaele law [12]:

$$\tau_c = \mu[tr(\dot{\gamma}^2)]^{\frac{m-1}{2}}\dot{\gamma} \tag{2}$$

where μ is the consistency index, $\dot{\gamma}$ is the rate of strain tensor, m is the flow index [13]. Newtonian fluids are described by a special case of the Eq. (2) with the flow index $m = 1$. Pseudoplastic (shear thinning) fluids are described by the flow index $0 \leq m < 1$. Dilatant (shear thickening) fluids are described by the flow index $m > 1$ [12-13].

By analogy with [14-15], the general three-dimensional problem, Eq. (1-2), was reduced to a one-dimensional problem by averaging physical quantities over the jet cross-section. A slender jet approximation was utilized by neglecting the tangential curvature of the jet surface. It was assumed that the flow of the jet depends primarily on the rate of stretching $\dot{\gamma}_{zz}$. The other components of the rate of strain tensor $\dot{\gamma}$ were neglected. The procedure of averaging lead to the problem formulation in terms of volumetric jet flow rate, Q, linear surface charge density, $S = \oint_L \sigma_e dl$ where σ_e is surface charge density, and jet radius, R. The reduced one-dimensional governing equations are:

$$\frac{d}{dz}\left[\frac{\rho Q^2}{2\pi R^2} + 2\pi R\sigma_s - \frac{S^2}{8\varepsilon\pi} - \pi R^2\mu\left[\frac{d}{dz}\left(\frac{Q}{\pi R^2}\right)\right]^m\right] = ES \tag{3}$$

$$\frac{dQ}{dz} = 0 \qquad \frac{d}{dz}(SQR^{-2}) = 0$$

3. Asymptotic analysis

Integration of the conservation equations in (3) gives two algebraic equations $Q = Q_o$ and $S = \pi I_o Q_o^{-1} R^2$, where Q_o is volumetric flow rate and I_o is electric current. In this paper, Q_o and I_o were considered external parameters of the problem. Variation of the jet radius was then described by an ordinary differential equation:

$$\frac{d}{dz}\left[\frac{\rho Q_o^2}{2\pi R^2} + 2\pi\sigma_s R - \frac{\pi}{8\varepsilon}\left(\frac{I_o}{Q_o}\right)^2 R^4 - \pi\mu R^2\left(\frac{d}{dz}\left(\frac{Q_o}{\pi R^2}\right)\right)^m\right] = \frac{\pi EI_o}{Q_o}R^2. \tag{4}$$

General closed-form solution of the Eq. (4) is not available. An asymptotic solution was found by integrating a simplified equation that was derived from the Eq. (4) assuming that the jet radius is small and monotonically decreasing function along the jet. Using the change of variables

$$\xi = R_o^2/R^2 \text{ and } s = z/z_o + s_o,$$

where the normalization parameters are characteristic jet radius, R_o and characteristic jet length $z_o = \rho Q_o^3/(2\pi^2 EI_o R_o^4)$, and s_o is a constant defined by the initial conditions, and integrating once the resulting equation, the following differential equation was obtained:

$$\beta\left(\frac{d\xi}{ds}\right)^m + s - \xi^2 = 0 \tag{5}$$

where $\beta = 4\pi^{(4-m)}\mu\rho^{-2}EI_o Q_o^{(m-5)}R_o^{8-2m}$ is a dimensionless parameter.

Using the parametric integration method [14], the Eq. (5) was transformed into:

$$\frac{ds}{dt} = \frac{m\beta t^{m-1}}{2\xi t - 1} \qquad \frac{d\xi}{dt} = \frac{m\beta t^m}{2\xi t - 1} \qquad 2\xi t \neq 1 \tag{6}$$

where t is the parameter. For large distances, $s \gg 1$, the solution can be reduced to a power-law approximation:

$$\frac{R}{R_o} \sim \left(\frac{z}{z_o}\right)^{-\alpha} \tag{7}$$

where exponent α depends on the flow index m. For $m > 2$, the exponent is $\alpha = m/2(m-2)$. For $m < 2$, it is $\alpha = 1/4$. The asymptotic exponent $\alpha = 1/4$ was obtained and experimentally verified in [17] for Newtonian fluids. Our results extend the applicability of the eq. (7) for shear thickening and dilatant fluids which rheological properties described by the Eq.(2).

The characteristic jet radius, R_o, was determined by jet stability analysis. Kinetics of thermal fluctuations of the jet flow parameters was considered. It was found that in order for thermal fluctuation level to be finite, the following inequality must be satisfied:

178

$$R \leq 2\left(\varepsilon\sigma_s Q_o^2/I_o^2\right)^{1/3} \tag{8}$$

The characteristic jet radius $R_o = 2\left(\varepsilon\sigma_s Q_o^2/I_o^2\right)^{1/3}$ can be interpreted as an initial jet radius. For a typical electrospinning process with parameters $Q_o \sim 0.1mm^3/s$, $I_o \sim 100nA$, and $\sigma_s \approx 50 \div 100mN/m$, this characteristic radius is $R_o \approx 50 \div 120\mu m$.

4. Conclusions

A model describing the variation of radius of a weakly conductive viscous polymer jet was developed. An asymptotic analysis of the jet radius decay was performed. It was shown that the power-law asymptote, originally obtained for the Newtonian fluids [17], is applicable to a wide class of fluids described by the nonlinear Oswald-deWaele law. The asymptotic exponent depended on the flow index in the nonlinear constitutive relation.

The work was supported by the National Science Foundation under Grant No. 9813098.

5. References

[1] Strutt J W (baron Rayleigh) 1882 *Phil. Mag. and J.* **144** 184-86; *The Theory of Sound* (London: Macmillan)
[2] Macky W 1931 *Proc.Royal.Soc.* **A133** 565-87
[3] Taylor G and Van Dyke M 1969 *Proc. Roy. Soc.* **A313** 453-75
[4] Taylor G 1964 *Proc. Roy. Soc.* A280 383-97
[5] Cloupeau M and Prunet-Foch B 1994 *J. of Aer. Sci.* **25** 1021-36
[6] Bailey A G *Electrostatic Spraying of Liquids* (New York: Willey)
[7] Doshi J and Reneker D 1995 *J. of Electrostat.* **35** 151-60
[8] Reneker D and Chun I 1996 *Nanotechnology* **7** 216-23
[9] Melcher J and Taylor G 1969 *Ann. Rev. of Fl. Mech.* **1** 111-46
[10] Groot de S R and Suttorp L G *Foundation of Electrodynamic* (Amsterdam: North-Holland Publ.Co.)
[11] Mestel A *1996 J. of Fl. Mech.* **312** 311-26; 1994 *J.of Fl. Mech.* **274** 93-113
[12] Darby R *Viscoelastic Fluids* (New York: Marcel Deccer)
[13] Byron B *Dynamics of Polymeric Liquids* (New York: Wiley)
[14] Yarin A *Free Liquid Jets and Films: Hydrodynamics and Rheology* (New York: Longman)
[15] Melcher J and Warren E 1971 *J. of Fl. Mech.* **47**(1) 127-43
[16] Kamke E *Differentialgleichungen reeller Funktionen* (Leipzig: Akad. Ver. Becker & Erler)
[17] Kirichenko V, Petryanov-Sokolov I, Suprun N and Shutov A 1986 *Sov. Phys. Dokl.* **31** 611-5

Inst. Phys. Conf. Ser. No 163
Paper presented at the 10th Int. Conf., Cambridge, 28–31 March 1999

179

Experimental study of a jet of conducting liquid subjected to a radial DC field

Heliodoro González†[1], Pierre Atten‡ and Khalil Chouman‡

† Depto. de Física Aplicada. E.S.I. Universidad de Sevilla. Camino de los Descubrimientos s/n, 41092-Sevilla. Spain

‡Laboratoire d'Electrostatique et de Matériaux Diélectriques, CNRS Av. des Martyrs BP 166, 38042 Grenoble Cedex 9, France

Abstract. The evolution of a conducting viscous ink jet surrounded by air at rest and subjected to a DC radial electric field is studied experimentally. EHD stimulation of one particular mode and measurements of the break-up length for different stimulation voltages allows us to characterize the growth of the perturbations. A dispersion relation for axisymmetric deformations obtained from a temporal linear stability analysis that includes the Kelvin-Helmholtz instability is compared with the experimental results.

1. Introduction

Charged jets are frequently found in some devices like ink jet printers. In these applications it is of crucial importance to control the size and charge of the resulting drops. As the charging process is performed by means of electrostatic influence and an electric field is acting on a significant part of the jet length, it is interesting to study its effect on the growth rates of perturbations. Our physical model is a conducting liquid jet subjected to a voltage difference with respect to a cylindrical coaxial electrode. The forces that we take into account are viscous ones and pressure effects at the free surface, including capillarity, the influence of a surrounding gas at rest and the electrostatic pressure. The relevant stability mechanisms are the classical capillary instability, related to a decrease in surface area for long wavelength axisymmetric perturbations, the point effect, responsible for destabilization of small wavelengths due to an increase in electrostatic pressure at the crests of the deformations, an opposite (stabilizing) effect, important for long wavelengths, caused by an increase of the local field strength at the valleys of the deformations, and finally the Kelvin-Helmholtz instability, originating in the relative velocity of the jet with respect to the surrounding air.

Previous theoretical works on this subject were carried out by Melcher (1963) and Saville (1970), for inviscid and viscous liquids respectively, although the effect of the air was not considered. However, to our knowledge, no experimental validation of their resulting stability spectrum, i.e., growth rate measurements for some representative wavenumbers, are available in the literature. The aim of the present work is to propose an improved dispersion relation, including the effect of the air, and to show experimental evidence of its validity. To this end, EHD stimulation has revealed to be a very efficient mechanism to create a perturbation of specific wavenumber and of magnitude much higher than the noise.

[1] E-mail: helio@cica.es. Also at Depto. de Electrónica y Electromagnetismo. U. Sevilla.

2. Dispersion relation

The system is specified by a set of geometrical and physical parameters: the radius of the jet, a and of the outer electrode, b; the jet velocity v; the liquid and gas densities, ρ and ρ_g, respectively; the surface tension σ, and finally the viscosity of the liquid, μ, (assuming that of the gas to be negligible).

The general formulation of the jet dynamics includes the Navier-Stokes equation for the liquid, the Euler equation for the surrounding gas and the continuity equation for both fluids, assumed to be incompressible. At the free surface we impose a kinematic condition that relates the normal velocities of the two fluids with the motion of the surface itself, and the stress balance between the hydrodynamic and capillary pressure terms, the electrostatic pressure and the viscous stress. The electrostatic pressure evaluation requires to solve the Laplace equation for the electric potential, along with boundary conditions (fixed constant potentials at the electrodes). Regularity conditions at the symmetry axis and the impenetrability condition on the outer electrode need also be used. If the velocity of the jet is high enough with respect to the typical velocity of propagation of perturbations it is usual to consider the jet as an infinite column and to perform a temporal rather than a spatial stability analysis; in that case a last condition concerning the velocity profile at the nozzle is disregarded.

A linear, modal stability analysis using cylindrical coordinates examines a small amplitude perturbation of the free surface with a dependence $\exp(\Omega t + im\theta + ikz)$, where $\Omega = \gamma + i\omega$ is complex in general, γ being the growth factor and ω the oscillation frequency, and k the wavenumber. We use nondimensional variables by taking the radius of the jet, a, and the capillary time, $t_c = \sqrt{\rho a^3/\sigma}$, as reference values for length and time, respectively. The other variables such as velocity, pressure, etc. have the same dependence with t, θ and z.

The resolution of the mathematical problem leads to an independent dispersion relation for each azimuthal mode characterized by the modal number $m = 0, 1, 2, \ldots$, giving Ω as a function of k. Here we restrict ourselves to present that corresponding to axisymmetric perturbations ($m = 0$) with the additional simplification $b >> a$:

$$\bar{\Omega}^2 \frac{I_0(k)}{k\,I_0'(k)} + 4C\,\bar{\Omega} \left[\frac{k\,I_0(k)}{I_0'(k)} - \frac{1}{2} + \frac{Ck^2}{\bar{\Omega}} \left(\frac{k\,I_0(k)}{I_0'(k)} - \frac{k_v\,I_0(k_v)}{I_0'(k_v)} \right) \right] = 1 - k^2 - \chi \left(1 + k\frac{K_0'(k)}{K_0(k)} \right) + \bar{\rho}\Omega^2 \frac{K_0(k)}{k\,K_0'(k)},$$

with $k_v^2 = k^2 + \bar{\Omega}/C$ and $\bar{\Omega} \equiv \Omega + ik\sqrt{We}$, where some nondimensional groups must be defined: i) the Weber number, $We = \rho v^2 a/\sigma$; ii) the Ohnesorge number $C = \mu/\sqrt{\rho\sigma a}$; iii) the electric number $\chi = \varepsilon_0 E_0^2 a/\sigma$ (with $E_0 = V_0/[a\ln(b/a)]$), and iv) the density ratio $\bar{\rho} = \rho_g/\rho$. It is not difficult to give a general dispersion relation for arbitrary m and ratio b/a, but in the experimental conditions that we are going to describe in the next section, the outer electrode is very far from the jet and the electric number is not high enough to make dominant an asymmetric mode ($m \neq 0$), as described in Melcher (1963) and Saville (1970).

3. Experimental set-up.

The experimental set-up of figure 1 has been described in Spohn et al (1993) and Barbet (1997). There is an hydraulic part including a constant pressure liquid reservoir, a circuit with thermal bath and a nozzle from which a 0.22 mm radius jet is formed with constant velocity of 5 m/s. Along its vertical path there is first the EHD stimulation region, where one k-mode is selected by means of a periodic electrostatic pressure produced by a thin blade electrode normal to the grounded jet and whose edge is very close to it; an AC voltage is applied to the stimulation electrode. In the second zone (the amplification zone) the jet is subjected to an almost uniform radial DC electric field by means of two long plane electrodes placed symmetrically, each one 5.5 mm apart from the jet. Both regions are shielded by a grounded metallic sheet with a small hole allowing the jet to pass through. The operating conditions are always such that the amplification region up to the break-up point is about 10 cm. The jet is monitored by means of a stroboscopic lamp and a CCD videocamera. The stroboscopic frequency is the same as the EHD stimulation frequency in order to obtain a fixed image. A phase shift electronic device allows us to choose any particular instant. Jet and electrodes form a block that can be vertically translated so that the distance of the breaking point to the nozzle, the break-up length L_b, is measured with a resolution

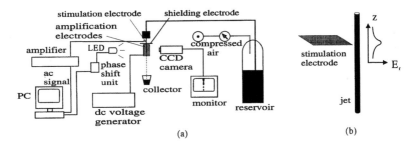

Figure 1. (a) Schematic drawing of the experimental set-up. (b) Stimulation electrode and qualitative distribution of the maximum radial field at the surface.

of the order of 10 μm. The whole system is mounted on an heavy optical rail with no other special insulation against vibrations.

The electrode connected to an AC voltage source $V_s \cos \omega_s t$ is responsible for the wavenumber selection. Its distance to the jet is between $a/2$ and a, so that only a small portion of the liquid surface is subjected to a strong electrostatic pressure with frequency $2\omega_s$. Although the initial deformation is small, there is an important momentum transfer to the mode whose wavelength is $\lambda = \pi v/\omega_s$. The voltage amplitude must be of the order of 1 kV to effectively make this mode to emerge from natural noise (the break-up length is typically reduced by 50%). External vibrations do not influence appreciably the jet dynamics as the stroboscopic image is very stable. After stimulation, the jet enters the amplification region and the amplitude of the selected mode remains small during most part of the amplification. If we observe a given section of the jet in a framework moving with the jet velocity, from linear theory the radius growth is exponential: $r(t) = a + \xi_0 e^{\gamma t}$, where ξ_0 may be considered as the initial amplitude induced on the free surface by the EHD stimulation. It is shown experimentally in Crowley (1986), Spohn et al (1993) and Barbet (1997), and theoretically justified, that the initial amplitude is proportional to the electrostatic pressure, i.e., $\xi_0 = AV_s^2$, with A depending on the geometry and fluid properties. Consequently, $L_b = (v/\gamma) \ln(a/\xi_0) = (v/\gamma)[\ln(a/A) - 2 \ln V_s]$. As v may be easily measured from the wavelength of the perturbation and the imposed frequency, the last equation gives a simple and accurate way to measure γ by merely changing V_s.

Two plane electrodes are actually used, instead of cylindrical ones, to facilitate the visualization of the jet. In spite of what could be expected, the electric field produced on the free surface is almost uniform with respect to the azimuthal coordinate θ. A numerical computation of the electrostatic pressure in our experimental conditions (minimum relative distance $b/a = 25$ and relative width of the electrodes $l/a = 44$) shows a maximum deviation of 0.3% from the mean pressure. However, this mean pressure is 0.857 times that of cylindrical configuration with the same value b for the radius of the outer electrode. We include this geometrical factor in the evaluation of χ. An imperfect centering of the system does not alter very much neither the uniformity of the pressure (less than 1% for a lateral displacement of the order of a) nor the pressure mean value.

4. Results

We used ink supplied by the group TOXOT ($\rho = 1160 \, \text{kg/m}^3$, $\sigma = 0.057$ N/m and $\mu = 16.25$ cps at 28°C). Several DC voltages were applied in the amplification zone for selected wavenumbers. The measured growth factors along with the corresponding theoretical predictions are shown in figure 2, classified in four representative stability curves with fixed DC voltage ($V = 0, 1840, 2800$ and 3740 V). The nondimensional numbers used to compute the curves are $We = 160$, $C = 0.130$ and $\chi = 0, 0.183, 0.425$ and 0.758 respectively (taking into account the geometrical factor). The theoretical predictions that include viscosity and the effect of the surrounding air at rest are in good agreement with the experimental values of the growth factor, although slightly underestimated in most cases. The theory also correctly predicts an almost linear dependence of γ with χ for k fixed and a change in

182

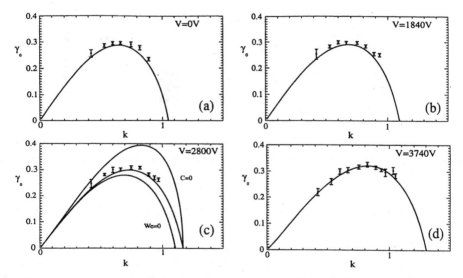

Figure 2. Growth factor γ of axisymmetric perturbations versus wavenumber k (both nondimensional) for several DC voltages. The points are experimental values. In (c) two theoretical curves labeled "$C = 0$" and "$We = 0$" show separately the effect of viscosity and surrounding air.

the sign of the slope near $k = 0.6$ (positive for smaller wavenumbers and vice versa). Note that the perturbations whose growth factors are very small are not efficiently stimulated in view of the limitation that air breakdown imposes on the initial amplitudes (giving too long break-up lengths). Figure 2(c) shows how the curves are modified by each physical mechanism: the curve "$C = 0$" omits the viscous damping and "$We=0$" does not include the Kelvin-Helmholtz instability. Only the combination of both effects give acceptable predictions.

5. Conclusions

The temporal linear stability theory of a conducting viscous liquid jet subjected to a radial DC electric field gives good predictions when the Kelvin-Helmholtz instability mechanism due to the effect of a surrounding gas at rest is included. As expected, the field intensity increases γ_0 for $k > 0.6$ and vice versa. The growth factor also increases with the jet velocity for all k, and finally, viscosity has a well-known damping effect upon unstable modes. In this experimental validation note the remarkable usefulness of the EHD stimulation method to select a particular mode from natural noise.

Acknowledgments

This work was carried out during the stay of one of the authors (H.G.) at the LEMD of Grenoble with financial support from the Junta de Andalucía (Spain).

References

Barbet B 1997 *Stimulation electrohydrodynamique et thermique de jets de liquide conducteur* (Grenoble: Thesis, LEMD-CNRS & UJF)

Crowley J M 1986 *IEEE Trans. on Ind. Appl.* **1A-22 6** 973-6

Melcher J R 1963 *Field-coupled surface waves* (Cambridge: M.I.T. Press)

Saville D A 1970 *Phys. Fluids* **13** 2987-94

Spohn A, Atten P, Soucemarianadin A and Dunand A 1993 *IS&Tth Congress on Advances and Printing Technology* 294-7

DISCUSSION - Section C - Atomisation and EHD

Title: Electrostatic Atomisation - Questions and Challenges
Speaker: A J Kelly

Question: S Cunningham
What is the life time of the formed crystal?

Reply: Given that charge is well known to be tenaciously bound to the droplet surface and is stable under droplet evaporation up to the point at which the Rayleigh limit is encountered the crystal will also be stable. In this regard, it is interesting to note that indirect evidence for crystallization was stumbled on, but not recognized, during the analysis of the Rayleigh bursting process, (cf. Roth D.G and Kelly, A J, 'Analysis of the disruption of Evaporating Charged Droplets', IEEE Trans. on Industry Applications, **1A-19**, 5, pp. 771-775, September/October 1983). At the time Roth and I were quite perplexed to find that the only way the modelling could be made to agree with data was for the surface charge to have zero mobility. Once it was assumed that charge was immobilized on the surface and could not freely move about, the theory replicated experiment. Obviously this is circumstantial evidence, but is consistent with the notion that the charge is locked into a stable framework encompassing the surface.

Question: Jen-Shih Chang
From the point of view of pollution prevention, will the charged spray combustion reduce thermal NO_x and VOCs due to the better mixing and reduced fuel injection velocity?

Reply: Electrostatic atomization provides, for the first time, true electronic control over both the droplet development process and droplet dispersal/mixing. This technology offers a new and as yet unexplored option for pollutant reduction. Whether it can be used to effectively control the production of pollutants (NO_x, SO_x, particulates and VOCs) is a moot point since no work has been done to explore the utility of this fuel preparation means. However, to cite one well defined and documented capability, the ability to maintain vigorous electrostatic atomization at all pressures down to ~20kPa should reduce, if not eliminate, VOC production during nozzle turn-on and decay transients, and post pulse nozzle dribble. In addition, the fact that we have been able to consistently burn Jet fuel with blue flames, in quiescent air, and at feed pressures of less than a bar, we feel is compelling evidence that electrostatic fuel preparation offers an ideal approach to pollutant control. Unfortunately, the engine and burner community has thus far failed to show any discernible interest in this option.

Question: W Balachandran
Did you study the charge crystal formation on the surface of a liquid jet?

How would this interfere with the dynamic surface tension?

Reply: No effort has been devoted to the question of surface charge crystallization on jet structures. There is every reason to suspect that crystallization is occurring since the energetics (eg. the ratio of charge electrostatic energy to thermal energy) are similar. The one major and complicating difference concerns the dynamics of the jet as it evolves to form droplets. As noted in my response to Dr Cunningham's question, circumstantial evidence indicates that crystallization plays a role in the droplet formation process associated with Rayleigh bursting. During the analysis of this process, (cf. Roth D.G. and Kelly, A.J., "Analysis of the Disruption of Evaporating Charged Droplets", IEEE Trans. On Industry Applications, **IA-19**,5, pp.771-775, September/October 1983) Roth and I were quite perplexed to find that the only way the modelling could be made to agree with data is for the surface charge to have zero mobility. Limited surface mobility, presumably due to locking of the surface charge in a crystallization process dominates the behaviour of Rayleigh burst droplet development, and by implication, will play a similarly important role in jet/droplet development and alter the "dynamic surface tension".

Title: Viscosity Effect on EHD Spraying of Liquids
Speaker: A Jaworek

Question: Kees Geerse
            What is the nozzle opening size?   Do you really have a cone-jet?

Reply: The nozzle used was a hypodermic needle 0.4 o.d. / 0.25 i.d. In the cone-jet mode, in a geometrically based definition, there is a well defined cone which gradually changes to a fine jet, that was observed by us in EHD spraying.

Title: Scaling of Droplet Size and Current Produced in the Cone-Jet Mode
Speaker: R P A Hartmann/Kees Geerse

Question: M K Mazumnder
            Could you please comment on the rate of evaporation of the stream of droplets produced by the cone-jet.

Reply: Depending on the liquid used, the rate of evaporation differs. Fast evaporating liquids, for example ethanol, will undergo Rayleigh explosions, i.e. the droplets very quickly become too small to keep their initial charge and a wide distribution of droplet sizes is obtained. To prevent this, different methods can be used - spray in a saturated atmosphere, so the evaporation rate decreases; use slower evaporating liquids. In some cases, evaporation is a profit, for example if there is something dissolved in the liquid, which has to become a particle itself, after evaporation, this particle is formed. To prevent Rayleigh explosions, (wide size distribution) the droplets have to be discharged with for example a corona needle.

Question: Zayed Huneiti
The axisymmetric mode of atomisation produces polydispersed droplets. When you derived the scaling law did you take the average size?

Reply: Axisymmetric break-up - monodisperse droplets first, drain droplets and satellites second. The scaling law is droplet-volume based, i.e. the diameter given by the scaling law is derived from the volume of the main droplet and the eventually occurring satellite. The break-up theory does not take into account the effect of satellites.

Title: Electrostatic Atomisation of Ultrafine Spray of Ceramic-based Solution
Speaker: P Miao

Question: S Gerard Jennings
Did you investigate the influence of the AC electric field on the deposition of the ceramic powder?

Reply: No, I did not investigate this influence.

Question: M K Mazumnder
Is there any advantage in using electrostatic atomization over an electrostatic spinning disc system which can be made to coat a large surface area easily?

Reply: Firstly, when electrostatic atomization is used to deposit ceramic particles on a surface, the liquid droplets which contain these particles are highly charged. This prevents the droplets from agglomerating. So the particle will be evenly distributed on the surface and the particle size distribution on surface will also be narrow. Secondly, particle size deposited on the surface can be controlled down to the original powder particle size before dispersion. Fine particles have the effect of lining the sides of the voids between particles which enable films to be deposited without flaws. Finally, the morphology and thickness of the film can be precisely controlled in the micrometer range. Although the electrostatic atomization technique (in cone-jet mode) using a single nozzle cannot be used to coat large surface areas like a spinning disc system, this limitation could be solved by proper design of a multi-nozzle system.

Title: Recent Advances in Enhancement of Heat Transfer and Mass Transport in Thermal Equipment with EHD
Speaker: J Seyed-Yagoobi

Question: T B Jones
Is it true that EHD-enhanced heat transfer still, after some 30+ years of work, suffers from the fact that geometries are designed with conventional

heat transfer in mind? Can this situation change?

Reply: Yes, this is true.   The key to achieving very high enhancement of heat transfer and the successful transfer of technology to industry is to design heat transfer surfaces with the EHD phenomenon in mind. This has not been done yet!

Title: Flow Electrification in High Power Transformers (Effect of Temperature Gradient and Charge Accumulation)
Speaker: G G Touchard

Question: J Seyed-Yagoobi
How does the charge relaxation time of the fluid compare with the travelling time of fluid within the tube?

Reply: On the same order of magnitude.

Hot Topic Discussion

Inst. Phys. Conf. Ser. No 163
Paper presented at the 10th Int. Conf., Cambridge, 28–31 March 1999

MECA Electrometer: Initial Calibration Experiments

M. Buehler, L-J. Cheng, O. Orient, M Thelen, R Gompf[1], J Bayliss[1], J Rauwerdink[1]

Jet propulsion Laboratory, California Institute of Technology, USA
[1]Kennedy Space Flight Center, National Aeronautics and Space Administration, USA

Abstract. The Mars '01 lander contains an electrometer designed to evaluate the electrostatic nature of the Martian regolith (soil) and atmosphere. In this paper the initial calibration and response of the triboelectric sensors were measured between –60°C and room temperature. The sensor has an electric field sensitivity of 35 kV/cm·V and room temperature drift of ~3 µV/sec.

1. Introduction

The electrometer is part of MECA (Mars Environmental Compatibility Assessment) project. The objective is to gain a better understanding of the hazards related to the human exploration of Mars. The electrometer will be built into the heel of the Mars '01 robot arm scoop as seen in Fig. 1. The electrometer operates over an 8-wire serial interface, is housed in a volume of ~50 cm³, consumes less than 250 mW, and weighs ~50 g.

As seen in Fig. 1, the instrument has four sensor types: (a) triboelectric field, (b) electric-field, (c) ion current, (d) temperature. The triboelectric field sensor array contains five insulating materials to determine material charging effects as the scoop is dragged through the Martian regolith. The insulating materials will be chosen after Earth-based tests in Mars simulant soils.

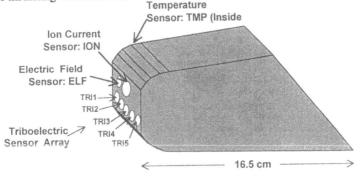

Figure 1. Electrometer sensor suite mounted in the heel of the Mars '01 scoop.

The engineering drawing of the scoop and the electrometer housing are shown in Fig. 2. During digging operation the electrometer is out of the way. After digging, the scoop is rotated so the electrometer head is pointing down toward the Martian soils allowing it to be rubbed against the Martian soil

190

Figure 2. Electrometer titanium housing attached to back of aluminum scoop.

The scoop is located at the end of the robotic arm on the Mars 2001 lander as depicted in Fig. 3. The electrometer is part of the MECA project which has a material patch experiment to determine the effects of dust adhesion, a wet chemistry laboratory with ion selective electrodes to characterize the ionic content of the soil, and microscopy station with optical and atomic force microscopes to determine particle size and hardness.

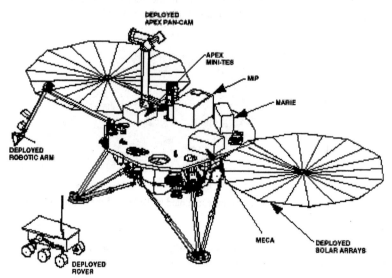

Figure 3. Overview of the Mars '01 lander. Some of the MECA experiments located in the box on the lander deck. The electrometer is located on the scoop on the robotic arm.

2. Operation

In the rubbing sequence, depicted in Fig. 4, the scoop is first lowered against the Martian soil. During the start of the traverse, the electrometer is zeroed by closing a switch which will be discussed later. After reaching the end of its traverse, the scoop is abruptly removed from the soil at which time the triboelectric sensor response, illustrated in the figure, is measured. The parameters and typical values for this operation are included in the figure caption.

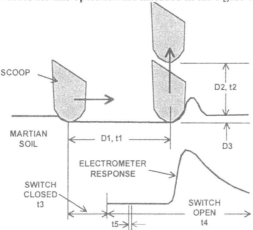

Figure 4. Operational scenario for the scoop where recommended operating parameters are: D1 = 10 cm is the traverse distance, D2 = 1 cm is the liftoff distance, D3 = 0.5 to 1 cm is the penetration depth, t1 = 10 s is the traverse time, t2 = 0.5 s is the liftoff time, t3 = 1 s is the switch close time, t4 = 19 s is the data acquisition time, and t5 = 0.1 s is the time between data points.

Physical insight into this process is illustrated in Fig. 5. As seen on the left, charge is generated triboelectrically across capacitor C3 as the insulator is rubbed on the Martian surface.

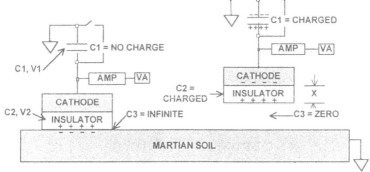

Figure 5. Charge distribution in the electrometer during rubbing (left) and after removal from the surface (right).

Since the charges are in close proximity across C3, no charge appears across capacitors C1 or C2. As the insulator is removed from the surface, the charges redistribute themselves across C1 and C2 according to the charge relationship Q1 = Q2 and provide the signal for the amplifier.

192

The sensitivity of the electric field developed across the insulator to the sensor output voltage is derived as follows. The charge relationship for the series capacitors, C1 and C2, is $Q1 = Q2$. Given that $Q1 = C1 \cdot V1$, $Q2 = C2 \cdot V2$, $C2 = \varepsilon A/X$, and $E2 = V2/X$ it follows that $E2 = (C1/\varepsilon A) \cdot V1$ or that the amplifier input sensitivity is $S1 = dE2/dV1 = C1/(\varepsilon A)$ where ε is the dielectric constant of the insulator and A is the area of the insulator. The amplifier output is related to it's input via the amplifier gain, G or $V1 = VA/G$. A theoretical value can be calculated for S1 from the component values. That is for $C1 = 1$ nF, $A = 0.316$ cm^2, and $\varepsilon = 8.85 \times 10^{-14}$ F/cm, the input sensitivity, S1, is 35.6 kV/cm·V. Thus, if $V1 = 1$ V, then the electric field, E2, is 35.6 kV/cm.

Experimental results for the ELF sensor, shown in Fig. 6, were taken at room temperature and low temperature where the sensitivity is 34.4 kV/cm·V at room temperature and 65.1 kV/cm·V at −66.2°C. This data can be analyzed using:

$$E2 = E20 + S1 \cdot V1 \tag{1}$$

where E20 is the electric field at $V1 = 0$. The voltage intercept, $V10 = -E20/S1$. Representing the temperature dependence of E20 and S1 by $E20 = E200 + KE(T - T0)$ and $S1 = S10 + KS(T - T0)$ where KE and KS are coefficients of the temperature dependent terms, E200 is the electric field at $V1 = 0$ and $T = 300K$, and S10 is the sensitivity at $T = 300K$. Combining the above equations leads to:

$$E2 = E200 + KE \cdot (T - T0) + S10 \cdot V1 + KS(T - T0)V1 \tag{2}$$

where T0 is 300K or 26.85°C. A least squares analysis of the data in Fig. 6 leads to E20 = -270.8 V/cm, KE = 2.85 V/cm·°C, S10 =34.14 kV/cm·V, and KS = -0.33 kV/cm·V·°C. Using Eq. 2, the electric field is determined given V1 and the temperature. Notice that S10 (34.14 kV/cm·V) determined from experimentation is close to the S1 value derived above using component values (35.6 kV/cm·V).

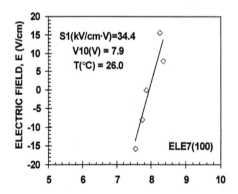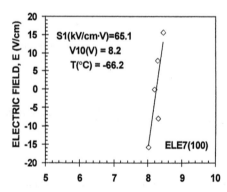

Figure 6. Room (left) and low (right) temperature response of the electric field monitor (ELF) where voltages are an average of 100 measurements. Signal averaging was necessary to increase the resolution of the 12-bit ADC from 2 mV/bit to 0.2 mV/bit.

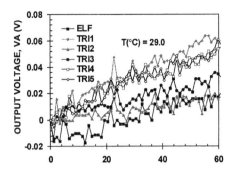 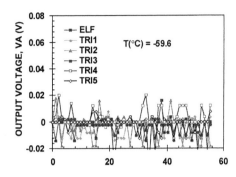

Figure 7. Room and low temperature voltage drift of the ELF and TRI sensors. At room temperature, the amplifier input voltage drift is between 1.25 and 4.5 µV/sec.

This electrometer is an induction field meter [1] operated in a direct current mode, where the operational amplifier input current charges C1. For the operational amplifiers used in this circuit the input current is about 2 fA at room temperature. At lower temperatures, this value gets much smaller. The results in Fig.7 have the expected behavior. At room temperature the drift relative to the amplifier output is about 3 µV/sec at room temperature (29.0°C) and essentially zero at the low temperature (-59.6°C). The data illustrates system noise which is related to the resolution of the 12-bit analog-to-digital converter which is 2 mV/bit.

3. Circuit Design

The electrical schematic of the electric field and triboelectric sensors is shown in Fig. 8. The design of the electric field sensor follows from the traditional electrometer [2]. The instrument is composed of a capacitive divider where C2 is the field sensing capacitor and C1 is the reference capacitor. The point between the capacitors is connected to the positive terminal of the first stage amplifier operated in the follower mode. The sensing electrode is protected by a driven guard that is connected to the negative terminal of the first stage amplifier. A second operational amplifier is added to provide additional amplification.

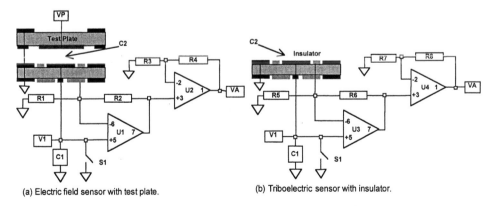

(a) Electric field sensor with test plate. (b) Triboelectric sensor with insulator.

194

Figure 8. Schematic circuit representations for the (a) Electric field sensor (ELE) and (b) Triboelectric sensor (TRI).

At the beginning of the measurements, C1 is discharged using the solid-state switch, S1 which has very low leakage. In the ELF sensor, C2 has an air dielectric; whereas, in the TRI sensor, C2 has an insulator dielectric which acquires charge during rubbing. The ELF sensor was calibrated by placing various potentials, VP, on the test plate and recording the response at the amplifier output, VA. A similar test plate was used to calibrate the TRI sensor.

4. Sensor Design

The cross section of the electrometer is shown in Fig. 9. The design principle was based on the apparatus used at the Kennedy Space Center for characterizing the electrostatic properties of materials. In that apparatus samples are clamped into a 15-cm diameter holder that grounds the periphery of the sample. Thus, after rubbing, samples are characterized by the rate of charge leakage to the ground ring. These features are replicated in miniature as seen in Fig. 9.

To characterize antistatic materials, an insulator sandwich is created as shown in Fig. 9. The antistatic film insulator is laid on top of a bulk insulator and clamped in place under the ground ring.

The assembly is designed so that the insulators can be replaced by removing the fasteners that consist of non magnetic stainless steel screws. The use of nonmagnetic screws is important since a major constituent of the Martian regolith is Hematite (Fe_2O_3) which is magnetic. To be in compliance with planetary protection, the cavity is sealed with epoxy.

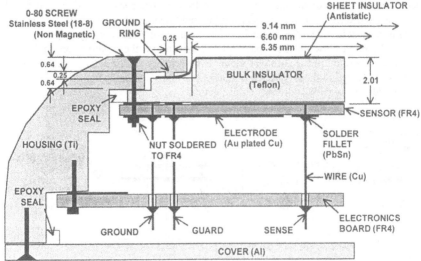

Figure 9. Partial cross section of the electrometer showing the sensor and insulator sandwich and their connection to the electronics board.

5. Experiments

Four different insulating materials were loaded into the titanium triboelectric sensor head and manually rubbed at room temperature with wool felt; the results are shown in Fig. 10. The falling response between 40 and 50 seconds represents the rubbing period. The curves then tend to zero after a period that exceeds 100 s. The large negative response is for the teflon which is to be expected for teflon rubbed on wool.

The results in Fig.10, can be related to the electric field using Eq. 2. For this analysis the amplifier output at VA = 0.4 V was chosen. First, V1 was determined using VA and the amplifier gain, G = 4. Then, Eq. 2 was evaluated using T = 27.5 °C and V1 = 0.1 V and the electric field calculated as E2 = -3.7 kV/cm.

The curves shown in Fig. 10 are encouraging in that they indicate that the electrometer signal is substantial, being in the 100 mV range. The signals are much larger than the millivolt signals measured during the calibration as depicted in Figs. 6 and 7. This means that the drift in the input voltage, as seen in Fig. 7, are not of concern for measurements taken at or below room temperature for times on the order of 10 seconds.

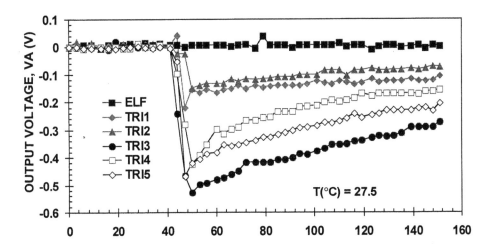

Figure 10. Response of triboelectric sensors after being rubbed with wool felt where TRI1 is ABS, TRI2 is polycarbonate, TRI3 is Teflon, TRI4 is Rulon-J, and TRI5 is Teflon.

6. Discussion

At this time, the flight units are under construction and the measurements presented here are from a prototype unit termed ELE7. The final selection of insulating materials will be done in the next few months. It is anticipated that dust cling to the insulator surface will affect the results in that the dust will compensate the triboelectrically induced charge in the insulator thus reducing the response. In the next few months, tests will be performed to quantify this effect and procedures sought to remove the dust.

References

[1]   J. A. Cross, *"Electrostatics: Principles, Problems and Applications"*, Adam Hilger (Bristol, UK)

[2]   *"Electrometer Measurements"*, Keithley Instruments (Cleveland OH, 1972)

Acknowledgements

The work described in this paper was performed by the Jet Propulsion Laboratory, California Institute of Technology, under a contract with the National Aeronautics and Space Administration. The authors are indebted to the managers who have encouraged this work. In particular from JPL, Michael Hecht, Lynne Cooper, and Joel Rademacher, from WVU, Tom Meloy, and from KSC Haesoo Kim and Rupert Lee.

In addition the authors appreciate the efforts of the MECA/Electrometer science advisor board that was convened at Electrostatics '99. The board has provided guidance and a number of important suggestions. The board members and their affiliations are: Peter Castle, University of Western Ontario, John Chubb, John Chubb Instrumentation, Michael Dyer, Du Pont, William Greason, University of Western Ontario, Paul Holdstock, British Textile Technology Group, Thomas Jones, University of Rochester, Z. Kucerovsky, University of Western Ontario, and David Swenson, EDS Association. File: Camb9516.doc.

Section D

Hazards

Inst. Phys. Conf. Ser. No 163
Paper presented at the 10th Int. Conf., Cambridge, 28–31 March 1999

Electrostatic Ignition Hazards Associated with Flammable Substances in the Form of Gases, Vapors, Mists and Dusts

M.Glor

Swiss Institute for the Promotion of Safety & Security, WKL-32.3.01, PO box, CH-4002 Basel, Switzerland

Abstract. The paper deals with the assessment of the electrostatic ignition hazards when flammable gases, liquids or powders are handled and processed in industry. It reviews the present state of knowledge in this field based on information available from literature, codes of practice and guidelines. The prerequisites for the ignition of a flammable atmosphere, the steps from charge separation to ignition of flammable atmospheres and the different types and incendivities of electrostatic discharges occurring in practice are outlined and discussed. In addition the electrostatic hazards related to the different types of FIBCs are reviewed.

1. Introduction

Build-up of static electricity is intrinsically related with most industrial processes. Static electricity may cause nuisance or damage and it may represent a fire or explosion hazard when flammable gases, liquids and powders are handleds. Thus, it is of great concern in chemical, plastics, pharmaceutical, foodstuff, printing and paint industries. The decisive factor for assessing the risk of accidents due to static electricity is the probability of coincidence in space and time between a flammable atmosphere and a high level of charge accumulation. The probability of a coincidence of this kind is greatest when the handling of a product gives rise both to high electrostatic charge densities and to a flammable atmosphere. This is especially true when dealing with flammable, non-conducting liquids such as fuels and apolar solvents or with flammable substances in powder form. Ignition sources may occur in practice due to human errors, technical malfunctions or they are intrinsically related to a process. In contrast to most types of other ignition sources static electricity belongs to all categories. Therefore, it is considered to be the ignition source least amenable to control.

Fires and explosions attributed to static electricity have ranged from filling a plastic bucket with toluene, washing of cargo tanks of 200,000 tons oil tankers to the transfer of combustible powder into large silos. Typical other accident reports include the filling of dryers with solvent wet powders and emptying intermediate flexible bulk containers filled with flammable powder.

The reports of phenomena due to static electricity date back to the 6th century BC when the ancient philosopher and scientist Thales of Milet observed the attraction and repulsion between amber and light particles. In the last centuries these phenomena have been shown on market places, in saloons and clubs for the amusement and amazement of people. With the industrialization and the frequent use of petroleum products in the present century the phenomena of static electricity became more frequent and obvious. Industry and academia started to investigate these phenomena. This research resulted in industry based (1-5), national and international (6-9) guidance to industry on safe manufacturing practices.

200

2. Systematic approach to the assessment of electrostatic ignition hazards

For every assessment of an electrostatic ignition hazard the scheme shown in Figure 1 is extremely helpful. In all cases where a fire or explosion in industrial environments is caused by static electricity, the sequence of events passes through the same stages. These stages are shown in Figure 1.

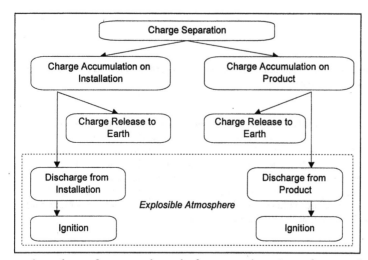

Fig.1: Scheme showing the different steps from charge build-up to ignition

Charge separation may occur if two surfaces which are in contact with each other are separated. If the mechanical separation process is fast enough compared to the mobility of the charge carriers, the surfaces are charged after separation. Apart from mechanical separation (so called tribo charging), there exist other charging mechanisms such as electrical induction, charge spraying, etc.

Surface contact, separation and movement involving poorly conducting materials are an intrinsic feature of many industrial processes. It must be kept in mind that in all these processes both surfaces in contact become charged after separation. This means that both, the equipment and the product may accumulate charge, as indicated in Figure 1. This fact has to be considered in every assessment of electrostatic hazards.

Charge separation does not, of itself, automatically lead to a hazardous situation. The amount of *charge accumulation* represents the decisive factor. This is determined by the rate of charge separation and charge dissipation. In practice, charge may accumulate on electrically insulated conductors, on insulating surfaces or on highly insulating products such as, e.g., hydrocarbon liquids or polymeric powders or on dust clouds and mists.

Fortunately *charge dissipation* occurs already at relatively high resistances to earth. The electric currents that usually occur in practice due to separation processes are very small. Typical values are 10^{-6} A or less. Under extreme conditions, values up to 10^{-4} A may be reached. For such low charging currents leakage of charge via resistances to earth in the range of 10^6 to 10^8 Ohm is sufficient to prevent hazardous level of charge accumulation.

If the accumulation of charges always grows higher, the resulting electric field in the air may reach its limit value. This limit value is also called the dielectric strength of air. Under normal conditions it amounts to about 3 MV/m. At this limit a so called *discharge* may occur. The total or only part of the energy stored in the charge accumulation may be released in such a discharge forming a hot discharge channel that may ignite a given flammable atmosphere.

The energy released in the discharge and the sensitivity of a given flammable atmosphere measured in terms of its minimum ignition energy determine whether *ignition* will occur or not. The assessment of the occurrence and incendivity of discharges in different situations in practice represents the most important but also the most difficult step in the analysis of electrostatic hazards. Because of the difficulty to predict the occurrence and incendivity of discharges in industry by the laws of plasma physics a more or less phenomenological approach is commonly used. The discharges occurring in practice are classified into different discharge types as outlined below. More details can be taken from different text books (10-12), codes of practice and guidelines (5-9).

3. Discharges - occurrence, incendivity and exclusion

Every assessment of an ignition probability is based on a comparison of the igniting power (incendivity) of an electrostatic discharge with the sensitivity of the flammable atmosphere. In a first approximation the incendivity of a discharge has so far been estimated by its total energy and the ignition probability assessed by comparing this total energy with the minimum ignition energy of a flammable atmosphere determined with a spark discharge from a capacitive circuit. The first problem which arises in such an approach is the calculation of the energy of the discharge which is far from being trivial for most discharges other than sparks. In addition, due to the complexity of the ignition mechanisms this approach is too simplified, since the incendivity of a discharge depends not only upon its total energy but also upon the energy distribution with respect to space and time. This spatial power density varies widely among the different discharge types. Due to these problems energy released in a discharge is best determined in ignition tests with flammable atmospheres. The energy determined in this way is called equivalent energy of the discharge and is defined as follows: A discharge has the equivalent energy W if it is just able to ignite an flammable mixture with a minimum ignition energy W, as determined with capacitor spark discharges.

3.1. Discharges from isolated conductors

Spark discharges occur between two conductors at different potential as soon as the electrical field in the gap reaches the breakdown value of about 3 MV/m at atmospheric conditions. In practice one of these conductors (including personnel) reaches a high potential since it is not connected to earth. Nearly the total energy stored in such systems (capacitors) is released in a single spark which generates a single discharge channel of rather high current density. Therefore, these discharges are rather incendive and their energy release can simply be calculated by the energy stored on the capacitor. Gases, vapors and dust clouds can be ignited by spark discharges

Theoretically spark discharges can easily be excluded by earthing all conductors. Practical experience shows however, that earthing of all conductors is difficult to ensure in practice. Earthing can often only be achieved by organizational measures (the operator must be aware that it is necessary to earth containers, etc.) and, if conducting and non-conducting materials are used for construction, the chance of overlooking an isolated conductor is high. Therefore, adequate training of personnel and the exclusive use of conducting material for the construction of apparatus and equipment are important prerequisites for the exclusion of spark discharges.

It is commonly agreed (5-9) that a resistance to earth of less than 10^6 Ohm is generally sufficient in the case of equipment and 10^8 Ohm in the case of personnel to avoid spark discharges. In the plant it is of course reasonable to insist on much lower values for the resistance to earth of equipment if earthing is achieved by metallic connections. If in these cases the resistance amounts to several orders of magnitude the connection to earth is defective and may exceed the critical value of 10^6 Ohm at any time.

3.2. Discharges from insulating surfaces, insulating liquids and insulating powders

If charges are arranged on non-conducting surfaces or within non conducting products they cannot be released in a single spark discharge. The mobility of the charges along the surface or through the volume is too low (if there exists any mobility) compared to the duration of a spark discharge. Under these circumstances three other discharge types may occur. Their occurrence depends on the geometrical arrangement of the charges and the surroundings.

If charges of one polarity are distributed on the surface or within the volume of non-conducting material, so called *corona* or *brush discharges* may occur as soon as an earthed conductive electrode is approached to the surface. The discharge results from the distortion of the electric field.

In practice, it can generally be assumed that only corona discharges occur if the radius of curvature of the electrode is below about 0.5 mm. Ignition of flammable atmospheres with a minimum ignition energy above 0.2 mJ by corona discharges has not to be expected. If the radius of curvature of the electrode is larger than 5 mm brush discharges are more likely to occur.

The characteristics and incendivity of brush discharges have been investigated by different authors (13-16). According to the present state of knowledge, brush discharges are very unlikely to ignite pure dust clouds containing no flammable gases and vapors. As a consequence, brush discharges must be excluded in areas where flammable atmospheres formed by flammable gases or vapors may be present.

Measures to prevent brush discharges are outlined in the literature (6-12). They are based on increasing the conductivity, limiting the extension of the non-conducting object or charge binding by image charges.

If the charges are not arranged in form of one single layer of one polarity on a non-conducting surface but in form of a double layer of charges of opposite polarity on the opposite surfaces of a non-conducting material in form of a sheet, *propagating brush discharges* may occur. They are initiated by an electrical short circuit of the two highly charged surfaces achieved either externally by the approach of connected electrodes to the two surfaces, or internally by a mechanical or electrical perforation of the dielectric layer. A physical description and treatment of such double layers of charges across dielectric materials and the formation of propagating brush discharges is found in ref. (12).

For a long time it has been assumed that these discharges can only occur if one surface of the insulating layer is in intimate contact with an earthed metallic plate. It can, however, be demonstrated that such discharges can also be generated by spraying charges of opposite polarity to the two surfaces of a dielectric sheet. Such a charging mechanism is observed, for example, when filling an insulating container with highly charged polymeric powder.

The energy released in propagating brush discharges is high enough to ignite most flammable gases and vapors and most combustible powders. Persons may suffer a serious shock when the human body acts as the initiating electrode for such a discharge.

Propagating brush discharges can be excluded by the use of conductive materials or insulating materials of low dielectric strength at all locations where the build-up of high surface charge densities may occur. If the breakdown voltage across a non-conducting layer or sheet is less than 4 kV or the surface charge density remains below about $2.7 \cdot 10^{-4} C/m^2$ propagating brush discharges have not to be expected

The phenomenon of discharges along the surface of highly charged bulked polymeric granules which nowadays is called *"cone discharges"* is known since about 20 years. First reports by Maurer and Blythe et al. (18,19) date back to 1979. These discharges are caused by the extremely high electrical fields resulting from the high space charge density which is generated when charged insulating particles are accumulated in silos or containers. Model calculations have shown that conditions necessary for the appearance of cone discharges do exist when filling silos with highly insulating powders or granules. The discharges may already be initiated at a low level of the charge to mass ratio of the incoming product (10^{-7}·C/kg in the case of a powder heap with radius 1 m) (20).

A comprehensive research project on the occurrence and incendivity of cone discharges sponsored by German and Swiss chemical industries and other institutions has been performed. The results are published in numerous publications (21-25,28) and in two final reports (26-27). Based on the resistivity, the minimum ignition energy and the particle size distribution of the product and the material, volume and geometry of the container or silo, a hazard assessment can be performed

Before anything was known about cone discharges, the charged dust cloud was considered to represent the major electrostatic hazard in silos. This speculation was based on observations during the eruption of volcanoes, where *lightning-like discharges* have been observed in the ash and dust clouds. However, so far no lightning like activity has been reported from industrial scale equipment. Systematic investigations with highly charged dust clouds in a $60m^3$ bunker showed however negative results (17). Thus, it can be concluded that lightning-like discharges are very unlikely to occur in industrial scale equipment.

4. Handling and processing of liquids

Extensive knowledge is nowadays available with respect to the ignition risks due to static electricity associated with the handling and processing of flammable liquids based on research and experience from the petroleum and chemical industries.

The hazards associated with charge build-up on packages, containers, apparatus and plant items have to be eliminated by an adequate choice of the material for construction and by earthing and bonding all conductive parts (see also spark, brush and propagating brush discharges). In addition earthing of the personnel via charge dissipative shoes and floors is mandatory.

The hazards associated with the charge on liquid products is given by the equilibrium resulting from charge separation, charge accumulation and charge dissipation (see Figure 1.) The parameters affecting this charge equilibrium state of the liquid are manifold. A summary is given in Table 1.

The most frequently applied measures to limit the charge accumulation on liquids to a safe level are the increase of the conductivity in case of liquids with a very poor conductivity such as e.g. the hydrocarbon liquids, or limiting the flow velocities of liquids in pipes. In special situations in industry it is, however, not possible to limit the charge accumulation on the liquid to a safe level. In these cases exclusion of the flammable atmosphere with inert gas blanketing is required. A typical example of this kind is the handling and processing of multiple phase liquid systems of poor conductivity.

It is strongly recommended to use guidance for the safe handling of flammable liquids in different industrial situations. This guidance can be taken from the comprehensive literature in this field (1-9) or in the form of expert advice.

204

Table 1: Parameters affecting the hazards associated with the charge on liquid products

Charge generation	Charge accumulation	Charge dissipation

process
• flowing
  • flow velocity
  • filters, valves, obstacles
  • turbulence
• filling
• emptying
• stirring
• mixing
• sampling
• atomization
• liquid jet cleaning
nature of liquid
• viscosity
• homogeneity
  • suspension
  • emulsion
  • multiple phase
	• quantity
	• volume
	• geometry
	• charge to mass ratio
	• presence of electrodes (intrusions such as level, temperature, etc. sensors influencing the electrical field)
		• conductivity of the liquid
		• conductivity of apparatus
		  • presence of liners or coatings in pipes, containers, etc.
		• electrical connection to earth

5. Handling and processing of powders

Most organic and polymeric powders are poor conductors. Thus they may become highly charged even when handled and processed in earthed metallic equipment. For a long period of time the ignition sensitivity of powders in the form of dust clouds with respect to electrostatic discharges has been underestimated. Therefore, in the past the guidance given in the guidelines with respect to powders was rather poor.

A lot of new knowledge has been collected during the last 10 to 15 years. The most important parameters in every hazard assessment for powders are:minimum ignition energy, resistivity of the powder in bulk, particle size distribution and content of flammable solvent

General guidance for the safe handling and processing of flammable powders can be taken from the new CENELEC technical report (8). If large amounts of powders with a minimum ignition energy less than 10mJ (measured in purely capacitive circuit) or powders in the presence of flammable gases or vapors are processed, expert advice is recommended.

Two special new topics, the use of FIBCs and the occurrence and incendivity of cone discharges, are shortly highlighted in this paper (see also section 3.2).

5.1. Flexible intermediate bulk containers (FIBCs)

FIBCs or Big Bags are now commonly used for the packaging and transport of powder materials. Their constructions, usually using a base fabric of insulating woven polypropylene may introduce an electrostatic ignition risk.

The initial work lead Maurer and co-workers in 1987 to propose a classification scheme for the different types of FIBCs (29). This essentially divided FIBCs into three types:

Type A which have no special safety precautions and are unsuitable for use with any potentially flammable material; *Type B* in which the FIBC wall fabric has a breakdown voltage of less than 4 kV in order to prevent the occurrence of propagating brush discharges. These can be used in the presence of potentially flammable dust atmospheres provided its ignition energy is greater than ca. 3 mJ. *Type C* which are constructed using a wall fabric which has a resistance to earth from any location on the FIBC, including the slings, of less than $10^8\Omega$. Such FIBCs, provided they are adequately earthed during use, are suitable for use in the presence of any potentially flammable atmosphere, both dust and gas or vapor.

In order to meet the requirements for type C, FIBCs are usually constructed either from a fully conductive fabric or from a non-conductive fabric containing interwoven conductive threads which are connected together and which then have to be connected to earth.

The need to ensure that type C FIBCs are adequately earthed during use and the difficulty that this presents in practice is illustrated by a range of incidents, described by Britton (30), which have occurred during powder handling operations involving FIBCs.

An alternative design of FIBC, commonly called *Type D* FIBC, involves the use of a woven polypropylene fabric containing interwoven conductive threads that are not connected together. It was believed that any charge build-up on such FIBCs would be dissipated by corona discharge from the threads and the low capacitance of the individual threads would prevent the occurrence of incendive discharges for all except the most sensitive flammable gas mixtures e.g. hydrogen/air. Laboratory and test site measurements have been performed with such Type D FIBCs by different authors (31-34). It could be demonstrated that with the new FIBCs it was no longer possible to ignite propane air mixtures by discharges resulting from the FIBC fabric. It was however still possible to charge unearthed objects or personnel close to the FIBC by electrical induction and corona to such a level that incendive sparks from these unearthed objects were observed.

6. Conclusions

Based on experience and industrial research the ignition hazards associated with static electricity can nowadays be assessed rather well. The present knowledge in this field is sufficient to prescribe safety measures for most common industrial processes. It must, however, be kept in mind that electrostatic phenomena are encountered so widely that it is not possible to cover all cases and that research in the field continues providing new information. In addition, new processes are developed, common processes are run under more hazardous conditions or conflicts of interests do not allow a standard solution. Thus, there are and there will be added in the future open questions concerning the assessment of electrostatic ignition hazards.

7. References

(1)   Schriftenreihe der ESCIS "Statische Elektrizität, Regeln für die betriebliche Sicherheit" Heft 2, Ausgabe 1997, zu beziehen bei Dr. R. J. Ott, Sektion Chemie, Abteilung Arbeitssicherheit, SUVA, Fluhmattstrasse 1, CH-6002 Luzern

(2)   API 2003 "Protection Against Ignition Arising Out of Static, Lightning and Stray Currents" 5th Ed. American Petroleum Institute (1991)

(3)   Model Code of Practice for the Petroleum Industry: Part 1. Electrical Safety Code. Institute of Petroleum, U.K. (1991)

(4)   Shell Safety Committee "Static Electricity- Technical and Safety Aspects" June 1988

(5)   ISSA Prevention Series No. 2017 (E) "Static Electricity, Ignition Hazards and Protection Measures" of International Section on the Prevention of Occupational Risks in the Chemical Industry of the International Social Security Association (ISSA). ISBN 92-843-1099-7, ISSN 1015-8022.

(6) Richtlinie für die Vermeidung von Zündgefahren infolge elektrostatischer Aufladungen, Richtlinien "Statische Elektrizität". Hauptverband der gewerblichen Berufsgenossenschaften, Zentralstelle für Unfallverhütung und Arbeitsmedizin, D-5205 Sankt Augustin, Fachausschuss Chemie, ZH 1/200,1989

(7) Code of Practice for Control of Undesirable Static Electricity, Part 1. General Considerations and Part 2. Recommendations for particular industrial situations. British Standards Institution London. British Standards Institutions BS 5958, 1981/82.

(8) Draft CENELEC report R044-001"Safety of machinery - Guidance and recommendations for the avoidance of hazards due to static electricity, elaborated by an European expert team forming WG5 of TC44X.

(9) NFPA 77 ,"Static Electricity" 1988 Edition, National Fire Protection Association USA

(10) G.Lüttgens und M.Glor, Statische Elektrizität begreifen und sicher beherrschen", Expert Verlag, Ehningen bei Böblingen, 1993.

(11) G.Lüttgens and M.Glor "Understanding and Controlling Static Electricity", Expert Publishing Company, Ehningen bei Böblingen, 1989.

(12) M.Glor, "Electrostatic Hazards in Powder Handling", Research Studies Press Ltd., Letchworth, Hertfordshire, England 1988.

(13) N.Gibson und F.C.Lloyd, British Journal of Appl. Phys. 16 (1965) 1619.

(14) M.Glor, J. Electrostat., 10 (1981) 327.

(15) K.Schwenzfeuer und M.Glor, J. Electrostat., 30 (1993) 115

(16) K.Schwenzfeuer and M.Glor, J. Electrostat. 40&41 (1997) 383

(17) P.Boschung et al., J. Electrostat., 3 (1977) 303

(18) B.Maurer, Chem. Ing. Tech., 51 (1979) 98

(19) A.R.Blythe and W.Reddish, Inst. Phys. Conf. Ser. No 48, p.107, Oxford 1979.

(20) M.Glor, J. Electrostat. 15 (1984) 223

(21) B.Maurer, M.Glor, G.Lüttgens and L.Post, J. Electrostat. 23 (1989) 25

(22) M.Glor, G.Lüttgens, B.Maurer and L.Post, J. Electrostat. 23 (1989) 35.

(23) M.Glor and B.Maurer, Proceedings of the International Conference "Sichere Handhabung brennbarer Stäube", Nürnberg, 1992, VDI Berichte Nr. 975 (1992) 207.

(24) M.Glor and B.Maurer, J. Electrostat. 30 (1993) 123

(25) A.Kuttler and M.Glor, J. Electrostat. 30 (1993) 285

(26) M.Glor and B.Maurer, Fortschritt-Berichte VDI, Reihe 3 Nr. 181 VDI-Verlag Düsseldorf 1989

(27) M.Glor and B.Maurer, Fortschritt-Berichte VDI, Reihe 3 Nr. 389 VDI-Verlag Düsseldorf 1995

(28) M.Glor and K.Schwenzfeuer, J. Electrostat. 40&41 (1997) 511

(29) B.Maurer, M.Glor, G.Lüttgens and L.Post, Inst. Phys. Conf. Ser. No 85, Sect. 3, Oxford 1987.

(30) L.G.Britton, Process Safety Progress, 12, 4 (1993) 241-250

(31) N. Wilson, Symp. on "Recent Developments in the Assessment of Electrostatic Hazards in Industry", IBC London 1989.

(32) M.A.Nelson, R.L.Rogers and B.P.Gilmartin, J. Electrostat. 30 (1993) 135-148.

(33) M.Glor, B.Maurer and R.Rogers, Proceedings of the Conference on Loss Prevention and Safety Promotion in the Process Industries, published by Elsevier Science B.V., Volume 1 (1995) 219

(34) K.Schwenzfeuer and M.Glor, The Journal of Powder/Bulk Solids Technology, June 1998, 29

Inst. Phys. Conf. Ser. No 163
Paper presented at the 10th Int. Conf., Cambridge, 28–31 March 1999

Large scale chemical plants: eliminating the electrostatic hazards

G R Astbury[1] and A J Harper[2]

[1]Zeneca Process Technology, P.O. Box 42, Blackley, Manchester. M9 8ZS
[2]Zeneca Process Technology, Earls Rd, Grangemouth, Stirlingshire. FK3 8XG

Abstract. This paper discusses a novel approach to avoiding electrostatic hazards arising from the treatment of effluent in a fine chemicals manufacturing site. Whilst methods of avoiding electrostatic hazards for small vessels and pumping small quantities through pipes are simple, the approach for orders of magnitude increases in scale found in treatment plants required novel methods which were solved by the application of both basic physics and chemical engineering methods.

1. Introduction

The volumetric flow of effluent from one of our fine chemicals sites is around 400 m^3hr^{-1}, peaking to about 1600 m^3hr^{-1} under storm conditions. The effluent is substantially aqueous and acidic in nature, but it is recognised that on multi-purpose chemical plants, immiscible, flammable solvents inevitably find their way into the site drains. Whilst the plants all have interceptors to minimise losses to drain, the effluent treatment plant has had to be designed to treat such solvents before discharge from the site. As many of the immiscible solvents used at the site are highly flammable and electrically insulating, the potential for the generation and accumulation of electrostatic charge on the immiscible phase was recognised and the effluent treatment plant was designed to accommodate this in a fashion that would allow safe operation.

2. Electrostatic hazards arising

The major electrostatic hazards arising in the effluent treatment plant occur from the materials of construction; the unavoidable presence of flammable, insulating, immiscible liquids; and the size of vessels and pipework. Where small quantities and small vessels are used, the occurrence of electrostatic hazards can be controlled relatively easily by the restriction of the velocity and choice of suitable materials of construction. The formation of electrostatically charged fine droplet mists can often be accepted in small vessels where the resulting electrostatic fields do not contain sufficient charge to produce an incendive spark discharge [1]. Also, where two immiscible liquids settle, the maximum depth of liquid is small enough to ensure that settling potentials are low enough to avoid spark discharges.

In the effluent treatment plant, restriction of velocities to low values would be impractical because of the pipe sizes involved. The tanks have to be made of corrosion resistant materials and are very large, so the use of exotic alloys is precluded on procurement and cost grounds.

Typically concrete tanks with a polymeric sheet lining are the most cost effective solution, but rely on adequate electro-conductivity of the polymer to dissipate the charge that may arise.

3.   Materials of construction

The materials of construction employed for the pipework and tanks were polymeric to provide chemical resistance against the acidic species present in the raw effluent. The major materials were glass-reinforced resins for pipework and smaller pumping tanks, and electro-conductive high density polyethylene sheet for lining the large concrete treatment tanks. Two resin systems used: polyester resin, and furane resin where the polyester system was inadequate.

The polyester resin system is amenable to the incorporation of fillers. Graphite powder was used both as a pigmenting filler for protection against ultraviolet light degradation and for electrical conductivity. Early test samples failed because either too much graphite was added, producing an adequately conductive but mechanically weak laminate; or too little was added resulting in high resistivity but excellent mechanical properties. The theoretical amount of graphite filler required to make a fully conducting matrix is far in excess of that actually needed when careful quality control of the mixing process is used. Eventually, the method of manufacture was refined so that reproducible results were obtained for each batch.

The furane resin system is not so amenable to the addition of filler and the requisite conductivity was achieved by the use of carbon-fibre surfacing tissue. The surface tissue was connected through the subsequent layers of resin and glass fibre by the use of woven carbon fibre "ties". This made a laminate which, although itself insulating, had conductive surfaces connected together through the laminate by the conductive "ties". This method of manufacture requires care that the number and spacing of the "ties" is sufficient to ensure that all parts have both conducting surfaces and through-connections between the surfaces.

The electroconductive polyethylene was a commercially obtained sheet material specifically designed for the installation in concrete tanks. One surface is fitted with a regular pattern of protrusions which are cast into the concrete to secure the sheet to the tank wall. The protrusions were also made of electro-conductive polyethylene, and allowed the sheet to be earthed through the concrete of the tank. The concrete is sufficiently conductive that the large bulk of it provided an adequate path for any charge to dissipate from the tank lining to earth.

4.   Pumping

Where an electrically insulating liquid is pumped as a continuous phase containing a dispersed material (for example, toluene contaminated with water), the liquid velocity requires restriction to about 1 m s^{-1} [2] to maintain a safe low level of charge. At the flow rates required for the effluent being pumped over 400 metres to the treatment plant, such a velocity would require very large pipes with an attendant high capital cost. As an alternative, small diameter pipes were used and the pumping velocity increased to the economic pipe velocity of about 5 m s^{-1}. This resulted in the potential to electrostatically charge any separate phase but, since the pipe was liquid filled, there was no flammable atmosphere present. Any spark discharge of the electrostatic charge would not result in the occurrence of a fire or explosion.

It was then necessary to ensure that any charge had decayed before the liquid reached a point where a flammable atmosphere could occur. The use of an inert atmosphere was precluded because of the difficulty of trying to maintain an inert atmosphere in 5 tanks each of 400 m^3 breathing at high rates, and the cost of the nitrogen required. In order to allow the charge to dissipate, "relaxation tanks" were provided at the end of the transfer pipes just prior to their entering the large tanks.

5. Relaxation tanks

The use of relaxation tanks has been proposed before [3] and this method was adopted to allow charge generated during pumping to dissipate. Whilst the concept of the relaxation tank was known and a residence time of three times the relaxation time or 100 seconds was suggested [4], it appeared that a method of determining a suitable tank geometry was unavailable, and much was left to the discretion of the designer. The required volume of the tank depends on the flowrate and diameter since any tank will have a flow velocity in it, which will tend to produce some charging. As a result of the low velocities suggested for two phase flow, an equation was derived to determine the required length of relaxation tank from the diameter and required flow.

If the relaxation time of the continuous insulating phase is taken to be τ, then 95% of the charge will have decayed in a time of 3τ. At a high pumping velocity of 5 m s^{-1}, the charge accumulated on the liquid will require this length of time to dissipate. Since the safe maximum velocity for a two-phase mixture passing into a partially filled tank is only 1 m s^{-1}, the velocity in the relaxation tank will have to be significantly below the 1 m s^{-1} to allow the charge to relax. According to Schön [5], the streaming current i_s is:

$$i_s = K v_t^2 d_t^2 \tag{1}$$

where K is a constant, v_t is the velocity in the tank in m s^{-1}, and d_t is the tank diameter in metres. The minimum tank volume would be given by:

$$V = 3\tau Q \tag{2}$$

where V is the tank volume in m^3, τ is the relaxation time, and Q is the flow-rate in m^{3}s^{-1}. However, this assumes that there is no charge generation within the tank. This implies that the velocity would be zero, and the diameter would then be infinite. Clearly this is impractical, so a finite diameter is required. If the streaming current in the tank is reduced to a fraction f of the streaming current at 1 m s^{-1} in the tank, then the effective relaxation would be 1-f. Therefore the relaxation tank would have to be increased in length by a factor of:

$$\frac{1}{1-f} \tag{3}$$

Since the fraction f of the streaming current is proportional to the (velocity)^2, Equations (2) and (3) can be combined to give a volume of tank with respect to the superficial velocity through the tank:

$$V = 3\tau Q \left(\frac{1}{1-v_t^2}\right) m^3 \tag{4}$$

Substitution of various velocities gives a tank diameter and hence length, and it is simple to select an appropriate shape and size of tank for the duty. In practice, the value of τ for the known worst case material used on the site was 15 seconds. A time of 75 seconds (5τ) was chosen as a compromise between 45 seconds (3τ) and 100 seconds for an unknown material.

210

6. Discharge to large effluent tanks

Once the raw effluent had been pumped close to the 400 m³ collection tanks, it had to flow by gravity into them for further processing. Clearly, the presence of electrostatically charged fine droplet mists would have to be prevented since the tanks could contain a flammable atmosphere. This precludes free-falling into the tanks. Although the obvious way to prevent such mists is to use a dip-pipe, the design of a dip-pipe requires care to avoid the occurrence of mechanical vibrations and pulsating flow [6]. At low rates, the velocity is so low in the dip-pipe that any immiscible phase could separate and accumulate inside the dip pipe. As the fall was over 4 metres, a significant layer of immiscible insulating liquid could accumulate, leading to the potential to generate hazardous levels of electrostatic charge [7]. At high rates the streaming current would become significant, so a compromise was to choose a velocity in the vertical section of the dip-pipe which was just above the minimum syphon velocity to ensure that any separate phase would be swept out by the flow [8]. By minimising the vertical length of the dip-pipe, the streaming current generated is kept below the infinite pipe current. Hence minimal charge is generated and transferred to the liquid within the large collection tanks.

7. Conclusions

7.1. The sheer scale of a large effluent treatment plant introduces potential electrostatic hazards which cannot be controlled by the normal precautions appropriate to small scale batch chemical plants.

7.2 Electro-conductive corrosion resistant plastics may be used without the introduction of an ignition risk from charge accumulated on the surface.

7.3 Great care is required in the manufacture of graphite-loaded glass reinforced plastic to ensure consistency from piece to piece, because of the potential variation in manual methods.

7.4. Relaxation tanks provide a method of allowing increased pipe flow velocities in smaller pipes, yet achieve adequate dissipation of charge from the liquid. A simple equation is given to determine an appropriate dimensions of tank for a given flow.

7.5. The avoidance of electrostatically charged fine-droplet mists by the use of dip-pipes requires care in their design to avoid the potential to accumulate a light phase within the pipe.

References

[1]   Post L, Glor M, Luttgens G, and Maurer B 1989 *J. Electrostatics* **23** 99-109
[2]   Gibson N and Lloyd F C 1970 *Chem. Eng. Sci.* 25 87-95
[3]   BSI 1991 *Control of undesirable static electricity* BS 5958: Part 1: 1991
[4]   Walmsley H L 1992 *J Electrostatics* **27** 1-200
[5]   Schön G 1965 *Handbuch der Raumexplosionen* ed H H Frytag (Weinheim Bergstr.: Verlag Chemie)
[6]   Simpson L L 1968 *Chemical Engineering* June 17 192-214
[7]   Howells P 1993 *Loss Prevention Bulletin* **114** (Institution of Chem. Engineers, London)
[8]   Hills P D 1983 *Chemical Engineering* September 5 111-4

Inst. Phys. Conf. Ser. No 163
Paper presented at the 10th Int. Conf., Cambridge, 28–31 March 1999
© 1999 IOP Publishing Ltd

Modelling the Electrostatic Ignition Hazards associated with the Cleaning of Tanks Containing Flammable Atmospheres

T J Williams[1] and R T Jones[2]

[1]Applied Electrostatics Research Group,[2]Wolfson Electrostatics, Department of Electrical Engineering, University of Southampton, Hampshire, SO17 1BJ, UK

Abstract. High-velocity jets of water or solvent are commonly used to clean tanks which have contained flammable liquids. This process creates a charged mist which can lead to the generation of an electrostatic spark or brush discharge that may ignite any flammable atmosphere present. Operators need to know the risk associated with their processes and modelling can help assess it. Numerical techniques are available to solve electrostatic problems for general tank geometries but tend to be expensive and difficult to use. This paper shows how Green's function can be used to solve Poisson's equation for a vessel of cylindrical geometry and finite length containing charged mist. For uniform space-charge density, an analytical solution can readily be implemented using cheaper proprietary software. An example from an industrial investigation is described in which the internal walls of a cylindrical vessel were washed with solvent jets.

1. Introduction

Metal tanks, used for the storage or processing of flammable liquids such as oils or paints, are often cleaned using high-velocity jets of water or solvent. It is well known that this procedure generates a charged mist within the tank which can lead to the ignition of any flammable atmosphere remaining, or that arising from solvent vapour, by causing an electrostatic discharge of spark or brush type. The ignition mechanism involved generally depends also on the presence of an object, conducting or non-conducting, within the tank as detailed by Tolson [1]. The object in question may be a solid fixture or even a moving slug of liquid (falling or projected).

The supertanker incidents of 1969 led to a spate of papers [2-11] which tried to explain either how the explosions occurred or how to avoid such incidents by making the cleaning operations safe. It was concluded that the explosions were probably triggered by sparks from free-falling water slugs resulting from the washing of oil tanks by jets of water [2,3,7].

Walmsley [10] showed how an induction-charged slug may be formed from a jet of water disintegrating (mainly by mechanical effects) in the presence of a strong electric field within a vessel, leading to a spark discharge at the wall as the slug subsequently impinges on it. He derived an approximate expression for avoiding incendiary sparks which requires the *jet safety factor* (dependent only on jet geometry to a first approximation) to be greater than the *material hazard coefficient* (see [13] for corrections to Eqns 13 and 14 in Walmsley's paper). The *material hazard coefficient* depends on the MIE (minimum ignition energy) of the flammable atmosphere and the maximum tank space potential (with the jet absent) which can be determined by modelling.

Explosion venting, containment or suppression systems can be installed or nitrogen inertion used to obviate such electrostatic hazards. These options tend to be expensive and nitrogen inertion, in particular, requires careful monitoring in view of containment problems.

The risks associated with industrial processes must therefore be established, especially if constraints (financial or otherwise) limit safety precautions. Modelling can be very helpful in assessing risk though indispensable is the on-site measurement of electrical parameters which may be required as input to the model (eg electric field strength or space charge density). Clearly, in order to assess the risk of potentially incendiary electrostatic discharges, it is important to be able to specify the space potential and electric field strength at points within a tank which contains a charged mist.

A simple analytical but approximate solution to Poisson's equation, which underlies the problem, can be determined by assuming an equivalent, spherical tank. At the other extreme, various sophisticated numerical techniques exist (eg the finite-element software OPERA 2D or 3D) which can provide an accurate solution in the case of general tank geometries but tend to be expensive and difficult to use (especially in cases where there is no rotational or translational symmetry). However, tanks often have cylindrical geometry for which analytical solutions, more appropriate than the spherical approximation mentioned above, can be determined even if the tank length is finite. The solutions generally involve special functions (Bessel functions in particular) which require a suitable mathematical software for implementation (eg Matlab). Such solutions, which are algebraic in nature, show clearly the dependence of the parameters involved and can often be posed in dimensionless form where they have more general applicability. The aim of this paper is to describe such an analytical solution.

2. Theory

For vessels of cylindrical geometry, Poisson's equation is naturally solved using cylindrical co-ordinates. This is straightforward for an infinitely long cylinder but Bessel functions arise if the length is finite. Analytical solutions can even be obtained for geometries containing an internal, concentric, cylindrical boundary (representing, for example, the central shaft of an agitator) and planar tank-ends. One such solution [4] involves standard Bessel functions but the roots of a transcendental equation must first be derived. A more convenient formulation involves modified Bessel functions as will now be described.

The potential per unit charge at the point (ρ,ϕ,z) inside a grounded, conducting, hollow cylindrical ring (of outer radius a, inner radius d and length L) due to a point charge q at position (b,α,c) inside the ring (ie the Green's function) may be expressed as follows [12] where $k = m\pi/L$ and $\delta_n^0 = 1$ if $n = 0$ and 0 otherwise.

$$V_-/q = \frac{-1}{\pi\varepsilon_0 L}\sum_{m=1}^{\infty}\sum_{n=0}^{\infty}(2-\delta_n^0)\frac{R_n(k,a,b)R_n(k,d,\rho)}{R_n(k,d,a)}\sin(kc)\sin(kz)\cos(n[\phi-\alpha]) \tag{1}$$

This is valid for the radial region $d \le \rho \le b$ (except when $\phi = \alpha$ and $z = c$ if $\rho = b$). Interchanging a and d in Eqn 1, and reversing the sign, gives the equivalent expression V_+/q for the radial region $b \le \rho \le a$ (except when $\phi = \alpha$ and $z = c$ if $\rho = b$). The function R_n, in Eqn 1, is defined in terms of the modified Bessel functions I_n and K_n of order n:

$$R_n(k,s,\rho) = K_n(ks) I_n(k\rho) - I_n(ks) K_n(k\rho) \tag{2}$$

By the superposition principle, the potential at the point (ρ,ϕ,z), due to a space charge density $P(b,\alpha,c)$ inside the ring, may be determined by integrating over all points where space charge is present. This may be expressed as shown in Eqn 3 where the space charge in an infinitesimal volume element $d\Omega$ is $P(b,\alpha,c)d\Omega$ and $d\Omega = bd\alpha dcdb$ for cylindrical co-ordinates. This is valid for the general space charge density P. If the density is uniform, so that P is constant within the ring, the integration is straightforward. The α integration, from 0 to 2π, has

$$V(\rho,\phi,z) = \int_0^{2\pi} \int_0^L \left[\int_d^\rho \left(\frac{V}{q}\right) P(b,c,\alpha)\, b\, db + \int_\rho^a \left(\frac{V_+}{q}\right) P(b,c,\alpha)\, b\, db \right] dc\, d\alpha \qquad (3)$$

value 2π if $n = 0$ and zero if $n = 1,2,3,...$ which shows that the n-summation in Eqn 1 degenerates to the $n = 0$ term only when substituted into Eqn 3 and the integration performed (this is consistent with the potential being rotationally symmetric, ie independent of α). Integration of the term $\sin(kc)$ with respect to c, between the limits 0 and L, is also straightforward and leads to a value of $2/k$ when m is odd and zero when m is even (where $k = m\pi/L$). With these values for the integrals, Eqn 3 may be expressed as:

$$V = -\frac{4P}{L\varepsilon_0} \sum_{\substack{m=1 \\ m\ odd}}^{\infty} \frac{\sin(kz)}{k} \left[\int_d^\rho \frac{R_0(k,d,b)R_0(k,a,\rho)}{R_0(k,d,a)}\, b\, db + \int_\rho^a \frac{R_0(k,a,b)R_0(k,d,\rho)}{R_0(k,d,a)}\, b\, db \right] \qquad (4)$$

To proceed, it is necessary to consider the following standard integrals (Eqns 5 and 6) and identity (Eqn 7) concerning modified Bessel functions:

$$\int_0^r b I_0(kb)\, db = \frac{r}{k} I_1(kr), \quad \int_0^r b K_0(kb)\, db = \frac{1}{k^2} - \left(\frac{r}{k}\right) K_1(kr), \quad K_0(x)I_1(x) + I_0(x)K_1(x) = \frac{1}{x} \quad (5), (6), (7)$$

Using Eqns 2, 5, and 6 in Eqn 4, performing much algebra and simplifying using Eqn 7, it can be shown that the potential, in dimensionless form, is given by:

$$\left(\frac{\varepsilon_0}{Pa^2}\right) V = -4 \left(\frac{a}{L}\right) \sum_{\substack{m=1 \\ m\ odd}}^{\infty} \frac{\sin(kz)}{(ka)^3} \left[1 - \left\{ \frac{I_0(k\rho)[K_0(kd) - K_0(ka)] + K_0(k\rho)[I_0(ka) - I_0(kd)]}{K_0(kd)I_0(ka) - I_0(kd)K_0(ka)} \right\} \right] \qquad (8)$$

The dimensionless radial and axial electric fields can be found by differentiating Eqn 8 partially with respect to ρ and z respectively and reversing the sign. Taking the limit as $d \to 0$ in Eqn 8 gives the dimensionless potential for a tank with no inner electrode.

The model was checked by evaluating the electric displacement (summing the series over odd values of m from 1 to 39) at all points on the interior surfaces of an annular tank ($d = 0.05$ m, $a = 1.2$ m, $L = 3$ m) and performing the surface integral numerically to evaluate the charge induced on the walls by an enclosed space charge of constant density (-10 nC/m^3). By Gauss's theorem, the total induced charge must be equal in magnitude but opposite in polarity to the total space charge enclosed. The total induced charge was calculated to be 133.77 nC, 1.3 % less than the actual magnitude of charge enclosed 135.48 nC. Further, the normal electric field values at the interior surfaces agreed with those determined using OPERA 2D (axisymmetric).

In some cases, Eqn 3 can be integrated analytically for non-uniform space charge density $P(b,\alpha,c)$ but the solution is more complicated. Such solutions are best restricted to axisymmetric space charge densities with variation only in the axial (c) and radial (b) parameters. The radial variable tends to be more restrictive as this is involved in the Bessel functions.

3. Industrial Example

A practical investigation was performed by Wolfson Electrostatics (University of Southampton) to assess the electrostatic hazards in washing a cylindrical vessel, used for blending paints, with

214

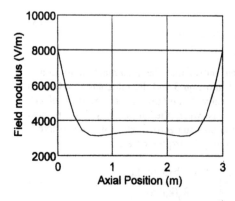 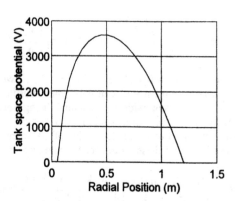

Figures 1a - Tank space potential versus radial position for an axial location of 1.5 m & 1b - Electric field modulus versus axial position for a radial location of 0.7 m (field mill positioned here flush with tank lid); uniform space charge density 169 nC/m³, tank outer radius 1.2 m, tank inner radius 0.05 m, tank height 3 m (series summed over odd values of m from 1 to 39).

solvent jets. The type of solvent and the amount of contamination in it were varied. The greatest value of space charge density in the solvent mist, assumed uniform, was determined to be 169 nC/m³. This was based on the electric field reading (8 kV/m from Fig 1a) given by a field mill, positioned flush with the tank lid, and its relationship to space charge density predicted by the model. The corresponding maximum space potential was calculated to be 3.6 kV (see Fig 1b), too small, according to Walmsley's hazard criterion [10], for an induction-charged slug to produce an incendiary spark. The maximum electric field strength in the tank, 40.6 kV/m (half way up the inner cylindrical wall), is too low to produce a brush discharge.

Acknowledgement

The authors are grateful to Industrial Coatings Akzo Nobel Ltd. (formerly Courtaulds Coatings Ltd.) for their support and permission to publish.

References

[1]   Tolson P 1989 *J. Electrostatics* **23** 89-98
[2]   van der Meer D 1971 *Proc. 3rd Conf. on Static Electrification (London) Inst. Phys. and Phys. Soc. Conf. Ser. 11* 153-157
[3]   van de Weerd J M 1971 *ibid.* 158-173
[4]   Smit W 1971 *ibid.* 178-183
[5]   Vos B 1971 *ibid.* 184-192
[6]   Hughes J F, Bright A W, Makin B and Parker I F 1973 *J. Phys. D: Appl. Phys.* **6** 966-975
[7]   Bustin W M 1973 *American Petroleum Institute Statics Research Program: Part 1*
[8]   Jones M R O and Bond J 1984 *Chem. Eng. Res. Des.* **62** 327-333
[9]   Jones M R O and Bond J 1985 *Chem. Eng. Res. Des.* **63** 383-389
[10] Walmsley H L 1987 *J. Phys. D: Appl. Phys.* **20** 329-339
[11] Post L, Glor M, Luttgens G and Maurer B 1989 *J. Electrostatics* **23** 99-109
[12] Smythe W R 1968 *Static and Dynamic Electricity* 3rd ed. McGraw-Hill New York
[13] Jones R T, Williams T J and Abu Sharkh S 1997 *J. Electrostatics* **40&41** 449-454

Inst. Phys. Conf. Ser. No 163
Paper presented at the 10th Int. Conf., Cambridge, 28–31 March 1999

Fuzzy Logic in Electrostatics - Hazard Assessment

E Balog, I Berta

Technical University of Budapest, Department of High Voltage Engineering and Equipment, Budapest, 1521 Budapest, Hungary

Abstract. Nowadays the use of fuzzy logic to handle uncertainty becomes more and more popular in scientific life. In this paper new and further developed method of fuzzy logic based automated diagnosis is presented to handle uncertainty in industrial electrostatics. An electrostatic precipitation process controlled by an operator in a cement kiln is modelled. In the system several symptoms are observed and with the help of a fuzzy expert system the relative existence of the possible faults may be determined. With automated fuzzy diagnostic, the fault existence of the observed system may be determined instantly and instructions can be received for the repairing. By monitoring the tendencies of the system state, predictions can be generated in order to prevent the whole breakdown. The diagnostic system reacts faster than the operator and also the detection of measuring equipment failure becomes possible.

1. Introduction

In the electrostatic gascleaning processes, one of the most important requirements is the continuous and effective operation of the precipitator (ESP). Only a relatively small decrease in efficiency could mean a huge amount of emitted dust contaminating the environment. In case of large industrial installations it is very advantageous to be able to determine the possible faults of the system from the symptoms before they occur. Thus the interruptions of the precipitation process may be effectively decreased and the maintenance can be optimally scheduled.

Fault analysis of large systems is a very complex task for operators. Despite the controlled process can never be known in details, the operator has to make proper decisions if a fault occurs. A mistake can be extremely dangerous and expensive. The use of fuzzy logic based automated diagnostics in the field of electrostatic precipitation helps operators to make proper decisions in order to prevent the environment effectively.

2. Functional areas of fuzzy expert systems

Expert systems are knowledge-based systems, they rely not only on data as an input, but also on expert knowledge. Table 1 shows some functions for which expert systems have been used. In the list "Monitoring and Control of Systems" are not involved, because this area is normally treated under "Fuzzy Logic Control".

Table 1

Category	Problem Description
1. Classification	Diagnosis, evaluation or interpretation of system or object characteristics
2. Debugging and repair	Prescribing remedies for malfunctions and/or executing plans to administer them
3. Design	Configuring objects and systems under constraints
4. Instruction and training	Diagnosing, debugging and repairing strident behaviour
5. Planning and decision support	Designing actions and supplying aids for better decision making primarily in the area of management
6. Prediction	Inferring likely consequences of given situations or scenarios

In the first category *diagnosis* means the determination of malfunctions or other system characteristics from observable indicators or symptoms. This function of fuzzy expert systems is applied for industrial electrostatics. The representation of the knowledge base (KB) is by means of *rules*, which are of the form "if a set of conditions are satisfied then a set of consequences can be produced". Applying rules is a convenient way to express domain knowledge as rulebases can easily be augmented by simply adding more rules [8]. The other categories may also be very useful in electrostatics, but their exact way of application is just under development.

3. Description of the model

In Figure 1 the model of an electrostatic precipitation process in a cement kiln can be seen. The raw material coming out from the kiln are grinded by the raw mill. At the same time hot flue gases from the kiln are used for drying the raw material in the mill. In case the raw mill is stopped, flue gases are lead directly to the ESP and it needs to be cooled down in the conditioning tower. This is achieved by the injection of atomized water into the tower by the

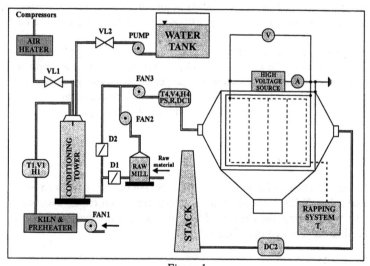

Figure 1

action of compressed air. By controlling the temperature and the flow rate of the compressed air and water, the appropriate conditions in the tower can be adjusted. D1 and D2 are the dampers allowing the control of flue gas flow. The ESP consists of two fields, both of them are energised by a high voltage power supply. With the help of the rapping system the precipitated dust is removed from the collecting plates of the ESP and the dust is collected. After the precipitation process, the cleaned gas enters the stack [4].

In order to be able to monitor the whole process, several parameters have to be measured. Before the conditioning tower the temperature and the flow rate of the flue gas are detected. Just before entering the ESP the temperature, the humidity, the particle size, the flow rate, the dust concentration in the gas and the resistance of the dust are measured. In the ESP the measured values are the voltage connected to the corona electrodes and the corona current. The rapping off-time is a pre-set value, thus it is known. Before the gas coming out from the ESP enters the stack, its temperature, flow rate and dust concentration are measured [1].

4.   About fuzzy logic

Fuzzy set theory is originated by Lofti A. Zadeh, and its purpose is to have a mathematical tool for working with generalities and imprecision. In contrast with ordinary set theory a fuzzy set is a class of objects with a continuum of grades of membership. The characteristic value may be found anywhere in the interval of zero to one. In this manner, an element can be considered a strong member (characteristic value is close to one), a partial member (characteristic value is around 0.5), or a weak member (characteristic value is close to zero) of a given set. In case of ordinary sets elements may have values of zero or one for non-membership or membership. As an example, the fuzzy set A may be defined as follows:

$$\mu_A(x_i)_{x \in X} = (0.5, 0.2, 1, 0.7, 0, 0.8) \tag{1}$$

where x_i represents the elements of A and the numbers between brackets are the characteristic values of each element. Equation (1) expresses that x_1 (the first element of set A) belongs to A according to a 0.5 degree, x_2 (the second element of A) belongs to A according to a 0.2 degree, etc.

5.   Fuzzy logic based diagnostics

Let us suppose that the previously described process is controlled by an operator. In order to reduce the risk of mistaken actuation an aid have to be provided for the operator. In other industrial fields scientists tried to put forward theories about such methods which help operators in decision making [5],[7]. The idea of the expert system is simple: a database that contains the relations between the symptom and fault combinations have to be created and in case a symptom is observed, the appropriate instructions about the troubleshooting may be received. It is similar to a trouble-shooting guide in a user's manual.

5.1. *Fundamentals*

The first step for the application of this method is the identification of possible symptoms and faults in the system. Symptoms are usually abnormal sounds, phenomena and parameter values, when one or more physical quantities approach or exceed given boundary values. In symbolic expert systems the parameter domain of the examined physical quantities are divided into small parts and the connections between a given part and the degree of existence of a fault requires a large database. In order to reduce the size of this database fuzzy logic is applied.

The observability of a symptom can be characterised by a fuzzy membership function $\mu_B(j)$, where set B denotes the array of symptoms. This function is represented in a co-ordinate system, where the horizontal axis represents the physical quantity, while the vertical one means the relative truth that the jth symptom B_j is observed (e.g. temperature is too high, etc.). After defining the symptoms the relations between the possible symptoms and faults must be examined. This can be represented by the relationship matrix R. Its element $R_{i,j} = 1$ if the ith fault (A_i) is connected to the jth symptom (B_j), 0 otherwise. In the previous expression the array of possible faults is denoted by set A. The strengths of the connections are stored in matrix T (KB), where $0 \leq T_{i,j} \leq 1$ is the relative truth that A_i implies B_j.

The task is to place bounds on the fault array at a given symptom combination, thereby narrowing the range of possibilities. It can be stated logically that the relative truth that a given symptom is observable will not exceed the relative truth that a related fault exists. If the observability of B_j is μ_{Bj} and the degree of existence of the related A_i is μ_{Ai}, the previous statement means that $\mu_{Bj} \leq \mu_{At}$ or in general for every A_i :

$$\mu_{Bj} \leq \max \left[\min \left(R_{i,j}, \mu_{Ai}\right)\right] \qquad (2)$$

If A_i is not related to B_j the minimum is 0, otherwise the value of the right side expression in (2) will be equal to the maximum existence of the related faults. From the definition of the Lukasiewicz implication, an element of T can be written as in (3) and from that equation μ_{Ai} can be expressed as in (4).

$$T_{i,j} = \min \left[1, 1 - \mu_{Ai} + \mu_{Bi}\right] \qquad (3)$$
$$\mu_{Ai} \leq \min \left[1, 1 - T_{i,j} + \mu_{Bj}\right] \qquad (4)$$

In formula (4) the inequality comes from (3) when $\mu_{Ai} < \mu_{Bj}$. For the purpose of diagnosis, the symptoms μ_{Bj} are known, the observation relations $T_{i,j}$ are understood thus the causes can be determined. This may be achieved by creating a lower bound for a fault combination from (2) and an upper bound from (4), so it is possible to determine which faults are the most likely responsible for the symptoms[6].

Table 2

Possible Faults				Symptoms				
F1	Failure of FAN1	F9	Failure in RAW MILL	S1	T1 is high	S9	V4 is high	S17	DC2 is high
F2	Failure of PREHEATER	F10	Failure of FAN2	S2	T1 is low	S10	V4 is low	S18	Amp. is low
F3	Failure of COMPRESSORS	F11	Failure of FAN3	S3	V1 is high	S11	PS is high		
F4	Failure of WATER PUMP	F12	Rapping System problem	S4	V1 is low	S12	PS is low		
F5	Failure of DAMP1	F13	Back Corona Discharge	S5	T4 is high	S13	R is high		
F6	Failure of DAMP2	F14	Failure of AIR HEATER	S6	T4 is low	S14	R is low		
F7	Failure of VALVE1	F15	Voltage controlling probl.	S7	H4 is high	S15	DC1 is high		
F8	Failure of VALVE2			S8	H4 is low	S16	DC1is low		

There are two types of fuzzy based diagnostics. The first is the off-line one, when we want to repair a system and we know the symptoms observed before the breakdown. The second type is called automated diagnostics, when the diagnostic system monitors the parameters continuously and displays the faults immediately. The process of automated diagnostics can be seen in Figure 3. For demonstration of its application in industrial electrostatics, a computer program has been developed.[3]

5.2. The expert system

First of all, the possible faults and the related symptoms must be identified. The list of possible faults and symptoms can be seen in Table 2. Symptoms are also defined by their membership functions. Secondly, R and T matrices have to be determined. This is the most difficult task, as the physical and hierarchical features of the whole system need to be known.

In the initial phase of generating the KB estimated values were applied and these values were "tuned" later: for a given fault combination we determined the symptoms by a simulation program based on Figure 3. These symptoms were the input parameters of the diagnostic program. The result of the fault analysis was compared to the initial faults and according to the alternations the database was modified.

If the degree of existence of the faults are known, it is possible to give instructions for the operator. The fault analyser selects the most significant faults and according to its output information the appropriate instruction can be selected from an instruction list. As a result, the calculations of the expert program provides a lower and an upper bound for the relative existence of each fault. To compare the various boundaries of faults and to decide which is the most relevant fault among them, it may be a rather complicated task. In order to help this decision making a simple transformation of the upper and lower boundaries were used. The upper and lower boundaries are replaced by their average value decreased by a value proportional to the difference between the two boundaries. With the help of this simple transformation the selection of the most relevant fault becomes much easier. It is important to note, that the diagnostic system is able to notice the failure of the measuring instruments at extreme input signal combinations.

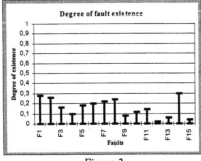

Figure 2

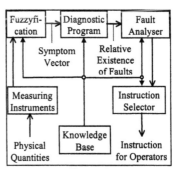

Figure 3

220

6. Results

Let us assume that the observability of each symptom equals zero ($\mu_{Bj} = [0,0,...,0]$). For this input vector, the fault diagnostic system provides a solution giving the upper and lower boundaries of the relative existence of the possible faults. The results can be seen in Figure 2. It can be seen from the solution that even in case of zero input vector, the existence of faults differ from zero. This means that there can be a well developed fault (e.g. F14) even before observing it. If the observability vector

$$\mu_{Bj} = [0,0,0.2,0.2,0.8,0.8,0.3,0.3,0,0,0,0,0.3,0.3,0,0,0.5,0.1], \qquad (5)$$

the solution is presented in Figure 4. The possible faults have several degree of existence in the interval [0,1]. In Figure 5, the results of Figure 4 are transformed as described in section 5.2. It shows clearly that the most dangerous fault is F14 which denotes the failure of air heater.

In certain observations the solution for the fault existence may not exist. If the KB does not contain contradictions this may be deduced for the malfunction of the measuring equipment. Thus with the help of automated diagnostics, the failure of the measuring equipment can also be detected.

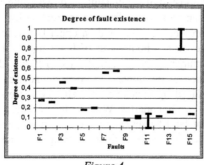

Figure 4

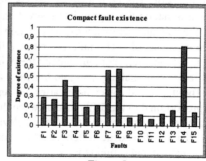

Figure 5

References

[1]   Böhm J 1982 *Electrostatic Precipitators Elsevier*
[2]   Chang J S, Crowley J M and Kelly A J 1995 *Handbook of Electrostatics Chapter 31. Static Electricity Hazards: Solid Surfaces and Gases* (Berta I) Marcel Dekker Inc. New York
[3]   Kiss I, Pula L, Balog E, Kóczy L T and Berta I. 1997 *Fuzzy logic in industrial electrostatics* 8th International Conference on Electrostatics Poitiers France
[4]   Reyes V, Nielsen A 1996 *Emission Control During Transition Periods Using an Expert Supervisory Control System* VI. ICESP Budapest
[5]   Tsukamoto Y and Terano T 1977 *Failure Diagnosis by Using of Fuzzy Logic, IEEE Proc. Decision & Control Vol 2*
[6]   Wells D J 1984 *The Diagnosis of Nuclear Power System Faults*
[7]   Wells D J 1983 *Associating Automated Diagnostics with Reliability Analysis* Presented on Winter Annual Meeting of the ASME, Boston
[8]   Zimmermann H J 1987 *Fuzzy Sets, Decision Making, and Expert Systems* Kluwer Academic Publishers

Inst. Phys. Conf. Ser. No 163
Paper presented at the 10th Int. Conf., Cambridge, 28–31 March 1999

Dust cloud ignition by electrostatic discharges and the mythology of Minimum Ignition Energy - do we know enough?

J M Smallwood

Electrostatic Solutions Ltd. 14 Courtland Gardens, Bassett, Southampton, Hants, UK.

Current practice assesses the electrostatic discharge ignition risk of a dust cloud by measurement of ignition sensitivity with a capacitive discharge circuit, expressed as the Minimum Ignition Energy. This implies that at lower discharge energies ignitions may not be obtained. A review of the literature demonstrates that this is misleading and ignition at lower energies may be obtainable by adding series resistance or inductance to the circuit. Some workers have indicated a reduction in discharge energy required for ignition to 1% or less of the zero ohm energy value, when 10-100 $k\Omega$ series resistance is inserted. Conversely, addition of series resistance reduces the incendivity of a capacitive discharge ignition (CDI) spark to gas mixtures.

Discharges from insulating surfaces are known to be less incendive to gas mixtures than CDI sparks due to their different space time characteristics. Nevertheless it is current practice to use gas mixtures to evaluate ignition risk from insulating surfaces for dust cloud hazard assessment.

There is a trend in modern practice to use static dissipative or conductive materials in the construction of items like FIBCs aimed at prevention of ESD ignition. These materials may have intermediate effective resistance values and could act like a capacitive discharge circuit with series resistance. The incendivity of a discharge from this type of circuit can be expected to be reduced for gas ignition but increased for dust cloud ignition.

This paper argues that conventional MIE values should be used with caution where a discharge may occur from an insulating or intermediate resistance source, as the incendivity of discharges from such materials to dust clouds has been inadequately studied. Where incendivity has been studied it has normally been using gas probe techniques, although gas mixtures have very different ignition characteristic compared to dust clouds. It is possible that the risk of dust cloud ignitions from insulators or materials of intermediate conductivity could be significantly underestimated using current MIE and incendivity evaluation methods.

1. Introduction

The risk of ignition of a flammable mixture by electrostatic discharge (ESD) can be considered dependent on two main factors[1], the spark ignition sensitivity of the flammable mixture, and the incendivity of the discharge (Gibson & Harper 1988). The idea is that if the incendivity of the spark exceeds the ignition sensitivity of the mixture, ignition is likely to occur. Incendivity and ignition sensitivity are usually described in terms of energy. The sensitivity of the mixture is expressed as the Minimum Ignition Energy (MIE).

2. Minimum ignition energy and discharge incendivity

In practice MIE is usually measured using a capacitive discharge ignition (CDI) circuit (Lewis & von Elbe 1951, Glor 1988). A typical example is the Hartman tube apparatus for measurement of MIE of dust clouds (British Standards Institution 1991). A capacitor is charged to a certain voltage and on breakdown of the spark gap, the stored electrical energy is dissipated in the spark. The MIE is taken as the $0.5\ C\ V^2$ energy stored in the capacitor prior to discharge. In practice no spark circuit has pure capacitance, and series resistance and inductance are also present. These are normally intentionally low, and the discharge waveform produced has high peak current (tens or hundreds of amps) and a damped oscillatory form.

Some workers have studied the effects of CDI circuit resistance and inductance on ignition of certain flammable materials. The effect of adding series resistance to the circuit is to increase discharge duration[2] (Rose & Priede 1959) and the reduce rate of energy dissipation. A significant amount of the stored energy is dissipated in the series resistance rather than the spark. For ignition of gas mixtures, adding series resistance increases the stored energy required for ignition, and hence the MIE value obtained.

In contrast, the effect of added series resistance on ignition of dust clouds can be to significantly reduce the stored energy required for ignition of dust clouds (Boyle & Llewellyn 1950). The MIE for granular aluminium and magnesium both decreased by a factor of about ten when series resistance of between 10^4 and 10^5 Ω were included in the circuit. Similar results were obtained by Line et al (1959) for lycopodium dust clouds. These results have been attributed to the ejection of dust particles from the ignition zone by the shock wave generated by the discharge, an effect which is reduced when the discharge is reduced in amplitude and lengthened in duration.

Inductance >1mH in the discharge circuit can also have the effect of reducing MIE (Line et al 1959, Glor 1988). The inductance also reduces the peak current and increases discharge duration[3], and can reduce the MIE measured for a dust cloud by more than a factor of ten.

The discharge from an insulating material or highly resistive material is rather different from a discharge from a metal object or CDI circuit (Gibson & Harper 1988, Gibson & Lloyd 1965). A brush discharge occurs, which is known to be less incendive to gas mixtures than the capacitive discharge. A conventional CDI circuit does not simulate this type of discharge well. Gibson & Lloyd 1965 proposed that the incendivity brush discharges

---

[1] In practice mixture ignition sensitivity and spark incendivity are dependent on a variety of other factors and may be very variable.
[2] For resistances above the value required for critically damped circuit conditions
[3] This is true for underdamped circuits giving oscillatory waveforms. For overdamped circuirs giving unidirectional waveforms increasing inductance decreases discharge duration.

should be characterised by an "equivalent energy", measured as the MIE of a gas mixture which can just be ignited by the discharge. They recognised that the difference in incendivity was due to the differences in spatial and temporal characteristics of brush and capacitive discharges.

3. Safety issues in current ignition test practice

Although the complex nature of ignition by electrostatic discharges has been recognised by many authors, some well established practices have arisen in ESD ignition hazard testing, which now should be re-examined as they could seriously erode the safety margins in operational circumstances. To summarise the factors giving rise to this situation;

- MIE of dust clouds and gas mixtures are generally measured using capacitive discharges with low circuit resistance and inductance.
- This type of CDI spark gives optimum ignition for gas mixtures. However, for dust clouds a factor of ten or more reduction of MIE can be obtained by altering the space-time characteristics of the discharge using added series resistance or inductance.
- Discharges from insulating surfaces have space-time characteristics which are not optimal and give reduced incendivity for gas ignition. Nevertheless it is current practice to evaluate their incendivity using ignition of gas mixtures.
- Incendivities of discharges from insulators evaluated using gas mixtures are then applied in assessment of ignition risk to dust clouds.
- There is little or no information available as to whether discharges from insulators are more, or less, incendive to dust clouds than CDI discharges. Although gas mixtures are less easily ignited, there is no reason to expect this is true of dust clouds.

We must conclude that incendivity data obtained using ignition of gas mixtures, combined with dust cloud MIE data obtained using CD discharges, could underestimate the ignition risk of discharges from insulators.

A typical example is in the current practice of investigating ignition risk in FIBCs using a gas probe. The discharge from the FIBC surface is made to pass through a flammable gas of known MIE. A judgement is made of the ignition risk arising from the FIBC design based on whether or not ignition can be made to occur.

A second cause for concern is the use of resistive materials in static dissipative equipment designed for ignition hazard avoidance. An example is a carbon fibre matrix woven into a cloth. In electrical circuit terms, a carbon fibre matrix of intermediate conductivity could be expected to resemble a CDI circuit with series resistance. Again, current practice would be to evaluate ignition risk using a flammable gas mixture, although a reduced incendivity to gas mixtures could be expected. In contrast, an order of magnitude increase in the incendivity of the discharge to a dust cloud could be expected.

Again we must conclude that incendivity data obtained using ignition of gas mixtures, combined with dust cloud MIE data obtained using CD discharges, could underestimate by an order of magnitude the ignition risk due to discharges from materials of intermediate resistance.

224

4. Conclusions

The MIE of flammable mixtures is usually measured as the stored energy in a capacitive discharge circuit which leads to ignition of the mixture. The circuit usually has minimal and undefined series resistance or inductance. This type of discharge has been shown to have high incendivity to gas mixtures. However, inclusion of 10^4-10^5 ohms series resistance, or >1mH inductance can increase the incendivity to dust clouds by an order of magnitude but reduces the incendivity to gas mixtures due to changes in the discharge space time characteristics.

Discharges from insulating surfaces, or conductors of intermediate resistivity, have different space time characteristics than CD discharges used in MIE test. Their incendivity to dust clouds has been little studied, and is usually evaluated using gas ignition. While the incendivity of these discharges to gas mixtures is reduced, their incendivity to dust clouds compared with CDI spark source is uncertain.

Thus CDI circuits give ESD in a form which ignites gas mixtures more easily than discharges from insulators or objects of intermediate resistance. However we have no reason to expect the same behaviour for dust cloud ignition.

It is therefore unsafe to rely on ignition of gas mixtures as an indicator of ignition risk to dust clouds from discharges from insulators or conductors of intermediate resistance. This concern is exacerbated by the measurement of MIE of dust clouds using discharges from CDI circuits with little series resistance or inductance, which are not optimised for incendivity to dust clouds.

References

British Standards Institution 1991 *Code of practice for the control of undesirable static electricity Pt1. General considerations.* BS5958: Part 1: 1991 pp 40-43

Boyle A R, Llewellyn , F J,. 1950 *The electrostatic ignitability of dust clouds and powders.* J. Soc. Chem Ind Trans. 69 pp. 173-181

Eckhoff RK. 1991. *Dust Explosions in the Process Industries.* Butterworth Heinemann ISBN 0 7506 1109 X

Gibson N., Lloyd FC. 1965 *Incendivity of discharges from electrostatically charged plastics.* Brit. J. App. Phys. 16, pp. 1619-1631

Gibson N, Harper D J. 1988 *Parameters for assessing electrostatic risk from non-conductors - a discussion.* J. Electrostatics 21 pp. 27-36

Glor M., 1988 *Electrostatic hazards in powder handling.* Res. Studies Press, ISBN 0 86380 071 8

Lewis B. L., von Elbe G. 1951 *Combustion, Flames and explosions of gases.* New York; Academic Press

Line L E, Rhodes H A, Gilmer T E. 1959 *The spark ignition of dust clouds.* J. Phys Chem. 63 290-294

Rose H. E., Priede T. 1959 *Ignition phenomena in hydrogen air mixtures.* 7th Symp. on Combustion 436-445

Inst. Phys. Conf. Ser. No 163
Paper presented at the 10th Int. Conf., Cambridge, 28–31 March 1999

The high resolution microcomputer system for recording of electrostatic discharges

L. Ptasiński, T. Żegleń and J. Gajda

Faculty of Electrical Engineering, Automatics, Informatics and Electronics,
University of Mining and Metallurgy, al. Mickiewicza 30, 30-059 Kraków, Poland

Abstract. The new system for recording of charge-time characteristics of ESD has been elaborated. In result of using new hardware and identification procedure the system can work in 8-channel mode and with much higher resolution. It is portable, with battery supply and may be used in practically all operating conditions.

1. Introduction

Electrostatic discharges occur in the form of pulse series. Measurement of charge transfer is particularly convenient to evaluate the ESD incendivity [1]. Basing on measurement of the charges transferred in pulses and time intervals between the pulses it is possible to determine the probability of ignition initiation in the continuous stationary processes [2].

The first microcomputer system for ESD recording has been presented in [3]. The system enabling the measurement of charge transferred in ESD was greatly improved. The main effort has been directed to increase the resolution and to create the possibility of simultaneous recording of amplitude-time courses of ESD in different places of the investigated site. Till now, the recording system worked in one-channel mode with 1 ms resolution [3].

2. Description of the system

The block diagram of the recording system is presented in Fig. 1. The pulse converters (1), (1')....for transforming charges of ESD pulses into voltage signals are the main parts of the measuring circuit. The separating units (2), (2')... separate galvanically the pulse converter from the data acquisition unit. The data acquisition system (3), of Iotech Co. WBK512 type, have 8 analog signal processing channels with highest sampling frequency of 1 MHz. The signals are sampled at selected frequency, processed into digital form and recorded in memory. For the data exchange with microcomputer notebook type AT-Pentium 133 MHz (4) the enhanced parallel port (EPP) is used which enables full advantage to be taken of its features. The acquisition system may work in different triggering modes. The system may be supplied from the mains or from the battery of unit WBK30A type (5). In the case of the battery supply the system does not require the galvanic separation units and can work for about 6 hours.

The pulse converters generate standarized voltage pulses of the waveform shown in Fig. 2. The amplitudes of the pulses are proportional to the charge value in the case of time intervals between successive pulses ≥ 500 μs (signals 1 and 2 on Fig. 2). This way the time response of

226

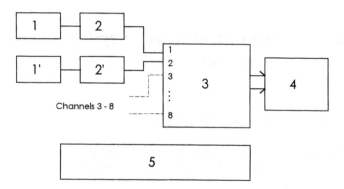

Figure 1. Block diagram of portable multichannel system for ESD recording.
1, 1′ - pulse charge converters; 2, 2′ - separating units; 3 – DSP-based multichannel portable acquisition system; 4 – notebook AT-Pentium 133 MHz; 5 – rechargeable battery module WBK30A.

the converter, practically determines the time resolution of the whole recording system. If the ESD pulses occur in time intervals shorter then approx. 500 μs (curve 3 on Fig.2) there is a problem of recognising and identifying parameters of multiple pulses.

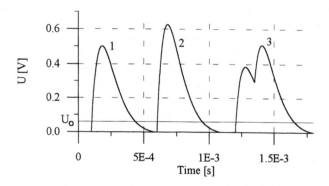

Figure 2. The output voltage waveforms of the converter for different time intervals between successive ESD pulses. U_o – selected triggering level; 1,2 – pulses with time interval 500 μs; 3 – two pulses with time interval 150 μs.

The whole process of pulses recording is controlled by the software written in PASCAL. There are two possible modes of the system operation, chosen depending of the type of the recorded pulses train.
- Each incoming pulse triggers the conversion and is recorded separately. For the time before the next pulse occurs the recording is switched off, and the time interval between two consecutive pulses is recorded separately on the basis of the real time clock readings. For such a mode of operation it is possible to record up to 1600 pulses and the time intervals between them in the range of (500μs ÷ ∞).
- After the first triggering the recording is continuous, and the results are later analyzed according to the algorithm described in 3. In this mode the recording time is limited to approx. 3,2s.

3. Identification process of recorded courses

An algorithm was sought which would allow the identification of parameters which characterise a single ESD, or parameters of each component pulse in a multiple ESD. The parameter estimation for all the ESD pulses included in the multiple one is performed consecutively for each pulse, beginning with the first one. A small (2-5) number of initial samples of the recorded pulse converter output form the estimation basis. The model describing the time response of the pulse charge converter for the single ESD pulse has two unknown coefficients. The graphical representation of the model response for such a case is presented in Fig.2 (curves 1 and 2). The unknown model coefficients are: t_Q - the time of the discharge occurrence, and Q_t - amount of the charge transferred during the discharge. The identification of the discharge parameters is equivalent to the both parameters value determination. In the case of multiple discharge identification this means finding the parameters value for each component pulse. From the model structure we can state that the model is linear taking the coefficient Q_t into consideration, and nonlinear for the t_Q coefficient. Identification process of the recorded courses is an iterative process of t_{Qk}, and Q_{tk} value determination, where k is the number of iteration equal to number of the component pulses in multiple ESD. Timing of the first pulse as well as the transferred charge is determined from the data with the so-called direct algorithm [4]. From the results i.e. t_{Qk} ESD pulse time and Q_{tk} charge estimation, using the pulse converter model, the pulse converter response to a single pulse with (t_{Qk}, Q_{tk}) parameters is calculated. Then, the calculated response is subtracted from the actual converter response recorded during the measurement. The subtraction result forms the basis for the iterative process of parameter estimation of the ESD pulse. The iterative process lasts until the subtraction results is below certain assumed level.

This iteration identification process allows for correct (at the assumed level of accurancy) estimation of parameters of the ESD series within the multiple ESD when component pulses in this series do not occur with time intervals shorter than 10µs.
The algorithms for identification of multiple ESD parameters and the results of their verification based on simulation are described in [4].

4. Exemplary measurement result

Fig.3 presents an example of the recorded time course multiple ESD (a) and charge-time characteristic of ESD pulses (b) for discharges that ignited an explosive mixture 22% hydrogen and air. The characteristic (b) was obtained from the course (a) using described identification process.

5. Summary

The new portable microcomputer system for ESD recording can work in two different modes, with the recording triggered by the consecutive pulses or with continuous recording.
The first mode enables recording up to 1600 pulses of charge transfer and the time intervals between consecutive ESD pulses with time resolution of 500µs without any time limitation. This mode is particularly useful for the investigation of the system where the ESD pulses are generated continuously. Example of this mode application is presented in [5].
In the second mode the recording time is limited to 3.2s.

228

Application of the algorithm mentioned previously made it possible to obtain the time resolution between pulses equal 10μs. The sampling frequency of 100kHz makes it possible to record simultaneously 8 independent trains of pulses in both modes.

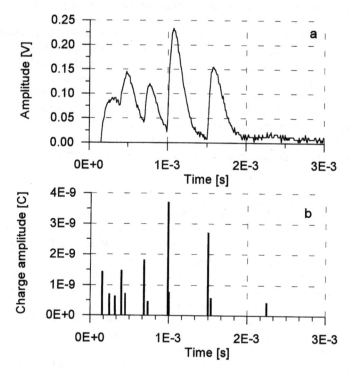

Figure 3. Recorded course of multiple ESD, obtained at sampling frequency 100 kHz (a) and result of the identification process (b)

References

[1] Glor M and Maurer B 1993 *J.Electrostatics* 30 123-134
[2] Ptasiński L. Żegleń T and Gajewski A 1996 *EOS/ESD Symposium Proceedings* 351-355
[3] Ptasiński L and Żegleń T 1995 *Inst. Phys. Conf.* Ser.No.143 349-353
[4] Gajda J 1999 *System Analysis-Modelling-Simulation (in print)*
[5] Ptasiński L. Żegleń T and Gajewski A 1999 *Electrostatics '99* (Cambridge Inst.Phys.Conf.)

Acknowledgements

The work presented in this paper was sponsored by the State Committee for Scientific Research, Warsaw (contract No: 8T10C01012).

Inst. Phys. Conf. Ser. No 163
Paper presented at the 10th Int. Conf., Cambridge, 28–31 March 1999
© *1999 IOP Publishing Ltd*

Flow electrification through porous media

T Paillat, E Moreau and G Touchard

LEA UMR 6609 CNRS, 40 avenue du Recteur Pineau, 86022 Poitiers, France

Abstract. This paper deals with streaming current generated by water flow through 4 geological porous media. These experimental measurements and a theoretical model allow to determine the space charge density ρ_W at the wall in these 4 cases and to compare these values to previous ones obtained with glass capillaries and glass porous samples. The results show that ρ_W is effectively the intrinsic parameter of the solid-liquid interface because it is not function of the geometry of the fluid flow paths but it only depends on the physico-chemical properties of the solid and the liquid.

1. Introduction

At the interface between a liquid and a solid, the specific adsorption process of some ionic species of the liquid by the solid induces an *electric double layer*. According to Stern model [1], there are two different regions inside the liquid : the *compact layer* so close to the wall that its electrical charges are not affected by the liquid flow, and the *diffuse layer*. When a liquid flows through a channel or a porous medium, it convects electric charges coming from the diffuse layer. Then a current called *streaming current* is generated. Inversely, when an electric field is applied on both sides of a porous medium, a part of the diffuse layer charges migrate and a flow may be induced by viscosity strength. This phenomenon is known as *electroosmosis*.

For example, the flow electrification phenomenon may be characterised at the oil-pressboard interface in high power transformers, where it is suspected to induce electrical discharges sometimes until the transformer destruction [2].

The electroosmosis and electrophoresis phenomena may be used for industrial applications which need the control of a flow : strengthening of clay soils [3] or removal of contaminants from polluted low-permeability soils [4] [5]. The nuclear waste storage represents an important challenge for the scientific community and this needs the knowledge of the hydraulic, electrical and ionic phenomena inside porous media. Furthermore, in order to use electrokinetic processes, one has to be able to predict the phenomenon as a function of the physico-chemical properties of the porous medium and the morphology of the fluid flow paths.

The ultimate goal of our work is to be able to determine the intensity of the phenomenon (the flow rate for electroosmosis and the convected space charge density for streaming current) from the analysis of the medium geometry. To achieve this the intrinsic physico-chemical parameter of the solid-liquid interface must be known. Touchard [6] showed that in the case of dielectric liquids this parameter was the space charge density ρ_w at the wall (*i.e.* at the solid-liquid interface). Recently we studied the streaming current generated by deionised water flow through glass capillaries of different radii [7] and through porous media of different pore sizes [8]. The results seemed to show that ρ_w did not depend on the radius of the fluid flow paths and that it was still the intrinsic physico-chemical parameter of the water-glass interface.

In this paper, the same experiments performed with several natural geological porous media are presented. This will allow us to discuss the values of ρ_w.

2. Theoretical model

The theoretical model has been accurately explained in a previous paper [8]. In order to be able to use the theoretical knowledge on electrokinetic phenomena in a porous medium, this has to be modelled. In the case of sintered glass, limestone and sandstone, it seems that the fluid flow path network may be represented by N parallel channels of constant radius r_0. Then, in the case of weak space charge density and laminar flow, the space charge density at the wall can be expressed as follows :

$$\rho_W = \frac{I}{D}\left(\frac{\alpha}{\delta_0}\right)^2 \frac{r_0^2}{8} \frac{I_0\left(\frac{\alpha r_0}{\delta_0}\right)}{I_2\left(\frac{\alpha r_0}{\delta_0}\right)} \qquad (1) \qquad \text{with } \delta_0 = \sqrt{\frac{\varepsilon_L D_0}{\sigma_L}} \qquad (2)$$

where δ_0 is the thickness of the diffuse layer (Debye length), σ_L the electric conductivity and ε_L the permitivity of the liquid, D_0 the mean ion diffusion coefficient, I the measured streaming current, D the experimental fluid flow rate, α a coefficient which takes into account the two diffusion coefficients of cations and anions inside the double layer, I_0 and I_2 the modified Bessel functions of zero and second order and r_0 the radius of the fluid flow paths given by :

$$r_0 = \sqrt[4]{\frac{D}{\Delta P} \frac{8\mu l}{\pi N}} \qquad (3)$$

ΔP being the pressure gradient applied on both sides of the porous medium, μ the fluid dynamic viscosity, l the sample length and N the number of fluid flow paths which can be determined by image analysis. In fact, the r_0 value corresponds to the radius of a capillary of length l which would induce, for a given pressure gradient, a flow rate equals to the flow rate generated in each channel of the porous medium.

A special equipment is used for hydraulic and electrical measurements in order to determine the streaming current and the fluid flow rate [8].

3. Results

Our previous works have been realised with 5 glass capillaries of different radii (from 10 μm to 100 μm) and 12 sintered glass samples of different pore sizes (from 1.65 μm to 10.8 μm). The main component of the solid phase was silica (SiO_2). The study presented in this paper uses 4 geological porous medium samples which are presented in Table 1.

Sample name	Rock name	Main component	Flow path radius r_0 (µm)
g1	Fontainebleau sandstone	SiO_2	3
g2	North sea sandstone	SiO_2	2.5
g3	Britain sandstone	SiO_2- FeS_2	0.16
c1	Limestone	$CaCO_3$	0.28

Table 1. Information about the 4 geological porous medium samples.

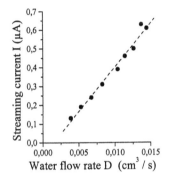

Figure 1. Streaming current versus water flow rate for Fontainebleau sandstone (g1).

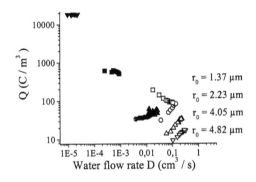

Figure 2. Convected space charge density Q versus water flow rate for 4 geological rocks *(solid symbols)* and 4 sintered glass samples *(open symbol)*. The r_0 values of the rock samples are precised on the figure.

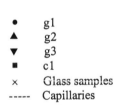

•	g1
▲	g2
▼	g3
■	c1
×	Glass samples
-----	Capillaries

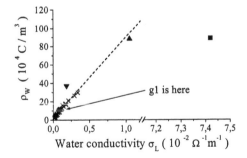

Figure 3. Space charge density ρ_W at the wall versus water conductivity σ_L.

Figure 1 is in good agreement with previously published results [2] [8] [9] [10] : it shows that the streaming current I is proportional to the liquid flow rate D. One can see that the current is positive. This signifies that the convected charges coming from the diffuse layer are positive charges. Figure 2 presents the convected space charge density Q versus water flow rate for the 4 geological porous media and for 4 sintered glass samples. It shows that Q is nearly constant for each sample whatever the flow rate. Furthermore, the main remark is that for a given solid-liquid interface (mainly silica-water in our case), the smaller

r_0 is, the greater Q is. Figure 3 shows the values obtained for the space charge density ρ_W at the wall as a function of the liquid conductivity σ_L for the 5 capillaries, with the 12 glass samples and with the 4 geological porous media. For the experiments performed with capillaries and glass samples, the water conductivity depends on its CO_2 concentration. For the geological porous media, the water conductivity before experiments is close to 15×10^{-4} Ω^{-1} m^{-1}. When the water flows through the path network, there is a chemical dissolution of the rock. This induces an increase of the fluid conductivity. Thus, the conductivity taken into account in computations is the equilibrium value, that is to say the value measured after experiments.

Results obtained with the capillaries *(dot line ----)* show that ρ_W increases when σ_L increases, independent of the r_0 values (which vary between 10 µm and 100 µm). This shows that ρ_W depends on the physico-chemical properties of the solid and the liquid, and does not depend on the channel geometry. The results obtained with the sintered glass samples *(cross symbol xxxx)* are similar. The results obtained with the geological media *(solid symbols ■ ● ▲ ▼)* depend on the main component of each rock. On one hand, the ρ_W values for Fontainebleau sandstone (g1) and North sea sandstone (g2), which are mainly composed of silica SiO_2, are equal to the values obtained with the silica-water interface. Britain sandstone (g3), composed of SiO_2 and FeS_2, has nearly the same behaviour. On the other hand, the results obtained with the limestone (c1) are strongly different because the physico-chemical properties of the limestone-water interface. Furthermore, they are similar to the results of Jounniaux and Pozzi with a Mériel limestone [10].

Consequently, these results confirm that ρ_W does not depend on the channel radius r_0, but only on the physico-chemical properties of the solid-liquid interface.

4. Conclusion

The goal of this work was to determine the values of the space charge density ρ_W at the wall induced by a water flow through 4 geological media : Fontainebleau sandstone, Britain sandstone, North sea sandstone and a limestone.

Streaming current measurements associated to a theoretical model allowed us to compute these values. They have been then compared to the values obtained previously with glass capillaries and glass samples, *i.e.* with the silica-water interface. It has been shown that ρ_W value is directly proportional to the silica content in the rock, and is independent of the radius of the fluid flow paths.

This study showed that the space charge density ρ_W at the wall was effectively the intrinsic parameter of the solid-liquid interface in the case of non-dielectric liquids.

References

[1] Stern O 1924 *Z. Electrochem.* **30** 508
[2] Paillat T, Moreau E, Touchard G 1998 *IEEE-CEIDP Annual Report* 182-187
[3] Lo K Y, Ho K S and I I Inculet 1992 *Can. Geotech. J.* **29** 599-608
[4] Moreau E, Grimaud P O and Touchard 1997 *Proc. ESA 25th Annual Meeting* 152-158.
[5] Probstein R F and Hicks R E, 1993 *Science* **260** 498-503
[6] Touchard G and Romat H 1981 *J of electrostatics* **10** 275-281
[7] Paillat T, Touchard G and Moreau E 1998 *Proc. ESA-IEJ Joint Symposium on Electrost.* 26-40
[8] Moreau E, Paillat T and Touchard G 1998 *Proc. ESA-IEJ Joint Symposium on Electrost.* 41-51
[9] Rice C L and Whitehead R 1965 *J. Phys. Chem.* **65** 4117-4123
[10] Jounniaux L and Pozzi J P 1995 *J. of Geoph. Research* **22** 485-488

Inst. Phys. Conf. Ser. No 163
Paper presented at the 10th Int. Conf., Cambridge, 28–31 March 1999

INCENDIVITY OF ELECTROSTATIC DISCHARGES IN DUST CLOUDS : THE MINIMUM IGNITION ENERGY PROBLEM

M. Bailey, P. Hooker, P. Caine

Zeneca Ltd., Manchester, U.K.

N. Gibson

Burgoyne Consultants Ltd., Ilkley, U.K.

Abstract. Minimum Ignition Energy of dust clouds is required to assess the electrostatic ignition risk. Recent studies are reported that indicate that the test methods in use to determine M.I.E. give markedly different values. The use of M.I.E. to define the incendivity of discharges from conductors and non-conductors is discussed and the value of data from present day tests is considered.

1. Introduction / the Problem

Assessment of electrostatic ignition risk in powder handling plants requires a decision as to whether a dust cloud can be ignited by spark, cone, corona, brush or propagating brush discharges. It is generally accepted that corona discharges from electrostatically charged bodies will not ignite dust clouds. The energies of propagating brush discharges can be several joules and can ignite a wide range of dust clouds. Safety is therefore achieved by avoiding the conditions under which such discharges would occur rather than consideration of the sensitivity of the dust cloud. The incendivity of spark discharges from conductors (e.g. metal plant) is assessed by comparing the total energy stored on the conductor with the Minimum Ignition Energy (MIE) of the dust cloud. Equivalent Energy (Ref. 1) has been used to assess the incendivity of brush discharges from plastic materials (Ref. 1) and of cone discharges from bulk powder (Ref. 2). This also requires a knowledge of the MIE of the dust cloud.

A valid measure of the Minimum Ignition Energy of dust clouds is required to assess the electrostatic ignition risk from spark, cone and brush discharges. However the distribution of energy in these discharges differs with respect to time and space. The total energy in a discharge does not necessarily define its ignition capability. MIE is determined by measuring the minimum energy released in a spark from an electrical circuit that just ignites the dust cloud. The measured MIE is a function of the circuit electrical characteristics. Siwek (Ref. 3) presenting data from three laboratories shows that inductance in the test circuit can decrease the MIE by 10-100 times (Figure 1). No one test circuit will simulate the different forms of electrostatic discharges. The problem is to define the test circuit most appropriate for characterising dust clouds with respect to their sensitivity to ignition by spark, cone, and brush discharges that can be used to assess hazard in industrial operations.

234

2. Sensitivity to Ignition by Spark Discharges

Spark discharges occur when static electricity accumulates on, and is released from, metal plant. The electrical characteristics of the test circuit should replicate those of unearthed metal plant.

Tests on a wide range of equipment in chemical plants (e.g. reactor, silo, pipework, tools, shovel valves, gas cylinder etc) show that the resistance is a few ohms and that with one exception the inductance is less than 3 μH. The exception was the internal metal coil in a flexible rubber connector (L = 220 μH). This indicates that the M.I.E. test circuit should be essentially capacitive. The measured MIE value depends not only upon the electrical characteristics of the test circuit but also on the geometry of the electrode gap, electrode size / shape and the discharge triggering mechanism. Data obtained by 12 test houses for 9 different powders is summarised in Table 1. This indicates MIE data is very dependent on test conditions. In industrial situations the "gap" across which a spark could occur cannot be controlled and it must be assumed that the discharge conditions producing maximum incendivity for a given energy value could be present. The test method used to assess electrostatic hazard from conductors should therefore be based on a capacitative discharge with an electrode configuration etc. that produces the lowest value of MIE.

Two types of equipment that are based on capacitative circuitry are in common use.

(a) Equipment based on BS 5958 (Ref. 4): sparks are released from the high voltage electrode (10-30kV) across a spark gap of not less than 2mm. The voltage is slowly increased until a discharge is produced and the energy in the spark is considered to be that on the capacitative circuit. Typical values of inductance is 12 μH.

(b) Kuhner Mike 3 equipment : this meets the draft European requirements (Ref. 5). It differs from the BS 5958 equipment in that the electrode is more pointed. The electrode gap is 6mm. A moving electrode system is used to trigger the sparks of 10mJ and above. For 1mJ and 3mJ sparks a high voltage relay is used to trigger the discharge at 15kV. The inductance can be 20 μH or 1020 μH. The former is used in hazard assessments.

MIE data obtained for 17 powders with the two test methods is shown in Figure 2. In no case did the BS 5958 test produce the lower MIE. With 4 powders the tests were in agreement. The Mike 3 test produced the lower MIE values with 13 powders. Most important the BS 5958 test failed to detect sensitive materials (i.e. MIE < 10mJ) in 6 cases.

It has been reported (Ref. 6) that when one powder was tested in 17 test houses using MIKE-3 equipment, 16 obtained the value 1 - 3 mJ and 1 the value < 1 mJ. This is good reproducibility.

It is concluded that the MIKE 3 equipment can provide data that can be reasonably used to assess the incendivity of sparks from insulated conductors. Although the test in BS5958 is based on capacitative circuit its electrode configuration etc is not such as to produce the minimum value of M.I.E. from such a circuit.

3. Sensitivity To Ignition By Brush and Cone Discharges

The characteristics of brush and cone discharges differ markedly in terms of the temporal and spatial distribution of energy from spark discharges. Comparison of the total energy in this type of discharge with the MIE determined using a capacitative spark circuit does not provide a measure of the incendivity of the discharges. To overcome this problem Gibson and Lloyd (Ref. 1) introduced the concept of Equivalent Energy. This states that if a brush or cone discharge just ignites a flammable atmosphere with an MIE of XmJ then it will not ignite a flammable atmosphere whose MIE exceeds XmJ.

At present the MIE value is that determined using a capacitative circuit. The essential pre-condition for the concept of Equivalent Energy to be valid is that the flammable atmospheres used in the ignition tests have similar combustion initiation characteristics to those for which the risk from brush and cone discharges is being assessed. This condition has been shown to be satisfied for common gases and vapours (Ref. 1). The effect of inductance on M.I.E. (Fig. 1) of dust clouds indicates that different dust clouds may not react in terms of combustion initiation to changes from spark to cone, spark to brush. Glor (Ref. 2) has overcome the problem for cone discharges by directly igniting dust clouds by them.

In the case of brush discharges, variations in M.I.E. over the range 1 - 10 mJ can markedly effect safety measures specified for a process. Three powders with low M.I.E. values have been tested with MIKE 3 apparatus using circuits with inductance of 20 µH and 1020 µH. The results are shown in Table 2. Narrow particle size bands were use to minimise the effect of the different settling times for particles of different sizes. The data indicates that, for dust clouds sensitive to ignition (i.e. M.I.E. < 10mJ), the effect of L may be present for some powders but it is small.

It is concluded that the equivalent energy concept can be used to give guidance on the incendivity of brush discharges but that further work is required on this topic to define safe limits.

Acknowledgement

The authors acknowledge valuable discussions with S. Newton, D. J. Harper, Zeneca Ltd.

References

[1]   N. Gibson and F.C. Lloyd. *Br. J. Applied Physics.* Vol 16. p.1619 (1965).

[2]   M. Glor, B. Maurer. *J. Electrostatics.* Vol. 30. p.123 (1993).

[3]   R. Siwek, C. Cesana. *Process Safety Progress.* Vol. 14, No. 2, p.107 (1995).

[4]   BS 5958. Part 1. British Standards Institution London U.K. (1991).

[5]   Draft European Standard *"Determination of Minimum Ignition Energy of Dust / Air Mixtures"* CENTC 305/WG1/SG:1.2 MIE (April 1998).

[6]   "Final Report on Calibration Round Robin" Adolf Kuhner AG. Switzerland (1998).

Table 1. Data from 12 Test Houses

Product	Minimum Ignition Energy (mJ)	
	Lowest	Highest
A	1 - 3	59 - 148
B	1 - 3	100 - 300
C	300 - 1000	2145 - 3890
D	100 - 300	3890 - 8700
E	10 - 100	> 12500
F	1 - 3	25 - 125
G	1 - 3	148 - 258
H	1 - 3	13 - 17
I	154 - 381	2500 -12500

Table 2 : Effect on MIE with changing Particle Size and Inductance

Particle size (μm)	Sample A		Sample B		Sample C	
	L=20μH	L=1020μH	L=20μH	L=1020μH	L=20μH	L=1020μH
150 - 250			>1000	>1000		
>212	3 - 10	3 - 10				
150 - 212	1 - 3	1 - 3				
125 - 150	1 - 3	1 - 3	30 - 100	10 - 30		
106 - 125	1 - 3	1 - 3	30 - 100	30 - 100	10 - 30	3 - 10
75 - 106	1 - 3	1 - 3	10 - 30	3 - 10	3 - 10	3 - 10
50 - 75	1 - 3	1 - 3	10 - 30	3 - 10	3 - 10	3 - 10
<50			10 - 30	3 - 10	3 - 10	3 - 10
<38					3 - 10	3 - 10

- All values are in mJ

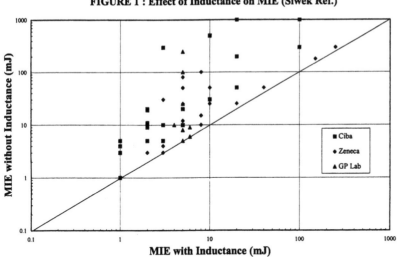

FIGURE 1 : Effect of Inductance on MIE (Siwek Ref.)

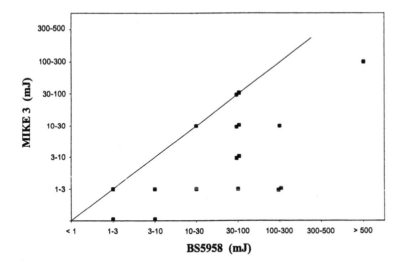

FIGURE 2: Comparison of BS5958 and MIKE 3 Test Methods

FIGURE 1. Effect of Inductance on MIK Growth Rate

DISCUSSION - Section D - Hazards

Title: Electrostatic Ignition hazards Associated with Flammable Substances in the form of Gases, Vapours, Mists and Dusts

Speaker: M Glor

Comment: John Chubb

Resistivity may be a useful parameter for 'continuous' operations but a charge relaxation measurement will be relevant in practice. It can be used with semi-continuous processes such as walking on a carpet. Upon stopping, charge relaxation competes with charge generation. I also suggest that charge relaxation (charge decay) measurements are a much easier way to assess insulating-type layer and powder materials. It is also worth noting that there are situations where neither charge decay nor resistivity necessarily indicate charge limitation processes. An example is the increase in body voltage on getting out of a car. In that, and such situations what needs to be measured is what is actually relevant i.e. body voltage. This emphasises the importance of assessing from the point-of-view of the end-user.

Comment: Istvan Berta

I think the most important problem with corona discharge is that corona is not a well defined physical phenomenon. A partial discharge on sharp conductors is usually called corona. If we speak about physical phenomena, corona can be formed of electron avalanches, streamers and leaders too. However in power engineering corona can be a rather large (long) discharge with rather high energy. In industrial electrostatics the energy of corona discharges is usually not concentrated in time and in space.

Question: J Smallwood

There is sometimes a conflict between the need to use plastics to line a plant and electrostatics hazard prevention - are conductive plastics finding application in this area?

Reply: Yes, these materials may be applied in classified areas. A problem can arise, however, from contamination by carbon black or other antistatic agents added to the polymers. If carbon black is used the conductivity is significantly increased and earthing is required as a consequence.

Title: Large Scale Chemical Plants: Eliminating the Electrostatic Hazards

Speaker: G R Astbury

Question: Martin Glor

(a) Have you performed any measurements in the large final tank? (b) How do you make sure that the small diameter pipes are always filled with liquid?

Reply: (a) No. (b) The small pipes were purged with water at high enough rate to ensure that any air was swept out. The end of the dip pipe was always left immersed in the liquid.

Section E

Applications and Processing

Inst. Phys. Conf. Ser. No 163
Paper presented at the 10th Int. Conf., Cambridge, 28–31 March 1999

Bipolar Spray Charging for Enhanced Deposition onto Nonconductive and Electrically Isolated Targets

S. Edward Law[1], Steven C. Cooper[2] and Whitney B. Law[1]

[1]Department of Biological and Agricultural Engineering, University of Georgia, Athens, GA 30602-4435, USA

[2]Electrostatic Spraying Systems, Inc., 67 Depot St., Watkinsville, GA 30677, USA

Abstract. Bipolar charging (*i.e.*, ±8.5 mC/kg @ 0-4 Hz) of conductive liquid being pneumatically sprayed from an electrostatic-induction nozzle has been investigated as a method for increasing droplet deposition onto four target types. For a grounded metal target cylinder and for a nonconductive cylinder with grounded substrate, there was no significant difference in the amount of deposition delivered by dc and by 1/8 Hz charged sprays; these two charging methods provided equally effective electrodeposition benefits averaging 8.4-fold significantly greater ($\propto = 0.01$) than uncharged spray. This experimental finding has important practical implications regarding improved design of portable human-carried electrostatic coating systems as well as reduced operational hazard.

1. Introduction

Electrostatic application of sprays and powders is a proven technology for dramatically increasing particulate deposition efficiency onto various target surfaces of commercial and environmental importance in industry and agriculture [1, 2]. Electric-field forces can routinely exceed by 20-50 fold the gravitational force on charged particulates and typically increase their mass deposition 2-8 fold as compared to similar uncharged particulates [3]. Certain type targets, however, present difficulty by being unamenable to conventional electrostatic-deposition processes; specifically included are nonconductive targets and conductive targets which are electrically isolated from the earth-potential reference of the electrostatic system. Examples are, respectively, plastic parts in manufacture, and in-flight objects (*e.g.*, pharmaceutical pills, airborne insect pests, food products and confectionaries, agronomic seeds, *etc.*) requiring coating with control agents, flavorings, fungicides, *etc.* In each case, the initial deposition of charged coating particles accumulates electric charge on the target in the absence of a conductive leakage path to earth. In addition to the deposited-charge accumulated using electrostatic-induction and triboelectric coating systems, ionized-field (*i.e.*, corona) systems exacerbate target-charge buildup by their inherent air-ion current. The elevation in target potential associated with charge buildup causes a detrimental limitation in electrodeposition benefit as subsequently approaching charged particulates are electrically repelled. Further problems result from inadvertent charge accumulation on nearby nonconductive and electrically isolated (*i.e.*, ungrounded) non-target surfaces including: a) possible incendiary electrical discharges originating from paint booths and curtains; and b) nuisance, and possibly hazardously distracting, discharges from operators of human-carried portable electrostatic coating systems.

244

Bipolar spray-charging strategies have earlier been investigated by Cooper and Law [4] to ameliorate another target-imposed problem limiting electrodeposition — namely, the deleterious electric discharges induced by excessive space-charge fields to flow from sharp edges and tips (e.g., plant leaf tips) of grounded, conductive targets. Alternating (e.g., 2-36 Hz) the droplet-charge polarity carried on successive spray-cloud sectors traveling through the nozzle-to-target region was experimentally shown to significantly diminish the spray cloud's overall space-charge electric field while still maintaining, to a degree, local deposition fields and beneficial charge transfers (Fig. 1) at internal and external target surfaces. For this present work it was hypothesized that excessive surface-charge buildups on nonconductive and electrically isolated targets undergoing electrostatic coating could likewise be controlled by appropriate bipolar spray-charging. The objective of this paper is to experimentally quantify spray deposition as a function of bipolar spray charging frequency (viz., seven values over the dc-4 Hz range), and compare results with similarly applied uncharged spray, for the following electrically differing target conditions: a) grounded conductive target; b) electrically isolated conductive target; c) nonconductive target; and d) nonconductive target surface overlying grounded substrate.

2. Experimental Analysis

Figure 2 illustrates the overall experimental setup for applying fluorescent-tracer-tagged aqueous sprays onto the various type deposition targets at specified bipolar spray-charging frequencies. Detailed descriptions of apparatus, procedures, and statistical design of the experiment follow.

2.1. Apparatus and Materials

2.1.1. Spray-charging. Spray liquid of 0.1 S/m conductivity was prepared by combining 1.5 g of fluorescent tracer particles (DayGlo Corp. - Blaze Orange GT15N), 0.1 g NaCl, and 1.0 mL of Triton X-100 non-ionic surfactant into 1000 mL of deionized water. The liquid was pneumatically atomized (207 kPa) at 0.93 mL/s into ~ 30 μm volume median diameter spray and inductively charged to ~ 8.5 mC/kg by an embedded-electrode spray-charging nozzle (Electrostatic Spraying Systems, Inc. - MaxCharge™ model) set midway its operational range at 650 V. Similar nozzles typically exhibit a 10-12 μC/kg·V linear spray-charging response vs. input voltage [5].

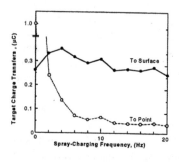

Fig. 1. Spray-charge transfers to grounded target-sphere and its insulating attached grounded discharge point as a function of square-wave bipolar spray-charging frequency [4].

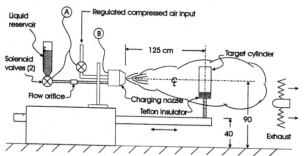

Fig. 2. Experimental setup for application of bipolar-charged sprays to cylindrical targets.
(Circled letters denote interconnections with electronic circuit of Fig. 3.)

2.1.2. *Bipolar voltage source.* For an 8 s standard spray-charging interval, the electronic circuit of Fig. 3 provided the nozzle's induction electrode with +650 Vdc or an integral number of square-wave bipolar voltage pulses of ± 650 V peak at specified frequencies; it also provided a synchronized 8 s energization to quick acting solenoid valves installed near the nozzle for gating air and spray-liquid flows. Upon push-switch command, valve opening was triggered by a falling edge of the square-wave signal. Following an 8 s delay, the solenoid valves closed to terminate a spraying event comprised of equal numbers of positive and negative bipolar spray-charging pulses. Actuation time was generally less than ~ 3% of pulse width for all frequencies.

2.1.3. *Target system.* Thin-wall cylinders (5.08 cm o.d. x 4.76 cm i.d. x 25.00 cm length) were vertically positioned perpendicular to the spray axis at 125 cm spacing from the nozzle face. Composition for conductive and nonconductive targets were, respectively, brass and polycarbonate (2.1 x 10^{14} ohm·m resistivity; 3.17 dielectric constant). A shrouded Teflon® insulator also of 5.08 cm dia. supported the targets at their lower end while a plastic cap covered their top end to prevent spray entry. When mounted, the 23.81 cm of cylinder length exposed to spray corresponded to 380 cm^2 target area.

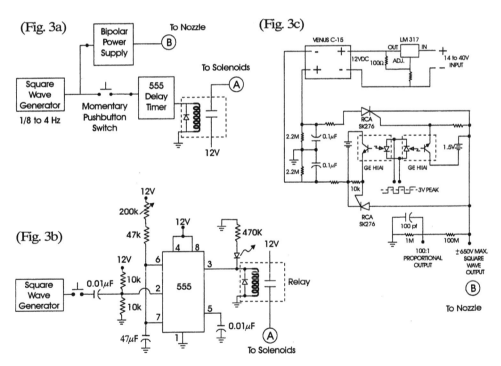

Fig. 3. Electronic controller for bipolar electrostatic spray system: a) block diagram of system; b) delay timer circuitry; c) variable frequency bipolar high-voltage power supply.

2.2. *Experimental Procedure*

2.2.1. *Fluorometric analysis.* Spray tracer deposited onto the targets was reclaimed into 300 mL of wash liquid (0.1% Triton X-100 surfactant in deionized water) by 15 min agitation in coaxial acrylic wash vessels [6]. Tracer deposition and its areal concentration (ng/cm^2) were subsequently quantified using a digital fluorometer (Turner model 450-005) exciting the tracer by quartz-halogen irradiation ($\lambda < 440$ nm) and measuring emission ($\lambda > 535$ nm) to include the tracer's 602 nm dominant fluorescence. Calibration standards established a linear (R^2=0.998) fluorometer response of: [(Fluorometer reading) = (0.686)(μg/L tracer concentration) - (18.238)]. Triplicate liquid samples from the wash cell for each respective sprayed target were fluorometrically analyzed, and daily calibration checks on freshly prepared 1500 μg/L tracer standard corrected the data for very slight instrument drift over time.

2.2.2. *Experimental design.* Using a randomized complete block statistical design, all 32 treatment combinations of the (4 target types) x (8 frequency settings) were completed within a given day; six such day-replications were made. Sprays were applied to the four earlier described targets at the following eight spray-charging levels: uncharged, 0 (*i.e.*, dc), 1/8, 1/4, 1/2, 1, 2 and 4 Hz. Experimental data were analyzed by SAS General Linear Models procedure using a standard analysis of variance followed by Tukey's Multiple Range Tests (TMRT) to identify significant differences in mean values of deposition achieved by the various spray treatments.

3. Results and Discussion

Table 1 summarizes the mean values for areal concentration (*i.e.*, deposition density, ng/cm^2) of tracer deposited onto specific type targets by each spray-charging treatment; the accompanying TMRT results indicate which means significantly differ one from another at the $\propto = 0.05$ level. Figure 4 illustrates the trends in deposition density exhibited as a function of spray-charging frequency, while Fig. 5 presents these results after having been normalized respectively to the uncharged deposition value for each type target.

Table 1. Mean values of spray-tracer deposition density onto four target types for eight conditions of spray charging, (ng/cm^2).

Spray-Charging Condition	Target Type			
	Conductive – Grounded	Conductive - Isolated	Nonconductive	Nonconductive with Grounded Substrate
Unch.	250.2 (1.0) [D]	268.1 (1.0) [A]	208.4 (1.0) [A]	181.8 (1.0) [D]
0 Hz	1908.7 (7.6) [A]	240.0 (0.9) [A]	133.6 (0.6) [A]	1290.7 (7.1) [AB]
1/8 Hz	1979.2 (7.9) [A]	322.0 (1.2) [A]	197.7 (0.9) [A]	1477.1 (8.1) [A]
1/4 Hz	1445.0 (5.8) [B]	325.5 (1.2) [A]	227.6 (1.1) [A]	1214.8 (6.7) [AB]
1/2 Hz	1191.7 (4.8) [B]	335.5 (1.3) [A]	304.3 (1.5) [A]	948.1 (5.2) [BC]
1 Hz	755.3 (3.0) [C]	403.3 (1.5) [A]	313.2 (1.5) [A]	668.1 (3.7) [C]
2 Hz	451.1 (1.8) [CD]	346.8 (1.3) [A]	254.3 (1.2) [A]	299.1 (1.6) [D]
4 Hz	321.0 (1.3) [D]	317.1 (1.2) [A]	304.3 (1.5) [A]	277.9 (1.5) [D]

[1] Numbers in parentheses are ratios of deposition as normalized to uncharged deposition for each target type.

[2] Means followed by same letter are not significantly different by TMRT at α=0.05.

As seen, deposition onto both the grounded brass target and the nonconductive target with grounded substrate was dramatically increased by spray charging at dc, 1/8 Hz and 1/4 Hz. There was no significant difference ($\propto$ = 0.10) in the amount of deposition onto the brass target delivered by dc and by 1/8 Hz charged sprays; these two charging methods provided equally effective electrodeposition benefits averaging 8.4-fold significantly greater ($\propto$ = 0.01) than uncharged spray. On the nonconductive target with grounded substrate, deposition delivered by the dc, 1/8 Hz and 1/4 Hz charged sprays could not be declared significantly different ($\propto$ = 0.10) one from another; they delivered an average 8.1-fold significantly greater ($\propto$ = 0.01) deposit than did uncharged spray. For both type grounded targets, deposition by all charged-spray treatments was significantly ($\propto$ = 0.05) greater than by uncharged spray, except for the 2 Hz and 4 Hz charged sprays which deposited amounts of spray not significantly different ($\propto$ = 0.10) than by uncharged spray.

For the isolated brass target and the nonconductive target, no significant differences ($\propto$ = 0.10) could be declared among the mean values of deposition delivered by the eight spray treatments. Because of the appreciable experimental variability encountered with these targets, an additional statistical test for lack of fit in linear regression was made across the dc to 1 Hz frequency range for deposition onto the nonconductive target; this verified that no significant lack of fit ($\propto$ = 0.10) can be declared. Furthermore, a test of the positive slope of regression confirmed it to be significantly different from zero for this dc to 1 Hz range, thus providing rationale for concentrating additional experimental analysis of bipolar spray-charging there.

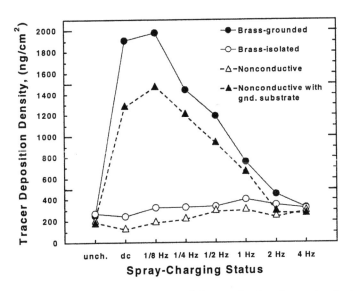

Fig. 4. Deposition onto four type target cylinders as a function of spray-charging status.

248

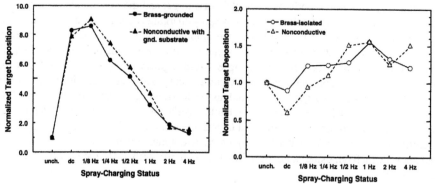

Fig. 5. Deposition values normalized to their respective uncharged-spray value.

4. Conclusion

This work experimentally documents that, as compared with conventional unipolar charged spray, a bipolar spray-charging strategy using relatively low square-wave frequencies of 1/8 Hz and 1/4 Hz will deposit equal amounts of surface coating, respectively, onto grounded metal targets and nonconductive targets having a grounded conductive substrate. As compared to similar uncharged spray, greater than 8-fold increases in deposition can be achieved at these two frequencies. Lesser, but still statistically significant, electrodeposition benefits can be had in the 1/2 Hz - 1 Hz frequency range; however, bipolar charging at 2 Hz and 4 Hz failed to provide deposition greater than uncharged spray, likely due to the turbulent intermixing and self-neutralization of the smaller bipolar sectors of the spray cloud in flight. Because there is zero net charge flux over time from a bipolar spray-charging nozzle to its target, improved engineering design options become possible to lessen grounding difficulties and charge-buildup problems on the operator of human-carried electrostatic coating systems operating in the 1/8 Hz to 1/4 Hz band. For nonconductive targets and electrically isolated targets, additional studies are required to identify optimal bipolar spray-charging frequencies and to better elucidate the role of target charge-relaxation.

Acknowledgements

This work was supported in part by funds provided by the Georgia Agricultural Experiment Stations. Appreciation is expressed to Mr. Patrick Harrell for assistance in fabrication of experimental apparatus and graphics preparation as well as to Mr. Michael Diaz for assistance in statistical analysis.

References

[1] Law SE 1987 *Rational Pesticide Use* (Cambridge: Cambridge University Press) 81-105.
[2] Bailey AG 1988 *Electrostatic Spraying of Liquids* (New York: Wiley & Sons).
[3] Law SE 1995 *Handbook of Electrostatic Processes* (New York: Marcel Dekker) 413-440.
[4] Cooper SC and Law SE 1987 *IEEE Trans. on Industry Applications* **23** 217-223.
[5] Law SE 1978 *Trans. of ASAE* **21** 1096-1104.
[6] Giles DK and Law SE 1985 *Trans. of ASAE* **28** 658-664.

Inst. Phys. Conf. Ser. No 163
Paper presented at the 10th Int. Conf., Cambridge, 28–31 March 1999
© *1999 IOP Publishing Ltd*

Marking with electrostatics

D A Hays

Wilson Center for Research and Technology, Xerox Corporation, 800 Phillips Rd., Webster, NY 14580 USA

Abstract. The early evolution of electrostatic marking is traced to the invention of xerography by Chester Carlson in 1938. Electrostatic marking technologies use an electrostatic force acting on pigmented toner particles to form images. Several classes of electrostatic marking physics can be described. The dominant class utilizes an image dependent electric field acting on charged, insulative toner particles. Electrostatic marking systems using different classes of marking physics can be grouped as either indirect or direct. The quality of images produced with electrostatic marking depends on the control of toner particle charge and adhesion. Triboelectricity is the dominant method for charging particles, in spite of a poor understanding of microscopic charging mechanisms and the enhancement of adhesion due to a non-uniform surface charge.

1. Introduction

Copying and printing technologies based on electrostatic marking have evolved to high levels of performance over the past 40 years. Xerography is the dominant electrostatic marking technology with about 50 billion dollars a year in worldwide revenue. This application of electrostatics represents a major innovation of the century. Electrostatic marking is based on a variety of phenomena, some of which lack a complete scientific understanding. In spite of this shortfall, continuous technology improvements have enabled high-quality, full color copying and printing at rates up to ~400 pages per minute.

Early discoveries of fundamental electrostatic phenomena provided a foundation for the invention of xerography in 1938. Although xerography has emerged to be the dominant electrostatic marking technology, several classes of electrostatic marking physics can be described. Furthermore, various electrostatic marking systems can be configured into different architectures. In spite of advances in electrostatic marking systems, challenges remain in controlling toner particle charge and adhesion.

2. Early discoveries in electrostatic marking

George Christoph Lichtenberg, a German physicist at Göttingen University, discovered in 1777 that an electrostatic pattern formed by a spark near an insulating material could be developed with powder [1]. The first studies in recording images with electrostatic marking principles were carried out much later in the 1920s and '30s by Paul Selenyi, a physicist at the Tungsram Laboratories in Budapest, Hungary. Powder was used to develop an electrostatic

250

image on an insulating material formed by an ion source scanned over the surface as the ion flow was modulated with a control electrode. This "electrographic" recording method enabled the first facsimile image transmission.

The work of Selenyi was influential in the invention of "electrophotography" (later to be called xerography) by U.S. physicist Chester Carlson on October 22, 1938. The invention of xerography represented the insight of using a photoconductive material to produce an electrostatic image that is developed with a charged, pigmented powder. Carlson built a prototype of a copying machine but was unsuccessful in attracting the interest of large corporations. In 1944, Carlson arranged an agreement with Battelle Memorial Institute to explore the process. Technology advances made at Battelle in xerographic materials and processes led to the introduction of the successful Xerox 914 plain paper copier in 1959 [2].

3. Classes of electrostatic marking physics

Electrostatic marking technologies utilize an electrostatic force acting on pigmented particles to form an image that is deposited on a medium such as paper. For discussion purposes, the electrostatic force acting on the particles can be approximated as

$$\vec{F} \cong Q\vec{E},$$
(1)

where Q is the charge on the particle and E is an applied electric field.

To form an image with charged particles, a variation (spatial or temporal) in the electrostatic force is required. For insulative particles, the variation in F may be decomposed into two parts as [3]

$$\delta\vec{F} = Q(\delta\vec{E}) + (\delta Q)\vec{E}.$$
(2)

The first term in Eq. (2) describes a class of electrostatics marking physics in which a variation in E acting on charged particles causes an image dependent variation in the electrostatic force. The xerographic process as illustrated in figure 1 represents a good example of the Q (δE) class, in which the spatial variation in δE is produced by image dependent light exposure of a precharged photoreceptor. An important attribute of this class is that a reversal of the electric field in non-image areas provides an electrostatic force directed away from the image receiver, as desired for preventing background development.

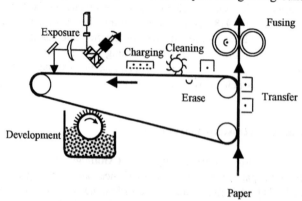

Figure 1. The six basic xerographic steps of charging, exposure, development, transfer, fusing and erase/cleaning are illustrated for a belt photoreceptor.

Another class of electrostatic marking physics is described by the second term in Eq. (2) in which an image dependent variation in the electrostatic force is obtained through a variation in the particle charge for a constant electric field. Although several imaging systems based on this electrostatic marking class have been proposed, no products have been brought to market to date.

If the pigmented particles are conducting and in contact with an electrode or other conductive particles when an electric field is applied, a charge will be induced on the particles that is proportional to the applied electric field. This represents a third class of electrostatic marking physics in which the variation in F becomes

$$\delta \vec{F} \propto \delta \vec{E}^{2}. \tag{3}$$

Note that the force is positive for either polarity of the electric field. A non-electrostatic force such as a magnetic force is usually needed to prevent particle deposition in the non-image areas.

4. Types of electrostatic marking systems

Figure 2 displays several types of electrostatic marking systems based on the $Q(\delta E)$ and δE^{2} classes of electrostatic marking. The system types can be grouped under headings of either "Indirect Printing with Latent Image" or "Direct Printing with Pigment". For indirect printing systems, a spatially varying latent electrostatic image ($Q(\delta E)$ class) is formed on a receiver that is subsequently developed with charged, pigmented particles. The electrostatic image can be formed by either photon exposure of a charged photoreceptor, as in xerography, or ion/electron deposition onto an electroreceptor (dielectric receiver), as in ionography.

For systems that utilize direct printing with pigmented material, an aperture array provides transient electric fields for image dependent control of charged, pigmented particle (or droplet) deposition onto a receiver. For these system types, the imaging and development occur virtually simultaneously. For systems utilizing charged particles, the transient electric fields provide an electrostatic force ($Q(\delta E)$ class) for image dependent control of pigmented particle deposition, as in the TonerJet® technology [4]. If the pigmented material is conducting, the transient electric field induces a charge on the material that provides an electrostatic force of the δE^{2} class. Examples of this system type include electrostatic ink jet printers and the recently announced Océ Direct Imaging technology [5].

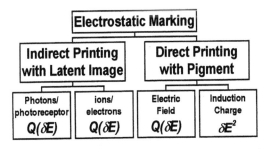

Figure 2. Types of electrostatic marking systems.

252

5. Challenges in controlling toner particle charge and adhesion

It should be evident that the image quality of electrostatic marking systems is particularly dependent on the ability to control the charge on toner particles [2]. Triboelectricity is the dominant method for charging the particles [6]. For two-component development systems, the particles are triboelectrically charged by mixing with larger carrier beads. For monocomponent development systems, the triboelectic charging is obtained with a compliant blade in rubbing contact with a toned roll [6]. A typical triboelectric charging level is 3 fC or 20,000 electronic charges for a 7 µm particle. (This is a charge imbalance of only 1 electron per 10^4 surface atoms!) The toner charge depends on a number of factors including the bead materials, toner resin and colorant, charge control agents, flow agents, relative humidity, and toner concentration. The optimum materials are chosen through empirical studies, since the microscopic mechanisms of triboelectricity are poorly understood.

The adhesion of charged particles must be sufficiently low to enable the movement of particles from one surface to another with an electrostic force. The adhesion depends on the particle's size, shape and surface roughness, toner materials, flow agents, charge level and the surface charge distribution [7]. An electrostatic component of the adhesion is obtained when the charge on the particles induces image charges in nearby materials. The adhesion is particularly high if the surface charge distribution is highly non-uniform due to triboelectric charging of irregularly shaped particles.

6. Summary

Xerography is the dominant electrostatic marking technology utilizing the $Q(\delta E)$ class of electrostatic marking physics. The technology has had a large impact on our society in facilitating the exchange and documentation of information. It is curious, however, that such a major innovation is based on the phenomenon of triboelectricity for charging toner particles, considering the fact that there is no clear understanding of the microscopic charging mechanisms.

References

[1] Carlson C F 1965 *Xerography and Related Processes*, edited by Dessauer J H and Clark H E 15-49 (New York: Focal Press)

[2] Schein L B 1996 *Electrophotography and Development Physics*, rev. 2nd ed. (Morgan Hill, CA, Laplacian Press)

[3] Schmidlin F W 1972 *IEEE Transactions on Electron Devices* **ED-19** 448-57

[4] Larson O 1997 *IS&T's NIP 13: International Conference on Digital Printing Technologies* 737-39 (Springfield, VA: IS&T)

[5] Geraedts J and Lenczowski S 1997 *IS&T's NIP 13: International Conference on Digital Printing Technologies* 728-31 (Springfield, VA: IS&T)

[6] Gruber R J and Julien P C *Handbook of Imaging Materials*, edited by Diamond A S 159-200 (New York, Basel, and London: Marcel Dekker, Inc.)

[7] Eklund E A, Wayman W H, Brillson L J and Hays D A 1995 *Electrostatics 1995* Inst. Phys. Conf. Ser. **143** 85-92 (Bristol and Philadelphia: IOP Publishing)

Inst. Phys. Conf. Ser. No 163
Paper presented at the 10th Int. Conf., Cambridge, 28–31 March 1999

Enhancement of Transfer Efficiency and Appearance in Automotive Powder Coating

M. K. Mazumder, R. A. Sims, D. L. Wankum, T. Chasser*, N. Grable, W. Chok, S. Robbins, W. Gao, and G. Tebbetts

University of Arkansas at Little Rock, Department of Applied Science, ETAS Bldg. 575, 2801 S. University, Little Rock, AR 72204, Tel: 501-569-8007, Fax: 501-569-8020, E-mail: mazumder@eivax.ualr.edu

*PPG Industries, 4325 Rosanna Drive, Allison Park, PA 15101, Tel: 412-492-5348, Fax: 412-492-5522

ABSTRACT Powder samples with volume median diameter ranging from 15-25 μm with different standard deviations were tested to examine how PSD influences the Q/M distribution, flow, and transfer efficiency of a powder, and finally the appearance of the films on polymer coated metal panels. Even with a minor difference of PSD within a narrow particle size range, a significant difference was observed in the flow, charging, transfer efficiency, and appearance characteristics of clearcoat powders.

1. Introduction

Transfer efficiency (TE) and Appearance (AP) in powder coating are both functions of physical and electrostatic properties of powders. The functions relating the variables are not the same for TE and AP, and when trying to simultaneously maximize both TE and AP, there are conflicts within these functions.

The appearance of a powder coated film depends upon the following factors: 1) physical properties of powder and the application process, 2) powder chemistry and melt rheology, and 3) the curing process. Previous studies [1] have shown that the appearance of the powder coated film was comparable to that of a solvent-based spray process when a narrow particle size distribution with d_{50} close to 10 μm was used. However, the improvement of appearance as particle size decreases is offset by poor performance in the dispersion and flowability of the powder. As particle size decreases, the powders become more cohesive and show more agglomeration, impact fusion, and high charge-to-mass ratio for a given applied voltage; consequently, there is an early onset of back corona.

As the thickness (t) of the powder layer increases during the deposition process, it affects the mechanisms that influence the appearance, often in a conflicting manner. As t increases, only large particles with high charge can deposit on the surface causing the top layer of the powder to become increasingly coarse as the thickness increases. Second, the deposition of charged powder on the target increases the electric field within the powder layer, and when the field exceeds the dielectric breakdown field, it leads to back corona. Third, a thick powder layer can, in spite of its coarse top layer and minor back corona effects, produce a very smooth film because of the enhanced melt rheology of a thick layer.

2. Appearance of the Powder Layer and Cured Film

2.1 The Product of (Q/M) of the Powder and the Thickness (t) of the Powder Layer Deposited on a Substrate

For a given dielectric constant of the deposited powder layer, the electric field strength E within the powder layer can be expressed as

$$E = \frac{(Q/M)\rho_a t}{\varepsilon_o \varepsilon_r} \tag{1}$$

where (Q/M) is the charge-to-mass ratio of the powder layer of thickness (t) deposited on the substrate, ρ_a is the packing density or the solid fraction of the deposited powder layer, ε_o is the dielectric constant of free space, and ε_r is the relative dielectric constant of the powder layer. For a powder layer thickness of 100 μm (cured film thickness of about 2mils) and the relative dielectric constant of the powder ($\varepsilon_r = 2$), the product (Q/M)t has an upper limit:

$$\frac{(Q/M)t\rho_a}{\varepsilon_o \varepsilon_r} \leq E_b , \tag{2}$$

where E_b is the dielectric breakdown field. The maximum Q/M must not then exceed 1.77 μC/g if the electrical field for breakdown for the powder layer is 10^7 V/m. The interstitial air will undergo breakdown within a powder layer when E exceeds E_b. If the powder is applied to an e-coated or primer coated panel, the electrical field for breakdown is approximately 10^7 V/m. However, for powder deposited on a bare aluminum panel, the interstitial space between the particles may cause breakdown as low as 3×10^6 V/m (breakdown electric field for air). Under this condition, the (Q/M) of the powder layer should not exceed 0.54 μC/g when t = 100μm. A safe practice will be to have the maximum Q/M of the powder not to exceed 0.54 μC/g for 100μm powder layer thickness for avoiding back corona. Therefore, (Q/M)t ≤ 177 for e-coated surfaces, and (Q/M)t ≤ 54 for bare metal surfaces when Q/M is expressed as μC/g and t is in μm. The minimum value of Q/M should be 0.3 μC/g, a suggested value for good adhesion of the powder layer based on previous studies.

2.2 Free Ion Current and Charge Decay

The ions deposited on the powder layer will increase the Q/M of the powder unless the charge decays to the substrate as rapidly as the build up occurs. The charge decay current is limited to only a few nano-amps across the powder layer because of its high resistivity ($10^{15}\Omega m$). The actual Q/M of the deposited powder is influenced by the initial charge and free ionic charge. Laboratory tests have shown that the Q/M of powder collected from test panels is as high as 6 μC/g, with initial Q/M less than 1 μC/g. Even though the free ions do not flow to ground across the powder layer, there is a corresponding image charge flow to the grounded substrate, creating a high-intensity electric field across the powder layer.

In a corona spray process, the maximum spray time depends upon the surface ion current density (J) due to the free ions that deposit on the surface of the powder layer. The typical value of the ion current, at a corona voltage under -60 kV, is approximately 2-6 microamps/m^2. The total amount of additional charge deposited by the free ions would be

$$Q_i = J \, T(s) \quad Coulomb/m^2 \tag{3}$$

where Q_i is surface charge density caused by the free ions.
The maximum time for spraying, therefore, will be

$$T(s) = (Q_{SM} - Q_c)/J \tag{4}$$

where Q_c is the surface charge of the particles acquired by corona charging and Q_{SM} is the maximum surface charge density before air breakdown occurs.

3. Experiments

Charge-to-mass ratio (Q/M) was measured using a corona gun for charging the powder, a grounded metal screen to collect free ions, and a Faraday cup with appropriate sampling techniques. To account for and eliminate the influence of free ions and back corona, measured values of Q_1/M_1, Q_2/M_2, Q_3/M_3... Q_n/M_n for a sampling time interval of T_1, T_2, T_3... T_n were plotted against the powder mass collected, M_1, M_2, M_3... M_n, respectively. When there is no free ion current and back corona, the slope of the plot Q/M versus mass of the powder collected is zero, showing that the Q/M is independent of sampling time. However, when free ion current is present (as in the case of a corona gun), Q/M changes with sampled mass m for powders of high electrical resistivity.

Experimental data show there is a linear relationship between Q/M and M of the collected charged powder samples. When we extrapolate the value of Q/M to M = 0, we obtain Q/M in the limit M close to zero, giving the initial value of Q/M in the absence of back corona. The slope of the curve depends upon the intensity of back corona caused by high resistvity of powder.

The resistivity of powder was determined by measuring the charge decay rate of powder deposited on a metal plate. The charge decay rate was measured by placing a powder coated test panel over a grounded metal substrate inside a chamber and plotting the decay of electric field with a field meter. The measured resistivity was greater than $10^{15} \Omega m$ for both powders.

The first pass transfer efficiency (FPTE) was measured using a 1m x 1m test panel and determining the mass of the powder collected on the panel as a function of the total amount of powder sprayed. Ratio of the mass of powder deposited on the test panel to the mass of powder sprayed provided the FPTE values.

The appearance was measured using a BYK Wave Scan Meter on a scale of GM Tension 0 to 20, where 20 represents best film appearance.

At least five experiments were conducted, and the average value was taken.

4. Results

The experimental results based on two powders are summarized in the following table. The two powders are of the same chemistry but were classified under different conditions so the mechanical properties are different for the two. POWDER A has larger particles with a narrow size distribution compared to the PSD of POWDER B, which has many more fines.

The narrow size distribution of POWDER A yielded a significantly better transfer efficiency than POWDER B. With a slightly reduced transfer efficiency, the additional fines in POWDER B yielded a much better appearance, especially with the thinner film compared to that of POWDER A.

Table 1 Summary of Powder Characteristics, Transfer Efficiency, and Appearance

		POWDER A			POWDER B		
PSD	d_{10} μm	8.56			4.38		
	d_{50} μm	22.43			15.13		
	d_{90} μm	39.20			33.53		
Q/M μC/g	-40 kV	-0.45			-0.55		
	-60 kV	-0.58			-1.12		
	-80 kV	-0.77			-1.20		
FPTE At –60kV	Nozzle	%	Std Dev		%	Std Dev	
	Fan	90.8	1.57		81.0	1.94	
	Cone	96.1	1.91		85.7	1.41	
Appearance at –60kV		Film Thickness mils	GM Tension	Std Dev	Film Thickness mils	GM Tension	Std Dev
		0.91	10.1	0.3	0.62	13.5	0.2
		1.37	15.4	0.5	1.40	17.8	0.5
		2.13	18.0	0.4	2.01	18.3	0.2

5. Conclusion

Experimental data show that, with appropriate adjustment of high voltage applied to the corona gun to control the free ion current and the thickness of the powder layer, it is possible to obtain both high FPTE and excellent film appearance with a proper choice of PSD. The achievement of high FPTE and excellent appearance using powder with d_{50} in the range 15 to 25μm is significant in automotive clear coat applications, since fine powder with d_{50} = 10μm is difficult to fluidize and disperse.

6. References

[1] Bailey A. G. 1998 *Journal of Electrostatics* **45** 85-120
[2] Sims R. A., Mazumder M. K., Liu X., Wankum D. L., Pettit P., and Chasser T. 1997 *Conference Record* IEEE Industrial Applications Society **Vol 3** 1697-1704
[3] Mazumder M. K., Wankum D. L., Sims R. A., Mountain J. R., Chen H., Pettit P., and Chasser T. 1997 *Journal of Electrostatics* **40-41** 369-374
[4] Chen H., Sims R. A., Mountain J. R., Burnside G., Reddy R. N., Mazumder M. K., and Gatlin B. 1996 *Journal of Particulate Science and Technology* **14-3** 239-254

Inst. Phys. Conf. Ser. No 163
Paper presented at the 10th Int. Conf., Cambridge, 28–31 March 1999
© *1999 IOP Publishing Ltd*

Factors influencing operational efficiency of an electrostatic solvent extraction system

M J Reeves[a]**, A G Bailey**[a]**, A G Howard**[b]** and C J Broan**[c]

[a]Dept. of Electrical Engineering, University of Southampton, Highfield SO17 1BJ, UK
[b]Dept. of Chemistry, University of Southampton, Highfield SO17 1BJ, UK
[c]Research and Technology, B709, BNFL, Springfields, Preston, Lancs. PR4 0XJ, UK
E-mail: mjr@soton.ac.uk. Fax: 44-(0)1703-593709.   Project sponsor: EPSRC

Abstract. A continuous electrostatic liquid-liquid solvent extraction process is being studied for the concentration of metals from dilute solution, as required for the recovery of toxic or valuable metals. Complementary extraction and stripping operations occur simultaneously in cells separated by a baffle. Aqueous crossover from the extraction cell to the stripping cell dilutes the product and must be minimised whilst maintaining high mass transfer of ions. Crossover increases with the electric field intensity, partly due to an increased number of smaller, penetrative droplets. It is evident that inlet droplets have net charges as they settle under gravity with rapid transverse oscillatory motion. Oscillation amplitude depends on both field strength and frequency, although oscillation frequency is lower than the applied frequency. At low frequencies, droplet-electrode contact commonly occurs, increasing the oscillation speed and encouraging crossover. This implies that charge has been exchanged at an insulated electrode. Oscillation is greatly reduced above 100 Hz, and crossover is decreased. Above 220 Hz, fine dispersions with small oscillation amplitudes result in lower aqueous crossover. The electric field across the baffle also influences aqueous crossover. The presence of a baffle field (above the critical field strength) causes aqueous crossover into the stripping side to increase with the stripping-side electric field.

1. Introduction

The electrostatic solvent extraction system described here is a development of a previously reported technique [1], [2], [3]. The process permits the enrichment of metal from dilute solution. It is potentially suitable as a pre-concentration step to electrowinning, and could process waste water from other hydrometallurgical operations. Extraction and stripping operations occur concurrently, breaking the chemical equilibrium limitation inherent in liquid-liquid extraction and allowing the residing organic extractant solution to be regenerated. The adjacent extraction and stripping chambers (Figure 1) receive respective aqueous feed and stripping droplets, delivered through non-immersed PTFE nozzles. The droplets in each cell pass between a pair of PTFE-insulated electrodes with high alternating voltages applied. Insulation restricts sparking, chemically isolates the electrodes and prevents bridging of droplets across a chamber through its non-wettability. Disintegration of inlet material into many small droplets raises the interfacial area and thus the mass transfer rate between the immiscible phases. Metal ions from the feed are extracted into the organic phase (1.5 l volume) and then reclaimed by stripping droplets.

A baffle, consisting of stacked elements of inverted-V cross section, divides the chambers and is positioned between wire-wound inner electrodes. This permits the organic phase to flow relatively unhindered throughout the unit whilst restricting the two aqueous phases from crossing compartments. The baffle is mounted on a support that divides settler zones, in which aqueous droplets coalesce out of the electric domain. Metal-depleted feed and aqueous concentrate outlet streams pass through siphon-breakers that allow settler height control. Two centre-grounded transformers with sinusoidal outputs are used to apply electric fields across the chambers and across the baffle region.

2. Electrostatic agitation

The application of an electric field polarises droplets and they deform into prolate spheroids with major axes parallel to the field. Stress increases with the field until there is zero net interfacial tension and disintegration occurs [4]. Below the critical field strength required for break-up, droplets periodically deform and relax under an alternating field. One of the droplet behavioural characteristics observed in this solvent extraction system is the near-horizontal translation of dispersed droplets from electrode to electrode in the alternating field, above the critical field strength, in addition to droplet deformation. These motions increase mass transfer, and achieving dispersions in this way uses less energy than mechanical agitation.

3. Experimental metal separation

1 mg/min nickel has been successfully processed at high efficiency using the unit. 10 vol% D2EHPA organic extractant in Isopar M diluent was contacted with a feed solution of about 100 mg/l metal with 8.0 and 1.5 g/l sodium acetate and acetic acid respectively. These added components benefit extraction by restricting the pH decrease when protons are released by D2EHPA during metal complexation. Acid stripping solution (pH 0) was used to recover nickel at a flow rate of about one-eleventh that of the feed. The stripping acid required a larger field strength {7 kV/cm (RMS)} than the feed {4 kV/cm (RMS)} for a similar degree of dispersion, indicating that the acid-organic interfacial tension was greater than the feed-organic interfacial tension. Figure 2 shows the approach to steady-state (when extraction and stripping rates become equal) of the concentrated nickel outlet.

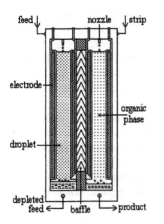

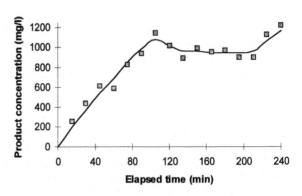

Figure 1: Electrostatic separator   Figure 2: Aqueous nickel separation

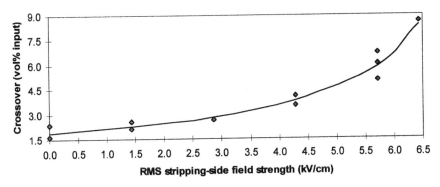

Figure 3: Effect of stripping-side field on aqueous crossover (110 Hz, 5 kV/cm RMS baffle field)

Atomic absorption spectrophotometry was used for the metal analyses. The steady-state concentration of metal in the organic phase was less than half of the feed metal concentration.

4. Experimental determination of aqueous crossover

After filling the unit with Isopar M, water was fed at a constant rate into the extraction chamber only. Under this simple mode of operation, crossover was defined as the volumetric percentage of water from the feed reaching the stripping side.

Raising the stripping field strength (Figure 3) increased crossover into the stripping chamber. The rate of crossover rise increased with field strength. However, without the baffle field present the data obtained was quite random. At zero stripping field, large coalesced droplets were seen falling out of the baffle into the stripping side. Generally, crossover increased with the field intensity on the extraction side, as did the probability of electrode-to-electrode droplet translations. There was a critical RMS field strength of about 3.5 kV/cm (Figure 4), below which crossover was negligible. From this value up to over 5 kV/cm (RMS) there was a transient region in which only a proportion of the droplets were shattered. From 5.5 to 6.5 kV/cm (RMS) the feed droplets were consistently disintegrated, with their sizes decreasing and crossover becoming increasingly more uniform over the baffle. At constant applied frequency there was little difference in results, irrespective of baffle field strength.

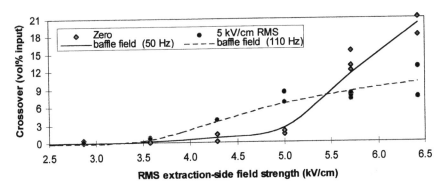

Figure 4: Effect of extraction-side field on aqueous crossover

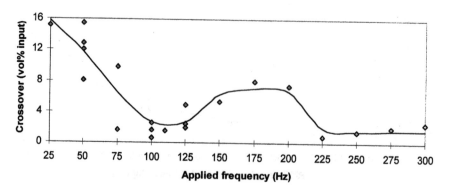

Figure 5: Effect of frequency on aqueous crossover

Regarding the effect of applied frequency on crossover, it is useful to consider the dispersion as a variable mixture of droplets of high transverse velocity that contact the electrodes, and orderly droplets that stay near the centre of the compartment. Without a baffle electric field, crossover generally decreases with an increase in frequency (Figure 5). At 25 to 50 Hz, nearly all the droplets hit the electrodes repeatedly on the way down to the settlers. Above 100 Hz, where crossover was at a minimum, most droplets were translating with an amplitude less than the distance between the electrodes. The 150 to 200 Hz plateau marked a significant change in droplet behaviour. Above 225 Hz, the dispersion was very fine and droplets did not deviate far from the centre of the chamber. Crossover was consistently low.

5. Conclusions

The applied frequency and electric field strength across a dispersion-filled chamber can be used to control droplet motion. Highly charged, high velocity and small diameter droplets that hit electrodes are encouraged by a low frequency, high strength field. This behaviour is beneficial for solute mass transfer, but detrimental to separation efficiency. It may be appropriate to opt for a higher frequency system that gives more orderly droplet behaviour to restrict crossover, but with a mass transfer penalty. The frequency of transverse droplet motion seems to increase with the applied electrical frequency, since droplet deviation from the centre of a compartment is reduced when the field changes direction more rapidly. Electrostatic forces of repulsion between recently fragmented droplets could be sufficient to hurl some of them into an electrode. For smaller transverse motions where droplets do not exchange charge with the electrodes, they must first have acquired a net charge, unless shape distortions are responsible for translation. Nozzle tips could accumulate induced charge in proximity to the field and charge dispensed droplets.

References

[1]   Gu Z-M, Xu M-X, Zhu L-Y & Jin L-R 1985 *Annu. Report of China Inst. of Atom. Energ.* 110-3.

[2]   Williams T J, Bailey A G & Broan C J 1997 *J. of Electrostatics* **40&41** 729-34.

[3]   Broan C, Hoskin S, Bailey A, Williams T & Reeves M 1998 *Int. Conf. on Process Innov. & Intens.*

[4]   Scott T C, Basaran O A & Byers C H 1990 *Ind. Eng. Chem. Res.* **29** 901-9.

Inst. Phys. Conf. Ser. No 163
Paper presented at the 10th Int. Conf., Cambridge, 28–31 March 1999

Electric Field Interaction with Debris Generated by Laser Ablation of Polyimide Films

C J Hayden, R Pethig and J P H Burt

Institute of Molecular and Biomolecular Electronics
University of Wales, Bangor, Gwynedd LL57 1UT, U.K.

Abstract Laser ablation is a method of machining structures to micron resolution. This technique is being used to micro-machine Biofactory-on-a-chip devices for manipulating bio-particles. The laser ablation process removes material efficiently, but also generates debris that degrades the micro-structures produced. We have found that externally applied fields in the range 0.7 to 1 MVm^{-1} can be used to control debris deposition.

1. Introduction

Laser ablation is currently being used as a method of micro-machining biofactory-on-a-chip devices [1]. These devices utilise electric fields to generate dielectrophoresis forces for manipulating very small particles such as cells and bacteria. In order to do this the devices require electrode and fluidic channel features that are similar to the dimensions of cells, i.e. microns in size. Various materials are employed to form the required structures, including plastic insulator films, resists and conducting metal layers.

Constructing these devices is challenging and requires specialist equipment. The system used in our work consists of an Exitech Series 8000 microfabrication workstation with a Lambda Physik Compex 110 krypton-flouride excimer laser. The workstation allows the workpiece to be positioned both horizontally and vertically to an accuracy of 0.1 micron, under computer control. The laser generates 248nm light pulses that are guided to the workpiece via a beam profiling mask and projection lens that allow the incident laser beam to be shaped. The masks used are typically chrome on quartz, which are mounted in an X-Y motion stage that has a total movement of 300mm x 300mm. An attenuator provides for accurate control of the intensity of the laser and a homogenizer utilises a 6 x 6 lens array to average the total beam power. This provides a pulse energy variation of less than ±5% RMS in 86% of the total beam area. The maximum fluences that are attainable at the workpiece are around 3 to 10 J cm^{-2}, depending upon the magnification of the projection lens used.

This system allows various materials to be accurately machined using laser ablation. Typically, plastic films are used to either separate the electrical contacts (machined in gold films) or to allow channels to be formed for fluidic transport within the Biofactory devices. A common insulator used in such devices is polyimide, either in sheet form or spin coated onto underlying structures.

When polyimide is ablated, the incident photons of light cause the polymer to de-polymerise through photochemical bond breaking. This, combined with other destructive forces, such as induced shockwaves and localised heating, results in the production of predominately small (<4 atoms) products and the ejection of this debris at high (supersonic) velocities [2]. This debris (mainly elemental carbon) can degrade the quality of the micromachined structures. An example of this is shown in Fig 1.

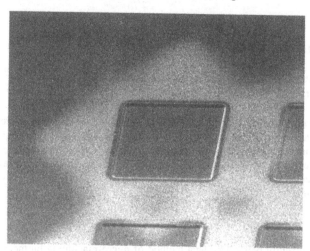

Fig 1. Scanning Electron Micrograph of a 50 micron square hole, ablated in a polyimide film. The debris can be seen to form a diamond pattern around the hole.

The debris generated can be seen as the pale, powder-like, material that surrounds or partially covers the ablated areas. In this report we describe how debris formation can be influenced by generating large, local, DC electric fields during the laser ablation process.

2. Experimental

Kapton HN® polyimide films, up to 100 micron thick, were mounted on glass substrates. Local electric fields were generated using 3.0mm ±0.01mm thick stainless steel electrodes (see Fig 2.) with a Fluke model 608B power supply capable of providing 6kV DC in 0.1V steps, with a maximum of 250mV ripple at full load.

To avoid electrical damage to the (earthed) laser workstation and associated systems, the electrodes were mounted in an insulated test area that in turn was mounted on a granite chuck. Glass apertures in the sides of the test area allowed use of the laser height measurement system and workpiece viewing cameras.

The same basic procedure was followed for all of the sample ablations. This consisted of a control test pattern being machined in the sample with no applied local electric field. The same test pattern was then machined in the polymer (at different sites) for different field strengths. The 'plumes' generated by the laser ablation process were photographed and estimated to extend about 1mm above the polyimide surface. The samples were then inspected using an optical microscope (OLYMPUS BX60). From this, images could be recorded directly to electronic format, using a video capture system. The samples were then gold-plated and examined using an ISI-40 scanning electron microscope.

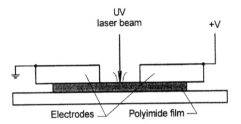

UV
laser beam   +V

Electrodes   Polyimide film

Fig. 2 Schematic of electrode setup used; inter-electrode spacing varied between 1 to 4 mm

3. Results

As shown in Fig. 3 the deposited debris pattern changed as the applied field was increased.

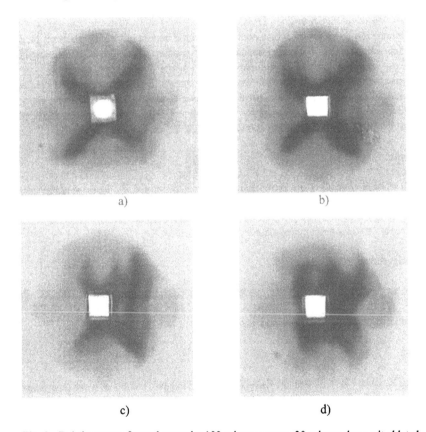

Fig. 3 Debris pattern formed around a 100 micron square, 30 micron deep, pit ablated in a polyimide film with a) zero b) 1kV c) 2kV d) 3kV applied to the electrodes shown in Fig. 2. See text for further details.

The debris can be seen to form 'petals' that surround the ablation area. With no applied field, the 'petals' are approximately symmetrical, and could result from the effect of shock waves thought to drive debris back onto the substrate [3]. As the applied field is increased

264

the debris 'petals' shift towards the right of the images until very little is deposited to the left of the ablated area. All images in Fig. 3 are oriented with the grounded electrode to the left and the positive electrode to the right of the ablated area. All of the sample ablations were in the middle of the inter-electrode gap (2.8mm), well away from the field fringing effects at the ends of the electrode strips. The electric field profiles were calculated using the Maxwell 2D field simulator (Ansoft, Pittsburgh). At mid-plume height (~0.5mm) the electric field at the centre of the ablated pit was found to be $0.35MVm^{-1}$, $0.68MVm^{-1}$ and $1.1MVm^{-1}$ for applied voltages of 1kV, 2kV and 3kV respectively.

4. Discussion

The results of Fig. 3 indicate that the trajectory of the ejected debris from the surface of the sample during ablation can be influenced by an external electric field. Because in all cases the debris particles are displaced towards the positive electrode, it can be deduced that either the particles themselves, or the associated plume of particles and gases carry a negative charge.

Although the particles of debris could be charged as a result of the ablation process producing free radicals (i.e. broken chemical bonds) another possible mechanism could be that of corona charging [4]. High voltage tests in normal atmospheric conditions on the electrode arrangement of Fig. 2 have shown that the threshold voltage for corona discharge occurs around 2kV (with complete electrical breakdown at around 3.5kV). Corona charging of the particles could therefore account for the non-linear relationship between the applied voltage and resulting debris deposition shown in Fig. 3.

5. Conclusion

The results of this investigation show that the debris generated by laser ablation can be affected by an applied DC electric field. Full control of this method will require a better understanding of the processes by which the debris particles are electrically charged. Corona charging appears to be a strong candidate for this process.

Acknowledgements

This work is funded by an EPSRC studentship (to C J Hayden) and a BBSRC grant (TO6337). We thank Prof. D M Taylor, A D Goater, M C Malpass, D Poirot, M S Talary and J A Tame for useful discussions and assistance.

References

[1] Pethig R, Burt J P H, Parton A, Rizvi N, Talary M S and Tame J A 1998 *J. Microtech. Microeng.* **8** 57-63

[2] Srinivasan R, Braren B and Dreyfus R W 1987 *J. Appl. Phys.* **61** 372-376

[3] Koren G and Oppenheim U P 1987 *Appl. Phys. B* **42** 41-44

[4] Taylor D M and Seeker P E 1994 *Industrial Electrostatics* Research Studies Press Ltd., Taunton. pp116-122.

Inst. Phys. Conf. Ser. No 163
Paper presented at the 10th Int. Conf., Cambridge, 28–31 March 1999

An experimental study on the possible use of the corona discharge in water treatment.

J Winkler[*], N Karpel Vel Leitner[**], H Romat[*]

[*] Laboratoire d'Etudes Aérodynamiques (U.M.R. 6609 C.N.R.S.)
Boulevard 3-téléport 2-BP179-86960 FUTUROSCOPE Cedex

[**] Laboratoire de Chimie de l'Eau et de l'Environnement (UPRES A 6008 C.N.R.S.)
ESIP, 40, Avenue du Recteur Pineau 86022 POITIERS Cedex - FRANCE

Abstract. We investigate the effect of the corona discharge on a layer of water. We study the evolution with time of a colouring (azocarmine B) mixed with the water and used to detect the creation of chemical radicals useful in the water treatment. The application of the corona discharge in the treatment of the water is studied.

1. Introduction

Nowadays the techniques used in water treatment are mainly based on oxidation processes. However, many substances remain refractory to the classical oxidants. Over the last decade, research has been done to implement new technologies called Advanced Oxidation Technologies (AOTs). These new technologies include electron beams irradiation and corona discharge, both having been recently used respectively in order to decompose organic compounds in water [1] and to remove volatile impurities from drinking water [2].
In this paper we relate the experiments done on water using the corona discharge technique. In the first part we describe the experimental device. Then we present the first experiment done with a grid used as an electrode brought to a high potential in order to obtain a corona discharge. The second experiment concerning the influence of the heights of the liquid and air layers is then presented with a new electrode system composed of needles. A configuration is chosen and the evolution with time of the decoloration of the solution is studied. Discussion on the results and the possibility of using this method in water treatment follows.

2. Experimental device

In brief we place a thin layer of liquid inside a cylinder and we impose a potential difference between two electrodes. In figure 1 we can see the different parts of the device : first the small cylindrical container whose internal radius and height are respectively 60mm and 30mm and into which the solution is poured and then the electrification system. This electrification system is itself composed of two parts, a dozen needles standing above the liquid and acting as a negative electrode and, below the liquid, the metallic base of the

266

cylinder which is grounded. The needles are embedded in a plain cylinder made of PTFE which fits exactly inside the other cylinder. All the needles are linked together to the power supply. A picoammeter gives the current. In order to limit the loss of electricity, the wall of the cylinder is made of PTFE (electrical insulator).

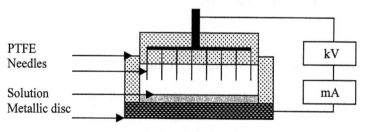

PTFE
Needles

Solution
Metallic disc

kV

mA

Fig. 1 Experimental device

The parameters that we can vary are the thickness of the solution, the distance between the surface of the liquid and the extremity of the needles, the electric potential and the concentration of the organic colouring azocarmine B which has been used because of its reactivity with radicals involved in the water treatment. The process of water treatment requires the presence of certain kinds of chemical radicals inside the water in order to destroy the organic compounds. The electrical species coming from the corona discharge are supposed to create those radicals whose presence will be detected by the decoloration of azocarmine B whose initial colour is red.

3. Preliminary experiments

In the first experiment we did, we had a metallic grid instead of the set of needles. Little holes had been driven in the grid in order to obtain tiny peaks on its surface. We poured a solution of 7.44 µmole/l of azocarmine B in pure water inside the cylinder, which gave a height of 8mm of liquid at rest. We placed the grid 5 mm above the surface of the liquid and applied the potential. We measured the current for different tensions and fixed it at 7000 V which is a value corresponding to a stable corona discharge between the grid and the counter electrode. With this potential, the solution of 7.44 µmole/l of azocarmine B was quite decoloured within 15 minutes and the current recorded was 0.4 mA. A spectrometric measurement of the concentration C of the solution gave a 70% reduction of the initial concentration C_0 : the electrical species involved in the mechanisms of the corona discharge have really reacted with the solution and directly or indirectly with azocarmine B.
We then changed the grid for the needles and did a few experiments in order to choose the most effective configuration. Varying the depth of the liquid hl, the distance ha between the extremities of the needles and the surface and the potential V, we finally concluded that the most effective configuration was the one with hl=4mm, ha=2mm and with potential V equal to 2.5kV. Indeed, in figure 2 (a) we can see that the three curves corresponding to ha = 2mm are below the two others, and among them the one corresponding to the smallest depth of water (hl=4mm) is the lowest one. In this configuration we preferred 2.5kV instead of 3kV

because the corona discharge gave rise to sparks of discharge at 3kV. (all these experiments were done with the same duration, 3mn).

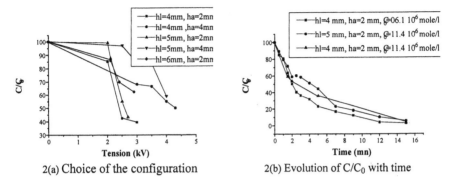

2(a) Choice of the configuration   2(b) Evolution of C/C_0 with time

<u>Figure 2</u> : Experiments on the decoloration of Azocarmine B

Another set of experiments were done for ha = 2mm and for different values of hl, varying V. The values of the current recorded showed that for ha = 2mm the current depended on the difference of potential and also that it did not change when hl varied. This is due to the difference of the order of magnitude of the resistivities of water and air. The global resistance of the total layer (air plus water) is more or less the resistance of the layer of air and therefore the current must not vary with the thickness of water. The same behaviour were recorded with ha equals to 4, 6, 8 and 10mm and in all the cases the current was obviously lower when the thickness of air increased. It was also the case for other thicknesses of water : the current does not depend on the thickness of the water whatever the case.

4. Evolution of the decoloration with time.

Following what we have just said above, we fixed the tension V to 2.5 kV, hl to 4 or 5mm, h_a to 2mm and we did experiments with two initial concentrations of colouring C_0 : 6.1 μmole/l and 11.4μmole/l. The results are given in figure 2 (b).
Whatever the thickness of water and the initial concentration of azocarmine B the evolution of the ratio C/C_0 is similar. The colouring has almost disappeared after 15 minutes.

5. Explanation and discussion

The corona discharge is a complex phenomenon creating ionic species whose characteristics depend on the medium where it occurs and on the potential applied to the electrode. In our needle-plate geometry as the potential increases we normally have first a pulse corona then a streamer corona then a glow corona and then a spark discharge [3]. In fact the phenomenon is a bit different depending on whether the potential is positive or negative. The objective of this article is not to study the mechanisms of the corona discharge, so we will not mention all the details of the differences between the two cases, Chang's publication [3] can help for more information on the subject. In our case, the potential was negative and because of the value of the potential fixed just below the potential of the breakdown, and also because of the glow that we observed continuously we can say, without having really investigated the phenomenon, that the experiments correspond to the glow corona discharge.

At this stage of our research we think that the decoloration of azocarmine B is due to the creation of radicals and active species like $OH°$, $H°$, O and O_3 which react in the water with the azocarmine B. According to Piskarev [4] who worked on similar problems, the active species would be first formed in the gas phase. The radicals originated in the gas would then undergo competitive reactions with themselves or with H_2O molecules or any substances in the vicinity of the surface layer of the water whose spatial extension is not more than 1mm. Among these species, the hydroxyl radical $OH°$ which is the basis of some AOTs is highly reactive with a wide class of compounds. This radical and also the dissolved ozone could be responsible for the removal of the colour during the treatment of the aqueous solution of azocarmine B. The attack of both ozone and $OH°$ radicals is generally directed to aromatic rings and conjugated double bonds and thus damages the chromophoric structures of the coloured compound. Owing to the poor selectivity of the active species generated towards organic compounds we can reasonably think that the process used in this work will act also with other many types of pollutants.

6. Conclusion

It was shown that a process based on corona effect allows the decoloration of an azo dye in aqueous solution. The use of this process involving highly active radical species like $OH°$ radicals merits being considered in the field of water treatment. However, numerous investigations are necessary for a better understanding of the parameters controlling the occurrence of the active species into the solution. The possibilities of optimised corona devices for the treatment of hazardous chemical wastes will then be considered with regard to the existing AOTs.

References

[1] Lubicki P., Cross J.D. and Jayaram S., Conference record of the ICDL. 12th International Conference on Conduction and Breakdown in Dielectric Liquids. Rome July 15-19 1996, p. 442-445.
[2] Al-Arainy A.A., Jarayam S. and Cross J.D., Conference record of the ICDL. 12th International Conference on Conduction and Breakdown in Dielectric Liquids. Rome July 15-19 1996, p.427-431.
[3] Chang J.S., Lawless A. and Yamamoto T., IEEE Transactions on Plasma Science, Vol. 19, n°6, December 1991, p. 1152-1166.
[4] Piskarev I.M., Sevat'yanov A.I., 6th International Frumkin Symposium Moscow 1995

Inst. Phys. Conf. Ser. No 163
Paper presented at the 10th Int. Conf., Cambridge, 28–31 March 1999

Physico-Dynamics in Electroatomiser Modelling

S. Viswanathan, H.C.Loy, K.K. Loh, and N.V. Ananthanarayanan
Department of Chemical and Environmental Engineering
10 Kent Ridge Crescent, Singapore 119260
Tel: (65) 8744309 Fax: (65) 8725483
E-mail: chesv@nus.edu.sg

and

W. Balachandran
Department of Manufacturing and Engineering Systems
Brunel University, Uxbridge, Middlesex, United Kingdom UB8 3PH
Tel: (01895) 203297 Fax: (01895) 812556
E-mail: emstwwb@brunel.ac.uk

Abstract
This paper focuses on the study of charged droplets in the presence of a strong hydrodynamic field by taking into account underlying physical phenomena associated with charged spray distribution. The foundation of the model was built around a Lagrangian particle-tracking scheme to simulate the formation of spray patterns for charged droplets. Various operating scenarios were simulated as a function of drop size, charge-to-mass ratio and supply voltage. This study indicated that the electric field has no effect on drop distribution in the strong hydrodynamic field for small drop sizes. In addition, charged liquid drops appear to spread better with increase in drop size and charge-to-mass ratio.

1. *Introduction*

The transformation of bulk liquid into sprays in a gaseous atmosphere is of importance in several industrial processes. Numerous spray devices have been developed and they are generally designated as atomisers or nozzles. Today, the use of electrostatically assisted nozzles is becoming popular in applications such as paint spraying, and wet scrubbers. In the case of wet scrubbers, electrical interaction mechanisms tend to play a major role along with impaction and diffusional scrubbing mechanisms. Pilat and Raemhild [1] have shown that the overall particle collection efficiencies vary from 25% (both particles and droplets uncharged) to 99.7% (both particles and droplets charged). The addition of electrostatic force gives rise to electrically augmented impact scrubbing, which is produced by an enhancement of relative velocity between droplet and particle [2]. This results in higher efficiency in the collection of sub-micron particles. It has also been shown that due to the enhancement of relative velocity between droplet and particle, the electro-charged water system would require a smaller liquid-to-gas ratio than a conventional type scrubber of equal collection efficiency. Although the use of such devices is becoming increasingly common, there is still a lack of fundamental understanding of physico-dynamics in electroatomiser modelling.

The objectives of this work are to:

1. Develop a model describing the dynamics and couplings of the physics driving the charged droplets.

2. Investigate the effects of drop diameter, charge-to-mass ratio and applied voltage on the spray pattern.

The focus is not on how drops are atomised or how charges are imparted, but on how the drops move and spread after leaving the nozzle under the influence of electric and hydrodynamic fields.

2. Problem Description

In wet scrubbers using electroatomisation systems, liquid is first atomised into tiny droplets using nozzles. The water droplets are then imparted with an electrical charge as they are released from the nozzle. The gas stream that is to be treated accelerates the charged droplets inside the scrubber. Based on review of literature [2, 3, 5, 7], the operating conditions were selected as given in Table 1.

S/No	Parameter	Value
1.	Nozzle orifice diameter	0.002 m
2.	Gas velocity	70 m/s
3.	Liquid flow rate	1.25×10^{-6} m^3/s
4.	Drop diameter	15, 30, 50, 70 μm
5.	Charge-to-mass ratio per drop	0.7, 1.4, 2.8 mC/kg
6.	Supply voltage	30, 60, 120 kV

Table 1: Operating Parameters

2.1 Solution Approach

As a first approach, the proposed model takes into account initial liquid momentum, hydrodynamic, gravitational and electric forces, and convective and diffusive mechanisms.

The foundation of the model is built around a Lagrangian particle-tracking scheme to simulate the formation of spray patterns for charged droplets. Steady-state spray patterns were computed using an iterative Particle Source in Cell (PSIC) approach, which represents momentum exchange between droplets and gas. Space charge on the surrounding electric field, caused by motions of charged particles, was computed as a natural extension of the PSIC method.

The solution approach was to consider a force balance (hydrodynamic, gravitational and electric forces) and mass conservation (continuity) with the assumptions that water drops are uniform in size; negligible interaction between the drops; an average space charge density for the spray cloud of drops; and drop movement in the axial direction by convection and in the lateral direction solely due to diffusion. The scrubber throat configuration and cell definitions used in the simulation are shown in Figure 1. For simplicity, a rectangular throat was considered. Symmetry along axial flow direction can be assumed since the nozzles are placed at the centre of the throat. Due to small nozzle to nozzle distance, negligible variation in drop concentration along the line of nozzle arrangement is expected. This makes the model two-dimensional and thus reduces the control volume for simulation.

2.2 Model Description

The process was modelled by considering the hydrodynamics effect along with the electrical force. The problem was first simplified by considering the water drops as being uncharged, and the solution was then extended to the case of charged drops.

2.2.1 Modeling for Uncharged Drops

The steady velocity profile for the water droplets can be written by making a force balance as [4]:

$$\frac{dv_D}{dx} = \frac{0.75C_{DN}\mu_G(v_G - v_D)}{D_d^2\rho_L v_D} + \frac{g}{v_D} \tag{1}$$

To solve for the drop concentration distribution in the x-y plane, a 2-dimensional mass balance (continuity equation) for a cell of drops is given by [4]:

$$\frac{\partial C_d}{\partial t} = -\frac{\partial}{\partial x}(v_x C_d) - \frac{\partial}{\partial y}(v_y C_d) + E_d\left(\frac{\partial^2 C_d}{\partial x^2} + \frac{\partial^2 C_d}{\partial y^2}\right) + Q_d \tag{2}$$

with labels: **bulk motion**, **diffusion**, **source**.

2.2.2 *Modeling for Charged Drops*

The solution for charged droplets takes into consideration the movement of charged drops in the presence of a strong hydrodynamic field. The presence of a space charge cloud of water drops sets up an electric field within the scrubber throat. As a result, these water drops experience an additional coulombic force [5]. The steady state velocity profile by making a force balance on the water droplets can now be written as[6],

$$\frac{dv_D}{dx} = \frac{0.75C_{DN}\mu_G(v_G - v_D)}{D_d^2\rho_L v_D} + \frac{g}{v_D} + \frac{\rho_m^q E}{v_D} \tag{3}$$

The continuity equation for the droplets remains unchanged. However, before computing the concentration profile the unknown electric field intensity E has to be determined first.

To determine electric field strength or intensity, the space charge potential V in the space charge cloud of drops, along the x-axis has to be determined. This can be achieved by solving the Poisson's Equation [5, 6] with appropriate boundary conditions as:

$$V = \frac{-\rho}{2\varepsilon_o\varepsilon_{rG}}x^2 + \left(\frac{\rho L_{TH}}{2\varepsilon_o\varepsilon_{rG}} - \frac{V_o}{L_{TH}}\right)x + V_o \tag{4}$$

where x = vertical distance from the origin.

Equation 4 contains an unknown term, the average space charge density ρ, that has to be calculated iteratively. An initial guess for the space charge density was calculated from the spray cloud formed by uncharged drops. This is therefore an overall average value for the entire spray cloud in the scrubber throat. With this initial value for average space charge density the electric field intensity can be calculated. This value is used to calculate the new velocity and concentration profile of charged droplets. Based on the computed concentration profile, a new average space charge density is calculated and compared with the previous value for convergence.

The criteria for convergence was set as $\dfrac{|\rho_{new} - \rho_{old}|}{\rho_{old}} \leq 0.1\%$

3. *Results and Discussion*

A number of simulation runs were conducted under various operating conditions as listed in Table 1.

3.1 *Effect of drop size and charge-to-mass ratio:*

To illustrate the effect of drop size on the charged spray pattern, the width of the spray plume is calculated at a throat length of 150 mm and plotted for different drop sizes in Figure 2. From the results, it is clear that for the same charge-to-mass ratio, the width of the spray reduces with increasing drop size. As the drop size increases the eddy diffusivity decreases. The electric field acting in the presence of a strong hydrodynamic field shows trends similar

272

to the uncharged case and is found to improve the spread only for large drop sizes. It can also be seen that an increase in charge-to-mass ratio does not have any effect on the spread for small drop sizes (below 30 μm). However the spread increased with increasing charge-to-mass ratio for drop sizes greater than 50 μm. This phenomenon could be explained by comparing the hydrodynamic force and the electric force. Figure 3 shows the simulation results for two widely varying operating conditions. For smaller drop sizes, the hydrodynamic force dominates over the electric force. However as the charge-mass ratio and drop size increases, the electric force becomes comparable to the hydrodynamic force. The electric force has a decelerating effect in the first half of the throat where the drag force is maximum, and an accelerating effect in the second half where the drag force is significantly lower. The combined effect of these factors decreases the rate of acceleration of the drops and improves the spread. Figure 4 shows a comparison of the dimensionless concentration contours in the entire throat for a typical charged and uncharged condition.

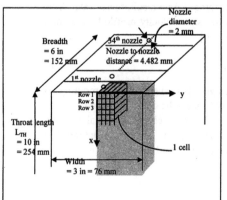

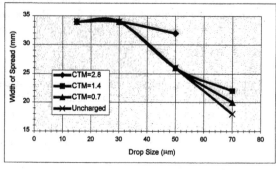

Figure1:Configuration of Scrubber Throat

Figure 2:Effect of Drop Size & Electric Charge on Spray Pattern

3.2 Effect of Supply Voltage

For the entire range of conditions simulated it was found that that the potential generated by the space charge cloud was much higher than the supply voltage. Hence, increasing the supply voltage has minimal effect on the electric field intensity and the electric force resulting in negligible variation in drop distribution.

However, this research is limited by the lack of an expression that can relate the charge-to-mass ratio to the supply voltage (among other parameters such as the drop size and liquid flow rate through the nozzle). A verification on the range and combination of operating conditions simulated, along with experimental data is required to substantiate the above findings.

4 Conclusions

A two-dimensional, steady-state, mathematical model was derived to simulate the flow pattern and behaviour of charged water drops accelerating in a high speed gas stream. The following conclusions can be drawn from the study:

1. The mathematical model provides reasonable insights into the underlying physical phenomena associated with the motion of charged spray.
2. The presence of an electric field has no effect on the droplet flow pattern in the strong hydrodynamic field for small drop sizes.
3. There is a better spread of charged liquid drops with the increase in drop size and charge-to-mass ratio.

Nomenclature

C_d	Drop concentration	drops/m^3
C_{DN}	Modified drag coefficient	-
D_d	Diameter of water drop	m
E	Electric field intensity	V/m
E_d	Eddy diffusivity of water drop in air	m^2/s
g	Gravitational acceleration	9.81 m/s^2
Q_d	Water drops source strength	drops/m^3-s
v_D	Velocity of water drop	m/s
v_G	Velocity of gas (air) at throat	70 m/s
v_x	Velocity component in the x-direction	m/s
v_y	Velocity component in the y-direction	m/s
V	Space charge potential	V
V_o	Supply voltage	V

Greek Symbols

Variable	Description	Value/Units
ε_0	Permittivity constant of vacuum	8.85 x 10^{-12} F/m
ε_{rG}	Relative dielectric constant of gas (air)	1.00054
μ_G	Viscosity of gas (air) at 30 ^{o}C	1.8464 x 10^{-5} N/m
ρ	Space charge density	C/m^3
ρ_G	Density of gas (air) at 30 ^{o}C	1.1768 kg/m^3
ρ_L	Density of liquid (water) at 30 ^{o}C	995.7 kg/m^3
ρ_m^q	Charge-to-mass ratio per drop	C/kg

References

1. Pilat M. J. & Raemhild G. A., Control of Particulate Emissions with U. W. Electrostatic Spray Scrubber, U. S. Environmental Protection Agency, Office of Res. & Dev., *Symp. on the Transfer & Utilisation of Particulate Control Technol., Vol. 3, Scrubbers, Advanced Technology and HTP Applications*, Report No. EPA-600/7-79-044C, PB-295-228, pp. 61-72, Feb 1979.
2. Aoki I., Sugahara T. & Matsuyama T., Collection Efficiency in Venturi Scrubbers by Utilising the Electro-charged Water for Scrubbing Liquid, *Technol. Rep., Kansai Univ., Osaka, Japan*, 1984, Vol. 25, pp. 91-96.
3. Anestos T. C., Sickles J. E. & Tepper R. M., Charge to Mass Distributions in Electrostatic Sprays, *IEEE Transactions on Industry Applications*, 1977, Vol. IA-13, No. 2, pp. 168-176.
4. Ananthanarayanan N. V. & Viswanathan S., Estimating Maximum Removal Efficiency in Venturi Scrubbers, *AIChE J*, 1998, Vol. 44, No. 11, pp. 2549.
5. Bailey A. G., *Electrostatic Spraying of Liquids*, Research Studies Press Ltd, 1988.
6. Loy H. C. & Loh K. K., Modeling of Electroatomisers, BEng. Thesis, National University of Singapore, Singapore, 1999.
7. Hu D. & Balachandran W., Non-wetting Electrostatic Assisted Nozzle for Spraying a Catalytic Melt, *Particulate Science & Technology*, 1997, Volume 15, pp. 1-12.

274

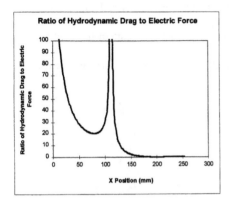

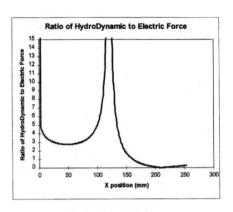

Drop Size = 15 μm;
Charge To Mass Ratio = 0.7 mC/kg;
Applied Voltage = 30 kV

Drop Size = 30 μm;
Charge To Mass Ratio = 2.8 mC/kg;
Applied Voltage = 60 kV

Figure 3: Effect of electric field in the presence of a strong hydrodynamic field
(GasVel = 70 m/s)

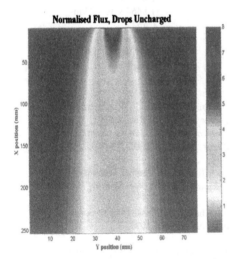

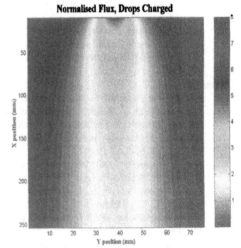

Drops Uncharged: Liquid distribution
under hydrodynamic force

Drops Charged: Liquid distribution
under hydrodynamic and electric fields
Charge To Mass Ratio = 2.8 mC/kg;
Applied Voltage = 30 kV

Figure 4: Comparison of liquid distribution contours for charged and uncharged drops
(GasVel =70m/s; Drop Size = 50μm)

Inst. Phys. Conf. Ser. No 163
Paper presented at the 10th Int. Conf., Cambridge, 28–31 March 1999

Cleaning of Dusty Gases by Pulse Corona in the Laboratory

M Nifuku [a), **E Kiss** [b), **I Jenei** [b), **M Horváth**[b) and **H Katoh** [a)

[a) National Institute for Resources and Environment
Onogawa 16-3, Tsukuba, Ibaraki 305-8569, Japan
[b) Dunaújváros Polytechnic of Miskolc University
H-2401 Dunaújváros, Táncsics Mihály str. 1/a, Hungary

Abstract. The authors have carried out a research to clean the exhaust gas by electrical discharge. The toxic gases were tried to be decomposed by pulse corona. A power supply produces electrical discharges with fast pulse rise time (the minimum is about 50 ns), high pulse peak voltage (maximum : about 80 kV) and the pulse repetition frequency 20 to 200 pulse/s. The dusts were investigated to receive highest electrical charge using the power supply mentioned above in order to evaluate the electrostatic precipitation property. The toxic gas were decomposed effectively with the pulse corona characteristics; pulse rise time less than about 100 ns, pulse peak voltage over about 25 kV and pulse frequency over about 100 pulse/s. The dust charge per mass were bigger at slower pulse rise time (a few hundred ns), 20 to 35 kV pulse peak voltage and higher pulse frequency (over 150 pulse/s) in the case of fly ash. The authors research indicated the possible application of pulse corona for dusty gas cleaning.

1. Introduction

Clean air is everybody's concern. Industrial and municipal wastes are increasing in our modern society. It is essential to process the wastes appropriate way in dealing with the increasing wastes, otherwise there is a chance where our environment could be polluted. Unfortunately, there are a lot of reports that air and ground water are already contaminated. Also, land for wastes disposal is becoming scarce now. Incineration will help solving this problem, because the volume of the wastes decreases drastically by incineration and the land area to fill the ashes will be reduced.

However, the incineration seems to accompany serious problems. Oile et al pointed out the discharge of dioxins from the incineration facilities in 1977 [1]. This might be related to the application of electrostatic precipitators in the facilities, because the electrostatic precipitators are being operated at high temperature (several hundreds degrees C) where the dioxins could be produced.

It will be a good idea to decompose component gases that might lead to the production of dioxins. This idea will contribute to suppress other toxic gases as well. Exhaust gases usually contain dusts. Therefore, it will be advantageous to decompose the polluting gases and to precipitate dusts at the same time. The authors have carried out the fundamental research in the laboratory. It was tried to decompose the gases effectively in the first place and then to suppress the dusts.

For this purpose, application of electrical discharge to dusty exhaust gas will be beneficial. Electrons produced by the electrical discharge will bombard neutral gas molecules and the molecules will be activated, leading to the new chemical reaction. Here, the energy on the electrons will play an important role to activate the molecules. This will depend on the discharging characteristics.

As this research being fundamental, the basic properties of particle charging were investigated to evelute the electrostatic precipitation characteristics.

There are many researches on the cleaning of exhaust gas. Various types reactors have been applied [2-6]. However, there are still many problems, such as influence of the electric discharge characteristics, suppression of dust in exhaust gas, etc., to be investigated.

The authors have tried to investigate the decomposition of toxic gases and the charging characteristics of dusts in the laboratory. It is tried to elucidate the influence of waveform on the gas decomposition and the dust precipitation.

2. Experimental method

Diluted sample gas was introduced into an electrical reactor after the reactor was vacuumed. A power supply produced pulse corona with predetermined discharge characteristics in the reactor. The relation between the electrical discharge characteristics and the gas decomposition ratio was investigated.

The experimental arrangements are shown in Figs. 1-3. The sample gas (volatile organic compound) was prepared in the glass container by regulating the water bath temperature and diluted by air. Thus the sample gas concentration was controlled. The evaporated gas was introduced into gas stream by using vacuum pump. The initial and the processed gas concentrations were measured by gas chromatograph. In the case of other gases (NO and SO_2), especially ordered gas concentration was used (N_2 dilution) or the gas was diluted by N_2 or NO_2. The gas concentration was measured by NOx and SOx analyzer. Fig. 4 shows the power supply. This apparatus produces various types of pulse corona, by changing pulse rise time, pulse peak voltage and pulse repetition rate. The sample gas used were benzene toluene, ethylbenzene, o-xylene, m-xylene, p-xylene, nitrogen monoxide and sulfur dioxide. A quadruple mass spectrometer was also used to identify intermediate products in the process of decomposition of the sample gases.

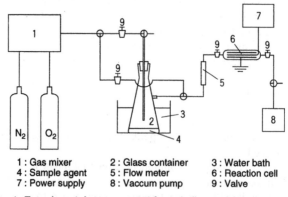

1 : Gas mixer   2 : Glass container   3 : Water bath
4 : Sample agent   5 : Flow meter   6 : Reaction cell
7 : Power supply   8 : Vaccum pump   9 : Valve

Fig. 1 Experimental arrangement for volatile organic compounds.

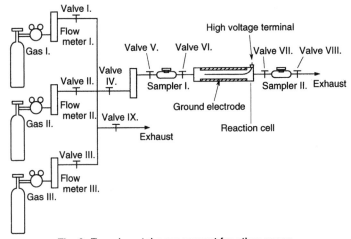

Fig. 2 Experimental arrangement for other gases.

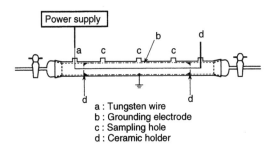

a : Tungsten wire
b : Grounding electrode
c : Sampling hole
d : Ceramic holder

Fig. 3 Gas reaction cell by electrical discharge.

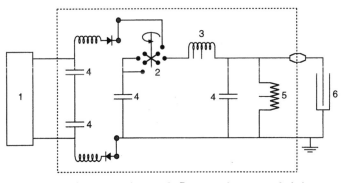

1 : High-voltage DC power supply   2 : Rotary spark gap   3 : Inductor
4 : Capacitor   5 : Resister   6 : Reaction cell

Fig. 4 Power supply (Manufactured by Masuda Research Inc).

Dust suppression by the electrical discharge was investigated by the assessment of electrostatic powder charge. Powder sample was supplied downward (vertically) from a feeder. The powder went through the electric field which was produced by the power supply. The powder was collected into a Faraday cage and the powder charge was measured. Fig. 5 shows the equipment. The power supply shown in Fig. 4 was also applied to charge powder sample. Here, the influence of electric discharge characteristics on the powder charge was investigated.

Fly ash was used as the powder sample. The resistivity was about 2×10^8 Ω·cm.

3. Results and Discussion

3.1. *Discharge characteristics of power supply*

The power supply shown in Fig. 4 produces various types of pulse corona. It is designed to produce the pulse with the minimum rise time about 20 ns, the maximum peak voltage about 90 kV and the maximum pulse repetition 200 pulse/s.

Using the reaction cell shown in Fig. 3, it was possible to apply about 45 kV (peak voltage). The half width value of voltage waveform applied in the experiment was around a few hundred ns. As the authors reported in another paper [7], the power supply produced the energy of the order of 10^{-15} joule/electron and the electron density was of the order of 10^{10} to 10^{11}/cm^3.

The power supply shown in Fig. 5 produced about the same pulse corona except the pulse peak voltage value. The maximum peak voltage produced was approximately 25 kV.

Based on these observation, the power supplies can be regarded as to provide enough energy for electron to proceed chemical reaction and powder charging.

3.2. *Decomposition of volatile organic compounds*

All the sample gases were decomposed effectively. The decomposition will be originated by the formation of radicals following the release or acceptance of electron into atoms. Actually, the decomposition was influenced by the pulse rise time, pulse peak voltage, pulse frequency, gas treatment time (residence time of gas in the electric discharge field), etc. as indicated by the authors' previous work (Figs. 6-9) [7].

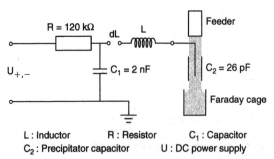

Fig. 5 Schematic diagram for powder charge measurement.

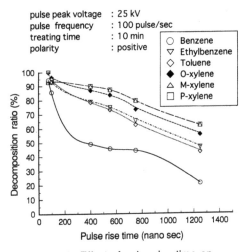

Fig. 6   Effect of pulse rise time on
gas decomposition.

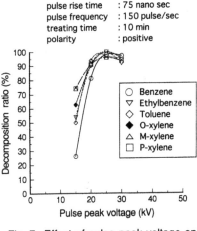

Fig. 7   Effect of pulse peak voltage on
gas decomposition.

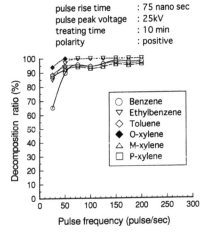

Fig. 8   Effect of pulse frequency on
gas decomposition.

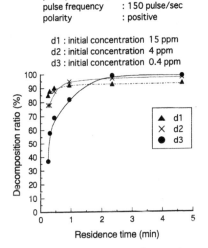

Fig. 9   Effect of residence time on benzene
decomposition in flow system

As for the relation between the pulse rise time and the decomposition ratio of the sample gases, the decomposition ratio was the highest (around 95 %) at the pulse rise time less than 100 ns. The ratio decreased with increase of the pulse rise time almost linearly. The decomposition ratio was about 40-65 % at the pulse rise time 1,250 ns (Fig. 6).

The decomposition ratio increased, reached peak and decreased slightly with the increase of the pulse peak voltage. The peak values of the decomposition ratio of the sample gases were around 95 %. The highest decomposition ratio was obtained at about 25 kV of the pulse peak voltage (Fig. 7).

280

Also, the decomposition ratio increased almost exponentially with the increase of the pulse frequency. The decomposition ratios reached plateau (around 95 %) at around 50 pulse/s of the pulse frequency (Fig. 8).

The decomposition ratio increased with the increase of the contact time of the pulse to the sample gas (Fig. 9). More chances of electron bombardment on gas molecules will be resulted in the longer contact time.

The sample gases were transformed into CO_2, N_2O and H_2O through aliphatic alcohols, acetic acid, formic acid, benzaldehyde, phenol, etc.

From the observation mentioned above, it could be concluded that the electrons produced by pulse corona had enough energy to decompose the volatile organic compounds. Also, the number of electron produced by the electrical discharge was enough.

3.3. Decomposition of NO and SO_2

NO and SO_2 gases were also decomposed satisfactorily. The influences of the pulse rise time, pulse peak voltage and the pulse frequency on the decomposition are shown in Figs. 10-13. In the figures, l_{rc} is the length of the reaction cell, D the inside diameter of the reaction cell, f the flow rate of gas, C the concentration of the sample gas, V the peak discharge voltage, t_{re} the residence time of the sample gas in the reaction cell, d the diameter of the discharge electrode (tungsten wire), L and L_2 the inductance in the discharge circuit, R_s the series resistance in the discharge circuit, R_p the parallel resistance in the discharge circuit and t_r the pulse rise time. The experiments were carried out in flow system.

The relation between the decomposition ratio and the discharge characteristics (pulse rise time, pulse peak voltage and pulse frequency) is approximately the same as the case of organic compounds.

In the case of VOCs, the sample was decomposed in the short contact time, but NO needed longer contact time for higher decomposition ratio. The shorter contact time will be requested for the effective application of this technique. This problem could be solved by finding more effective discharge properties and the authors are preparing that new project.

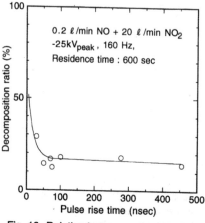

Fig. 10 Relation between pulse rise time and NO decomposition at negative polarity.

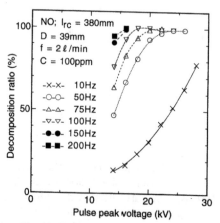

Fig. 11 Relation between pulse peak voltage and NO decomposition ratio at positive excitation.

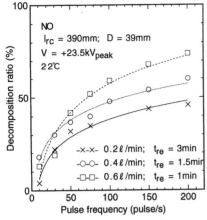

Fig. 12 Relation between pulse frequency
and NO decomposition ratio at
different residence times.

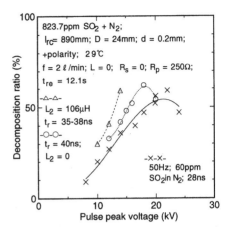

Fig. 13 Relation between pulse peak
voltage and SO₂ decomposition
ratio at positive pulse.

3.4. *Powder charge*

The powder sample received good amount of charge. Figs. 14-16 show the influence of the discharge characteristics on the charge per mass.

Fig. 14 shows the relation between the pulse rise time and the charge per mass. It is indicated that the charge per mass increases with the increase of pulse rise time, reaches peak and decreases. This is contrary to the case of gas decomposition. It is shown that the slower pulse rise time is required to give larger particle charge.

Fig. 15 shows the influence of pulse peak voltage on the charge per mass. The charge per mass increases, reaches peak and decreases slightly with the increases of the pulse peak voltage. In the case of particle charging, the charge per mass increase rate is much smaller than the case of gas decomposition. The pulse peak voltage value to produce the maximum charge per mass is about the same value (about 20-25 kV) as in the case of decomposition of organic volatile compounds.

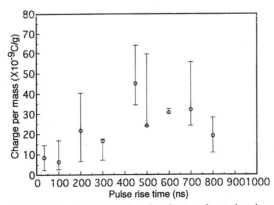

Fig. 14 Relation between pulse rise time and powder charge.
(Power supply: positive, Pulse frequency: 165 Hz,
Pulse peak voltage: 20 kV)

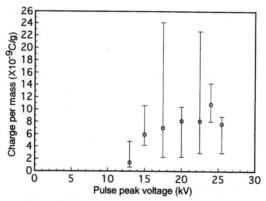

Fig. 15 Relation between pulse peak voltage and powder charge.
(Power supply: positive, Pulse frequency: 165 Hz,
Pulse rise time: 470 ns)

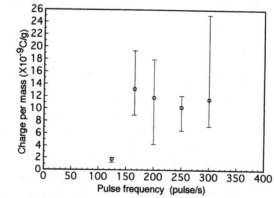

Fig. 16 Relation between pulse frequency and powder charge.
(Power supply: positive, Pulse peak voltage: 20 kV,
Pulse rise time: 470 ns)

Fig. 16 shows the relation between pulse frequency and powder charge. It is indicated that higher pulse frequency is required to give larger charge per mass comparing to the case of gas decomposition.

The experimental data in Figs. 14-16 are shown with positive polarity charging. The experimental data with negative polarity charging showed approximately the same results. However, the charge per mass was double or more in the case of negative polarity charging.

In this research, it was tried to clean the flue and exhaust gases (dusts and gas at the same time) using a power supply, as mentioned so far. For this purpose, the pulse charging applied in this research will be advantageous compared to traditional negative DC charging. As it is indicated by some researchers [8-10], the pulse energization is more advantageous than DC charging and intermittent energization. The corona discharge is more uniformly produced and the current density on collector of EP is more homogeneous than the case of traditional charging system. The space charge of ion produced around the discharge electrode will be spread spatially. The pulse energization can suppress back corona and no reentrainment of dusts will appear. Thus the pulse energization will materialize higher precipitation and will also have energy saving effect. [8-10].

However, the authors powder charge per mass is somewhat smaller (10^{-3}-10^{-1} order) than in-situ dust charge [10]. This will probably be because the in-situ power supply is more powerful than the authors power supply.

Comparing the results of gas decomposition mentioned in the sections of 3.2. and 3.3. with the results of powder charge mentioned here, the influence of the pulse rise time on the gas decomposition ratio and the charge per mass is different between the two. This might be because the production of ion will not be enough to charge particles when the pulse rise time is too fast. For this problem, the authors are now investigating other types of discharge which could be effective for both the gas decomposition and the particle charging.

The influence of pulse peak voltage and pulse frequency on gas decomposition and charge per mass is approximately the same between the two cases.

In practice, it will be requested to handle larger flow rates. For this purpose, the more effective discharge method will be necessary for the faster flow. If not, a multiplicity of small diameter reactors in parallel will be a solution.

4. Conclusions

It was shown that the flue and exhaust gases were decomposed satisfactorily by applying electric discharge. The decomposition was influenced by the characteristics of electric discharge.

The fast rise of pulse discharge (less than about 100 ns) was necessary for high decomposition of gas. Also, high pulse peak voltage (about 25 kV) and fast repetition of the pulse (more than about 60 pulse/s) were important for the higher decomposition.

The pulse energization of powder was also effective for electrostatic precipitation. The slower pulse rise time (about 450 ns) was necessary for higher charge per mass. Also, high pulse peak voltage (about 20 kV) and fast repetition of the pulse (more than about 150 pulse/s) were necessary for the higher charge per mass.

From these results, it could be concluded that the pulse discharge is a useful technique to clean dusty flue and exhaust gases.

References

[1] K Olie, P L Vermeulen and O Hutzinger 1977 *Chemosphere* **8** 455-459

[2] T Oda 1995 *Proceedings of the Institute of Electrostatics Japan* **4** 283-288

[3] A Mizuno 1995 *Proceedings of the Institute of Electrostatics Japan* **4** 289-295

[4] O Tokunaga 1995 *Proceedings of the Institute of Electrostatics Japan* **4** 296-300

[5] T Yamamoto 1995 *Proceedings of the Institute of Electrostatics Japan* **4** 301-305

[6] T Ohkubo and Y Nomoto 1995 Proceedings of the Institute of Electrostatics Japan **5** 369-374

[7] M Nifuku, M Horváth, J Bodnár, G Zhang, T Tanaka, E Kiss, G Woynárovich and H Katoh 1997 *Journal of Electrostatic* **40&41** 687-692

[8] Y Matsumoto and Y Nakayama 1985 *Proceedings of the Institute of Electrostatics Japan* **5** 315-323

[9] S Masuda 1990 *Proceeding of the Institute of Electrostatics Japan* **6** 441-442

[10] H Fujishima and K Tomimatsu 1990 *Proceedings of the Institute of Electrostatics Japan* **6** 443-450

Inst. Phys. Conf. Ser. No 163
Paper presented at the 10th Int. Conf., Cambridge, 28–31 March 1999
© *1999 IOP Publishing Ltd*

Streaming Current Charge versus Tribo Charge

R Baur, H –T Macholdt, E Michel

Clariant GmbH, Business Unit Pigments, G 834, D-65926 Frankfurt am Main, Germany

Abstract. In electrophotographic toners (electronic ink) so-called charge control agents (CCAs) are key ingredients which adjust the triboelectrically generated charge of the toner. Typically CCAs exist within the toner matrix as discrete particles in the nanometer range ($\ll 1$ µm). Two CCA cation systems with different counter-anions are synthesized and investigated with two different physical methods: 1. q/m-measurements to detect the tribocharge influence in powder toners and 2. electrokinetic streaming current potential measurements to detect the surface charge of the suspended CCA particles. The results of both measurements are compared.

It is shown that negative charge directing CCAs (q/m) show anionic surface charge while positive directing CCAs in toners exhibit cationic surface charge. In addition, the anion directs strongly the charge both in sign and magnitude.

1. Introduction:

The electrophotographic printing process became one of the most successful technical applications for electrostatic charging phenomena[1]. During the development step the latent image - existing as a charge screen on the printing drum (photoconductor) - is developed with oppositely charged toner. In the complex toner formulations CCAs direct the chargeability of the toner precisely in sign, magnitude and time-dependency. They are incorporated by blend or polymerization processes into the formulation (ca 1%-3%) together with the other ingredients. Various explanations for the charge-directing influence of CCAs are discussed in the literature[2] e.g. electron exchange models, proton/ion transfer, water bridging, etc. But until now no model exists which would allow to forecast the experimental charging properties of chemical compounds as efficient CCAs. At present, the charge-directing influence of CCAs in toners is most widely tested with the q/m- Blow Off method [3]. These toners are prepared according to conventional blend methods followed by milling and classifying process steps. In order to find new efficient CCAs, a better understanding of the basic influencing factors is necessary. In this paper conventional q/m-measurements are compared with surface charge measurements detected with an electrokinetic method called Streaming Current Detection (SCD). In toners for q/m measurements nearly all of the CCA particles are coated by the toner matrix while during SCD measurements the particles are surrounded by the liquid dispersion medium water. It will be discussed whether a parallel between both measurements results exist and how structural modification of the investigated CCAs influences these effects. This paper will not discuss recent developments like so-called Charge Stabilizers which act either together with positive or negative CCAs or even as external CCAs on the final polymer-based product [4].

2. Experimental

Test toner and q/m-detection are carried out according to previously descibed methods[3]. The CCA concentration is 1 % in a styrene/acylic resin (Dialec S 309, styrene/methacrylate copolymer 60:40, T_G = 60°C). The chemical substances are synthesized according to the literature [5].

The streaming current potential is registered with a commercially available SCD (Streaming Current Detector) unit (PCD 03 Particle Charge Detector; Muetek Munich Germany). The central measuring device consists of a PTFE-cell (figure1) with a PTFE piston inside, which moves periodically (4Hz) up and down. For detection a 1 % dispersion in water of the insoluble CCA is filled into the the measuring cell. After a short time (approx. 15 sec.) the system is in equilibrium and the SCD device registers the charge sign. The absolute surface charge value is determined by neutralizing the surface charge with an oppositely-charged titration agent. For negative surface charges poly(diallyldimethylammoniumchloride) and for positive charges poly(ethylensulfonate) sodium salt are used. The charged particles build with the oppositely charged titration agent a so-called interpolyelectrolyte complex which is more or less irreversible [6]. The titration agent consumption at the isoelectrical point (point of zero charge) in µl quantifies the charge magnitude.

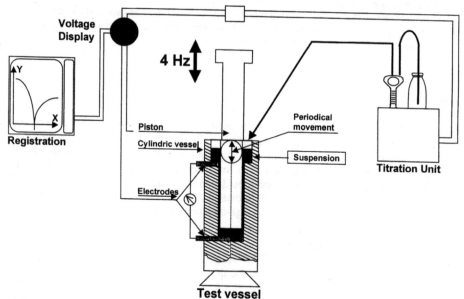

Figure 1. SCD titration unit with measuring cell and piston (mid), titration unit with charge neutralizing agent (right) and registration unit (left). The minimum of the curve defines the isoelectrical point of the suspended system. Y-axis is related to the detected potential, X to the µl consumption.

3. Streaming Potential of CCAs versus Tribo Charge of CCAs

In general, when two phases are placed in contact, a difference in the potential between them will be developed. This difference may be generated by different functional groups at the surface, orientation of dipoles, absorption, ionization, etc. Therefore at the interface of a liquid and suspended particle, separation of charge occurs. Theoretical models describe the structure of the interphase as a double layer with an adsorbed or rigid layer and a diffuse or mobile layer[7]. The charge on the surface of the solid phase is compensated by an opposite charge in the liquid created by e.g.counterions. The thickness of the layers depends e.g. on the ionic strength of the involved ions. The interface between both layers is called the shear plane and the potential existing at the shear plane is known as the zeta potential.

Streaming current measurements as described are executed according to a method called particle charge detection (PCD). The streaming potential is created when a liquid is forced to flow periodically relative to an (electrified) immobilized (wall-adsorbed) particle surface. The diffuse, mobile layer follows the movement. It causes an excess charge due to the liquid movement which could be measured. Advantage of this method is a quick and easy measuring in combination with less sensitivity to environmental influences. Exact detection of zeta potentials is complicated because of various experimental influences. Therefore the surface charge is detected more accurately by detecting the consumption of oppositely charged titration agents via interpolyelectrolyte complex building.

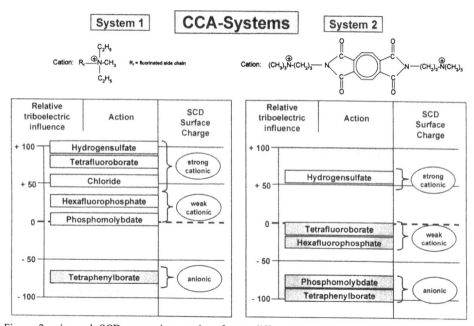

Figure 2. q/m and SCD measuring results of two different ammonium cation salt systems in comparison. Monocationic salts (system 1) left; biscationic (system 2) right. The detected maximum values are defined as 100 and the other values are set in relation.

Two ammonium salts are selected for the investigations. System 1 (figure 2: left) a mono-cation and System 2 (figure 2: right) a biscation with two positive charges in one cationic molecule part. In System 2 the two cation centres are separated by an aromatic amide spacer. Additionally the counter-anions for both cations are modified. The q/m measurements for simulating the triboelectric influence in hydophobic toners exhibit:

1. A strong influence on tribocharge both in sign and magnitude by the counter-anion. For both systems hydrogensulfate shows strong positive tribocharge influence while tetraphenylborate directs extreme negative. Similar behaviour is observed for the biscation.

288

2. Using the same counter-anions for the biscation system 2 the values shift to more negative values. For the biscation even the tetrafluoroborate and hexafluorophosphate anion show negative values

Surprisingly, SCD measurements in water based dispersion show a strong parallel.

1. The sign of the detected surface charge is identical with the sign of the triboelectric charging for all ionic systems.
2. The relative magnitude of the surface charge is also comparable. Strong positive friction charge influence shows high consumption of negative titration agent and therefore exhibits strong cationic surface charge. For the anionic systems the situation is vice versa.
3. Also the differences between the monocationic and biscationic system become observable.

Obviously there exists a strong parallel between the triboelectrical influence of the investigated CCAs incorporated in hydrophobic polymeric resin and the suspension of the pure salt particles. The counteranion influence could be explained in the sense that strong polarizing anions like BF_4^-, Cl^- or HSO_4^- show a positive charge-directing influence while less polarizing anions like $B(C_6H_5)_4^-$ act vice versa. If one cation involves two cationic centres separated by an aromatic amide spacer (system 2) the influence is reduced or shifted to more negative values. Tetraphenylborate is the anion of the weak acid $HB(C_6H_5)_4$ while the positive charge-directing anions correspond to strong acids. The acidity influence seems to be reduced in the biscation sytem. The results implicate that the acidity of the systems influences the triboelectric charge as well as the surface charge strongly. Several models for triboelectric charging are described. Ion or proton transfer seems to be the explanation for the salt systems tested. Further investigations with multicationic-centered ammonium salts or multianionic-centered counter anions seem reasonable to get a deeper understanding of these surprising results.

4. Conclusion:

The triboelectric charge influence of two CCA ammonium test systems in a polymeric toner matrix is compared with the surface charge detected by streaming current detection. System 1 includes only one cationic center; System 2 is based on two cationic centers. In addition, the counteranions for both cations are modified. The sign and relative magnitude of the final electrostatic toner charge and detected surface charge exhibit strict parallelism. It became evident that the acidity of the anions tested influences strongly the tribo charge behaviour as well as the surface charge of the pure particle. It seems possible that SCD-measurements are a simple new method to improve CCA developments.

5. References

[1] Schein L B 1992/1996 *Electrophotography and Development Physics (Springer/Laplacian)*
[2] Castle G S P 1997 *J Electrostatics* **40/41** 13-20
[3] Baur R and Macholdt H-T 1993 *J Electrostatics* 30 213-221
[4] a) *Products for the Non-Impact Industry,* Clariant GmbH, DP 7701 E 10.97 b) US 5,502,118. Castle G P Higashiyama Y Inculet I I Brown J D 1993 *J Electrostatics* 30 203-212
[5] EP 0506867/ DE 3837345
[6] Dubin P Bock J Davies R M Schulz D N *1994 Macromolecular Complexes in Chemistry and Biology* (Springer Verlag)
[7] Dörfler H-D 1994 *Grenzflächen und Kolloide* (VCH VerlagWeinheim)

Inst. Phys. Conf. Ser. No 163
Paper presented at the 10th Int. Conf., Cambridge, 28–31 March 1999
© 1999 IOP Publishing Ltd

Neutralisation of static surface charges by an ac ionizer in a nitrogen environment

J S Chang, K G Harasym, P C Looy, A A Berezin

Department of Engineering Physics, McMaster University, Hamilton, ON, Canada L8S 4M1

C G Noll

ITW Static Control & Air Products, 2257 North Penn Road, Hatfield, PA 19440-1998, USA

Abstract. Charge decay and residual potential (balance or offset voltage) were measured in gases with various ion mobilities. It was found that charge decay, especially for positive charges, occurs more quickly in nitrogen than in air. The residual potential on the probe is negative in pure nitrogen and increases towards positive values with the injection of small quantities of air to the ionizer. The fluctuations in the residual potential are generally less than 3 V peak-to-peak. As with air ionizers, the charge decay rate in nitrogen increases with superficial gas-flow rate. The results are consistent with a theory by Chang and Kodera.

1. Introduction

Static elimination is often necessary in manufacturing processes to eliminate hazards, increase production rates, and protect sensitive products and processes from electrostatic discharges. In recent years, nuclear (radioisotope), ultraviolet, soft x-ray, and corona discharge ionizers have been suggested for use in pure nitrogen environments [1] [2] [3] [4].

Nitrogen is used to provide an inert environment in many industries, and ionizers based on electrical corona are generally preferred in applications. The fundamental difficulty for corona in nitrogen environments has been known for many years [5]. The negative charge carriers formed in the discharge are free electrons and these do not readily attach to atomic or molecular species. The sensitivity of the negative carrier mobility to the concentration of electronegative gases effects the control of electrical static eliminators. The alternative technologies also produce positive ion and free electron pairs. The balance of these ionizers, however, is not easily controlled and the ionizers themselves introduce hazards to the work place. Electrical ionizers are preferred if methods can be found to establish balanced production of free electrons and positive ions.

The present work is part of experimental and theoretical efforts towards understanding the mechanisms and control of static elimination in nitrogen. In this paper attention is focused on the residual potential on a probe that arises from the differing mobilities of the positive and negative polarity charge carriers. The probe of interest is used conventionally for measurements of charge decay time and balance.

2. Theory for static elimination in a nitrogen environment

As noted above, ionization in nitrogen by all known methods results in the production of positive ions and free electrons. There is usually a preferential loss of free electrons to grounded components in the process environment and there is a fundamental bias given to charged objects that are to be neutralized. The interactions between bipolar ions and a charged object can be analyzed by a model proposed by Chang and Kodera [6]. This model was originally developed for arbitrary-shaped electrostatic probes in atmospheric plasmas. Here, the model will be briefly reviewed and used to estimate the ion deposition current and residual potential on a charged object — in our case a probe used to measure charge decay and residual potential.

We assume first that the ionizer produces a bipolar distribution of charge carriers. These carriers move in the vicinity of the probe, either conveyed by gas circulation or driven by electrical forces to walls or the probe itself. The flow of carriers constitutes a current near the probe. It is assumed that a current I_p reaches and is proportional to the charge on the probe Q_p. where $Q_p=C_pV_p$ and we have assumed the

$$I_p = \frac{dQ_p}{dt} = \frac{d(C_pV_p)}{dt} = (I_- - I_+) = C_p\frac{dV_p}{dt} \tag{1}$$

probe's capacitance C_p is constant. The currents of positive and negative ions (and free electrons) to the probe are then, respectively [7]

$$I_+ = \frac{eN_+D_+A}{L}\frac{\phi_p/\epsilon}{1-\exp(-\phi_p/\epsilon)} \tag{2}$$

$$I_- = \frac{eN_-D_-A}{L}\frac{\phi_p\exp(-\phi_p)}{1-\exp(-\phi_p)} \tag{3}$$

These expressions are written for the case where the Laplace potential $L/\lambda_D=0$ without gas flow. The quantity A is the surface area of the probe, L is the characteristic length, $\epsilon=T_-/T_+$ is the temperature ratio, D is the diffusion coefficient, N is the number density of ionic species, e is the electronic charge, k is the Boltzmann constant, λ_D is the Debye length, and $\phi_p=eV_p/kT_-$ is the dimensionless surface potential.

We obtain from Eqns. (1) to (3) estimates of the potential decay on the probe

$$\frac{V_p}{dt} = \frac{eAN_-D_-}{LC_p}[\frac{a}{\epsilon[1-\exp(-\phi_p/\epsilon)]} - \frac{\exp(-\phi_p)}{1-\exp(-\phi_b)} \tag{4}$$

and the condition on residual potential (offset voltage or balance) when $I_p=0$ yields

$$\frac{a}{\epsilon[1-\exp(-\phi_p/\epsilon)]} = \frac{\exp(-\phi_p)}{1-\exp(-\phi_p)} \tag{5}$$

where $a=N_+D_+/N_-D_-$ is the diffusion current ratio. For $\phi_p > 1$, Eqn. 5 can be simplified to

$$-\phi_p=\ln(a/\epsilon) \tag{6}$$

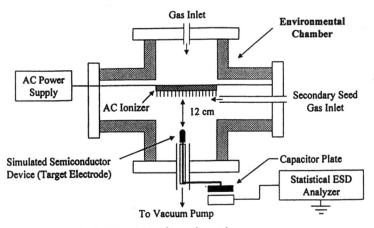

Fig. 1 Schematics of experimental apparatus.

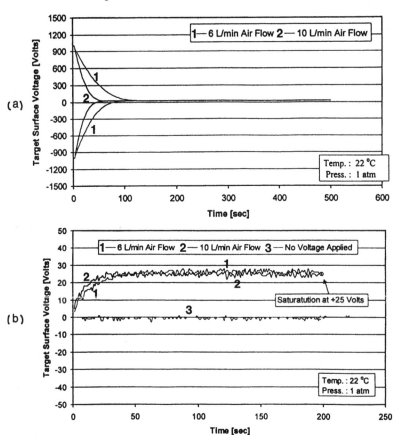

Fig. 2   Surface voltage/charge neutralisation characteristics for various dry air flow rate at applied voltage
V_{pp} = 14.4 kV a) Charge decay characteristics and b) Floating characteristics

or

$$V_p = \frac{kT_-}{e}\ln(\frac{N_+D_+}{N_-D_-}\frac{T_+}{T_-}) = \frac{kT_-}{e}\ln[\frac{N_+\mu_+}{N_-\mu_-}(\frac{T_+}{T_-})^2]$$ (7)

with $\mu = eD/kT$.

For corona discharge in air:

$kT_+/e \approx 0.1$ eV for positive ions (N_3O^+, NO_3^+, etc.),
$kT_-/e \approx 0.1$ eV for negative ions ($N_2O_2^-$, etc.),
$kT_e/e = 2\text{-}10$ eV ≈ 4 eV for electrons,
$D_- = 0.0329$ cm^2/s or $\mu_- = 1.414$ cm^2/Vs,
$D_+ = 0.0496$ (0.0512) cm^2/s or $\mu_+ = 0.743$ (0.718) cm^2/Vs for N_3O^+ (NO_3^+), and
$D_e = 2\times10^7\lambda_e$ $(T_e)^{\frac{1}{2}} \approx 2800$ cm^2/s for 4 eV.

Hence, in air for $N_+=N_-$, the residual potential in volts is

$$V_p \approx 0.1\ln(1.53N_+/N_-) \approx 0.042$$

The residual potential in nitrogen is negative and much larger in magnitude.

$$V_p \approx 4\ln(6\times10^{-7}N_+/N_-) \approx -57.31$$

The cases with gas flow [8] and space charges ($L/\lambda_D > 1$) [9] are expected to lead to greater residual potentials.

3. Experimental Apparatus

The experiments were conducted in a closed chamber that was maintained at atmospheric pressure and temperature. The arrangement of equipment is shown in Fig. 1. The ionizer consists of eighteen needle-type corona emitters supported by an insulation system within a grounded metal casing. It is operated with 14.4 kV peak-to-peak at 60 Hz on the high voltage wire. The corona emitters are capacitively coupled to the wire. This capacitive coupling limits carrier current flow from the emitters and aids in the production of bipolar corona ($N_+=N_-$). The carrier production, however, is not generally balanced by this method.

The superficial nitrogen flow rate in the chamber was from 6 to 10 /min. These gas flow levels approximate the still-gas case, yet maintain a dilute bipolar plasma for static elimination. Clean dry air (0 to 200 cc/min) was injected into the ionizer in several tests. Measurements were secured using a half-sphere, conductive probe located 12 cm downstream from the ionizer. The capacitance of the probe and measurement system C_p is approximately 300 pF.

4. Experimental Results

The charge neutralisation and residual potentials for various dry-air flow rates are reported in Fig. 2. The residual potential in the dry-air case is approximately +25 V and is largely determined by the geometry of the ionizer and chamber.

When clean dry air is injected to the emitter zone of the ionizer in a nitrogen environment, the residual potential progressively rises from negative values to the value observed in the air

293

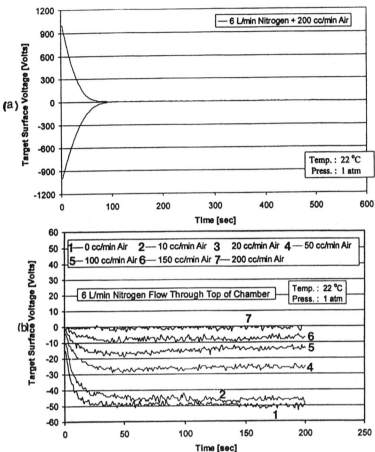

Fig 3   Surface voltage/charge neutralisation characteristics for various air seeded flow rate for pure nitrogen environments at V_{pp} = 14.4 kV. a)   Charge decay characteristics and b) Floating characteristics.

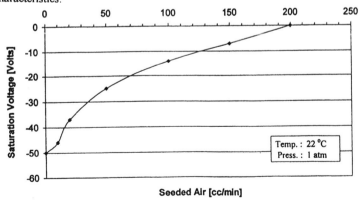

Fig. 4   Surface voltage/charge as a function of seeded air flow rate at V_{pp} = 14.4 kV in pure nitrogen environment.

environment. The effects of air addition are illustrated in Figs. 3 and 4. Figure 3a also shows that the potential fluctuation does not depend significantly on the rate of air injection. The amount of air required to stabilize ion production in the corona is approximately 200 cc/min, as shown in Fig. 4. This amount of air can be reduced by controlled injection about the emitter tips.

5. Concluding Remarks

It was found that charge decay, especially for positive charges, occurs more quickly in a nitrogen environment than in air. The residual potential on the probe is negative in a pure nitrogen environment and increases towards positive values with the injection of clean-dry-air to the ionizer. The fluctuations in the residual potential for air and nitrogen environments are less than 3 V peak-to-peak. As with conventional air ionizers, the charge decay rate in nitrogen increases with the superficial gas-flow rate. The results are consistent with predictions of a model where free electrons are assumed to be the dominant negative charge carriers. The residual potential arises from the differing mobilities for positive- and negative-polarity charge carriers, and is independent of the ionization source where equal carrier concentrations are generated.

Acknowledgement

This work is supported by ITW Static Control & Air Products and Natural Sciences and Engineering Research Council of Canada.

References

[1] Inaba H and Ohmi T *et al* 1992 *IEEE Trans. Semiconductor Manufacturing* **5**(4) 359-67
[2] Gosho Y and Yamada M *et al* 1990 *Jpn. J. Appl. Phys.* **29**(5) 950-1
[3] Inaba H and Ohmi T *et al* 1994 *J. Electrostatics* **33** 15-42
[4] Noll C G and Lawless P A *et al J. Electrostatics* **44**, 221-238
[5] Weissler G L 1943 *Phys. Rev.* **63**(3,4) 96-107
[6] Chang J-S and Kodera K 1985 *J. Geophysical Res.* **90**(D4) 5897-900
[7] Laframboise L G and Chang J-S 1977 *J. Aerosol Sci.* **8** 331-8
[8] Chang J-S and Kodera K 1978 *et al Proc. Soc. Atmos. Elect. Jpn.* **20** 58-66
[9] Chang J-S and Kodera K *1978 et al Conf. Rec. IEEE-IAS Mtg.* 38-40

Inst. Phys. Conf. Ser. No 163
Paper presented at the 10th Int. Conf., Cambridge, 28–31 March 1999

The influence of flow regime on electrostatic destabilisation of water-in-oil emulsions: A theoretical and experimental assessment.

N J Wayth[1,4], T J Williams[1], A G Bailey[1], M T Thew[2] and O Urdahl.[3]

[1] Department of Electrical Engineering, University of Southampton. SO17 1BJ.
[2] Department of Mechanical & Production Engineering, University of Bradford. BD7 1DP.
[3] Statoil, Veslefrikk Technical Department, 5020 Bergen, Norway.
[4] now at Greater Forties, BP Amoco plc, Dyce, Aberdeen. AB21 7PB.

Abstract. In this study the growth of water droplets in a water-in-oil emulsion under an applied electric field is considered. The significance of the flow regime, laminar or turbulent, has been investigated theoretically and experimentally with respect to droplet coalescence and break-up mechanisms. It is shown that whilst turbulent flow may enhance the coalescence of small droplets, it can also lead to disintegration of larger ones. Experimental assessment of suspended droplet size, performed on emulsions flowing through a purpose-built compact electrocoalescer test duct, showed good agreement with theoretical predictions of maximum droplet size under turbulent flow and electrostatic stresses. It is suggested that in this case the droplet break-up was primarily due to turbulent stresses in subsequent pipe-work rather than the combined turbulent and electrostatic stresses in the duct itself.

1. Introduction

The amount of water co-produced with crude oil tends to increase during the lifetime of a reservoir. Conventional separation is performed by large gravity/electrostatic treaters. Not only do these entail a large volume, mass and "footprint", but they are also sensitive to wave motion (if offshore) and offer little flexibility in coping with changes in production. This has led to an impetus for the development of more compact and flexible separation equipment. In previous publications [1 - 3] we described the development of a CEC (Compact Electrostatic Coalescer) for the separation of w/o (water-in-oil) emulsions. The purpose of such a device is to grow water droplets to larger diameters, allowing more efficient subsequent phase separation. Laboratory tests have been performed with three different test ducts, using distillate oils to simulate crude oil. Additionally, tests have been performed on a prototype CEC device using crude oil. To date the results have been very encouraging, with water droplet growth of over ten-fold recorded under some conditions.

Whilst much has been done previously to investigate the principles behind the electrostatic coalescence of water droplets in oil [4 - 8], there has been little attempt to isolate the effects of flow regime under an electrostatic field. This work seeks to demonstrate the different mechanisms applicable to water droplets under laminar and turbulent flow regimes. Maximum stable droplet sizes have been considered by combining the fluid and electrostatic stresses. The difference in collision rates between laminar and turbulent flow, with or without a superimposed electrostatic field, has also been investigated. The theoretical work is complemented by experimental tests performed using one of the laboratory test ducts.

2. Theoretical development

For maximum droplet size under turbulent flow the Hinze [9] expression is employed (ρ_c = oil density, ε = mean turbulent energy dissipation rate per unit mass and γ = interfacial tension):

$$C = 0.725 = D_{max\_t} \left[\rho_c / \gamma \right]^{0.6} \varepsilon^{0.4} \tag{1}$$

Considering now the electrostatic stresses acting on a droplet, maximum droplet size under an applied electrostatic field E_0, in a dispersion of continuous phase permittivity ε_c, is related to the dimensionless electrostatic Weber number. This has a critical value of $We_{crit} = 0.409$ according to Roskenkilde [10]:

$$We_{crit} = 0.409 = \left(D_{max\_e} \, \varepsilon_c E_0^2 \right) / \gamma \tag{2}$$

Assuming the two stresses given in equations 1 and 2 may be combined additively, an expression for overall maximum droplet size (D_{max_c}) may be derived:

$$\frac{\gamma}{d_{max\_c}} = \rho_c \frac{(D_{max\_c}\varepsilon)^{2/3}}{C^{5/3}} + \frac{\varepsilon_c E_0^2}{0.409} \tag{3}$$

Substituting $X = (D_{max\_c})^{1/3}$ allows equation 3 to be converted into a quintic. Although there is no analytical solution, *Matlab* software has been used to determine numerical solutions.

A similar correlation was developed for the combination of stresses from electrostatic forces and steady-state laminar flow, using a relationship developed by Rumscheidt and Mason [12] for maximum stable droplet size under laminar flow. However, it was found that under the conditions appropriate to this work the laminar shear played an insignificant role compared to that of the electrostatic stresses. Using relationships summarised by Pearson et al. [11] for turbulent and laminar collision mechanisms a *Matlab* program was developed to show droplet growth under these mechanisms. The turbulent collision mechanism was found to be far more effective than laminar shear though space restrictions preclude the presentation of results here.

3. Experimental procedure

Experimental work was performed in a duct of rectangular cross-section (100 mm wide x 20 mm high). The w/o emulsion system investigated in this work was formed using *Shell Tellus R5*, a hydraulic oil, as the continuous phase. This was used to simulate crude, for reasons of safety and practicality. Distilled water was used to form the dispersed phase of the emulsion. The resultant emulsion formed by mixing the two phases was equivalent in stability (at 25 °C) to a medium-stability crude oil at treatment temperature (60 °C). The rectangular CEC duct was mounted horizontally and consisted of 7 modules each 0.5 m long and possessing a pair of parallel plate electrodes mounted at the top and bottom.. Each electrode was insulated from the emulsion by acrylic of thickness 5 mm giving an electrical treatment length of 3.5 m, a flow gap of 20 mm and an electrode spacing of 30 mm. Each electrode pair was energized using a transformer operating at a frequency of 50 Hz. A full description of the experimental rig is reported elsewhere [1-3]. The present tests were performed at 1 and 2 vol% water cut. In order to investigate the effects of flow regime on coalescence performance a fixed electrical residence times of 3s was used. This was achieved by varying the number of electrodes energized and meant that one energized module was required for each 20 litre/min of flow. Flow rates up to 100 litre/min were used which gave turbulence in the duct beyond about 60 litre/min. Applied potentials of 0, 7.5, and 17.5 kV peak-to-peak were used, giving electric field strengths of up to 5.8 kV/cm. Coalescer performance was assessed by measuring water droplet size distributions at inlet and outlet using a previously developed technique [1-3] based on a *Malvern* Laser Diffraction Particle and Droplet Size Analyser. The characteristic diameters $D_{(v,0.10)}$, $D_{(v,0.50)}$ and $D_{(v,0.90)}$ were taken in order to indicate not only the median droplet size but the range of the size distribution as well. This also gave indication of turbulence-induced droplet break-up and the growth of very small droplets

4. Results

Inlet droplet VMD (volume median diameter ie. $D_{(v,0.50)}$) was found, at both 1 and 2 % water cut, to reduce from around 10 μm to 7 μm as flow rate increased. With no electric field the outlet VMD was virtually the same as that at the inlet, indicating no droplet growth due to laminar or turbulent mixing alone. When an electric field was applied to the emulsion, rapid growth of water droplets resulted. Considering first the results at a water content of 1 vol% (see Fig 1a), growth was noticeably improved at 17.5 kV compared with that at 7.5 kV. An increase in growth also resulted as the flow rate was increased, implying enhanced coalescence from turbulent mixing, although this reached a plateau at flow rates between 80 and 100 litre/min.

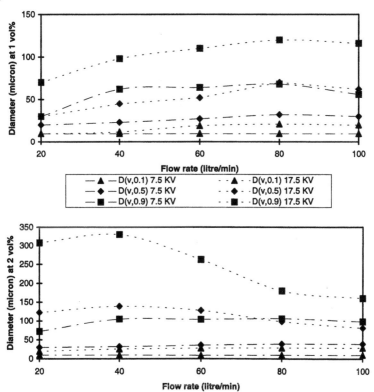

Figures 1a and 1b - Experimental Results at 1 and 2 vol% water respectively.

298

At 2 vol% water cut (see Fig 1b), droplet growth was significantly greater than at 1 vol%, again with 17.5 kV giving dramatic improvement over 7.5 kV. Here the $D_{(v,0.1)}$ reached 40 μm at the highest flow rate and potential. However, a clear reduction in $D_{(v,0.5)}$ and $D_{(v,0.9)}$ droplet size was obtained at flow rates of 60 litre/min and above, indicating the likelihood of droplet break-up. Figure 2 shows theoretical maximum droplet sizes against measured $D_{(v,0.9)}$ for these conditions. It can be seen that at flow rates of above 60 litre/min D_{max_c} in the duct, with an applied potential difference of 17.5 kV, is greater than the D_{max}, in the 1" pipe-work and transitional section downstream of the CEC duct. It may be inferred from this that the likely cause of droplet break-up, in this instance, is due to turbulence alone in the narrower downstream sections, rather than the combined electrostatic and turbulent stresses acting on the suspended water droplets within the CEC duct itself.

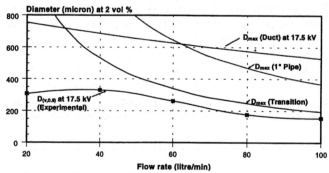

Figure 2 - Comparison between measured $D_{(v,0.9)}$ and predicted D_{max}. Predictions are only valid under turbulent flow (at approximately >20 litre/min for transition and pipe, and > 50 litre/min for duct).

References

[1]   Urdahl O, Williams T J, Bailey A G and Thew M T 1996 *Chem Eng Res.Des.* **74 (A2)** 158-165
[2]   Harpur I G, Wayth N J, Bailey A G, Thew M T, Williams T J and Urdahl O 1997 *J. Electrostatics* **40-41** 135-140
[3]   Urdahl O, Berry P, Wayth N, Williams T, Nordstad K, Bailey A and Thew M 1998 "The Development of a new Compact Electrostatic Coalescer Concept" *SPE Conference New Orleans* (Presentation cancelled due to hurricane Georges.)
[4]   Williams T J 1989 "The Resolution of Water-in-Oil Emulsions by the Application of an External Electric Field" *PhD Thesis Department of Electrical Engineering University of Southampton*
[5]   Williams T J and Bailey A G 1984 "Changes in the Size Distribution of a Water-in-Oil Emulsion Due to Electric Field Induced Coalescence" *IEEE-IAS-84 Annual Meeting Chicago Conf. Rec. IAS* **14** 1162
[6]   Waterman L C 1965 "Electrical Coalescers: Coalescence" *Chem. Eng. Prog.* **61(10)** 51-57
[7]   Urdahl O 1993 *PhD Thesis Dept. of Chemistry University of Bergen Norway*
[8]   Taylor S E 1988 "Investigations into the Electrical and Coalescence Behaviour of Water-in-Crude-Oil Emulsions in High Voltage Gradients" *Colloids and Surfaces* **29** 29-51
[9]   Hinze J O 1955 "Fundamentals of the Hydrodynamic Mechanism of Splitting in Dispersion Processes" *AIChe. J.* **1** 3
[10]   Rosenkilde C E 1969 *Proc. Roy. Soc.* **A312** 473
[11]   Pearson H J A, Valioulis L and List E J 1984 "Monte Carlo Simulation of Coagulation in Discrete Particle-Size Distributions. Part 1. Brownian Motion and Fluid Shearing" *J. Fluid Mech.* **143** 367-385
[12]   Rumscheidt F D and Mason S G 1961 *J. Colloid Sci.* **16** 238-261

Inst. Phys. Conf. Ser. No 163
Paper presented at the 10th Int. Conf., Cambridge, 28–31 March 1999

New Results in ESP Modelling

I. Kiss[1], J. Suda[2], N. Szedenik[1], I. Berta[1]

[1]Technical University of Budapest, Dept. of High Voltage Engineering and Equipment
Egry J. u. 18., 1111 Budapest, Hungary
[2]Technical University of Budapest, Department of Fluid Mechanics
Bertalan L. u. 4-6., 1111 Budapest, Hungary

Abstract. In the paper our further developed ESP model and its results are presented to simulate the movement of the dust particles in turbulent boundary layer type flow field. With the new development, the space charge distribution can be calculated more realistic, therefore the effect of the non uniformity of the electric field caused by the space charges can be taken into consideration more accurately.

1. Introduction

The paper proposes a better model for determining the transport of dust particles in turbulent flow in an ESP channel. A computer program has been developed with numerical straight forward marching method of finite differential forms (FDID). The turbulent boundary layer type of flow for the typical case of wire-smooth plate model-precipitator equipment has been calculated by solving numerically the turbulent boundary layer equation.

The results of the numerical calculations presented in this paper demonstrate both the effect of turbulent diffusion in the turbulent boundary layer on dust motion, on streamlines, and the effect of the particle-size and non uniform electric field including space charges due to the charged dust-phase and ionic clouds on the electric conditions in the channel.

2. Formulation of problem

Among different types of electrostatic precipitators, the plane-wire types are widely used in the industry. Therefore such type of precipitator was examined. We used the parameters of an existing model precipitator in our simulation model, proposing further experimental validation of the computer code. Horizontal cross-section of the model precipitator is outlined in Fig.1.

The typical wire-smooth plate type model ESP used for lab-scale experiments has seven discharge (corona) electrodes (DE) in one ESP channel. The grounded collecting electrodes (CE) are parallel smooth plates. Channel half width s=33 mm; precipitator length L_{CE}=300 mm; corona wire radius r_{DE}=0.5 mm; wire distance 2c=50 mm.

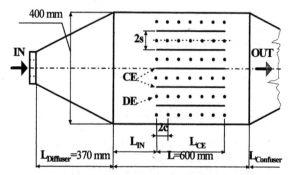

Fig. 1. Model-scale ESP cross-section.

In the literature a limited number of theoretical models are available to follow the particle transport in ESPs. One possibility is to compute the dust particle trajectories in the precipitator channel. This model is often called as Particle Tracking Model (PTM). The application of PTM with a turbulent transport model requires an appropriate setting of the time-dependent turbulent velocity field, see Riehle [7]. An another method for determining the turbulent transport process is the application of k-ε model in a fluid dynamic simulation program, proposed by Gallimberti [2] and Medlin et al. [6].

The turbulent transport of solid particles in ESPs is a highly complicated process because it involves inhomogeneity and changes of electrical and fluid dynamical conditions such as space charges of ionic clouds and dust-phase and the turbulent flow field. That is why the real process can only be approximated by using different simplifying assumptions describing this highly complex process.

The following major assumptions are made in developing the proposed model to study the particle transport of a plate-wire ESP:
1. The system is in steady-state operating condition.
2. Two-dimensional, boundary layer type turbulent flow field is in the ESP channel.
3. The streamwise diffusion can be neglected, there is no mixing in the streamwise direction.
4. The suspended dust-phase in the gas is a continuum phase.
5. Gravitational forces acting on the particles are neglected.
6. The dust particles are regarded as spherical, monodisperse particles.
7. There is no re-entrainment from the wall, and no dust layer on the collecting electrodes.
8. The effect of ionic or electric wind on the gas flow is not taken into consideration [4], but the effect of ionic and dust space charges can not be neglected.

3. Method of computation

3.1 *Electrostatic field*

The first task is the determination of the space charge. It consists of two parts, the first is the space charge deriving from ions and electrons (mainly ions, because their mobility are much smaller than the electrons) while the second one derives from the charged particles. The ion concentration can be calculated as in [3], using the modified Peek's law.

To model the charged dust particles, in the first step the Cochet formula [6] was used. The main idea is that the dust particles can collect only a limited amount of charge, the value of which is dependent on the features of the dust and the electric field. (In the future further analysis is planned according to [8] to determine the charge process when the dust is not monophase and contains not only spherical particles. It is an important remark, that the density of the dust is dependent on the x, y, z coordinates, so it is 3 dimensional.) In the calculation of the electrical field it can be taken into consideration with 3D space function as an input for the numerical field computation. To calculate the field strength and the potential distribution in the ESP the Poisson equation has to be solved. For this purpose the Boundary Element Method (BEM) was chosen. At a given point 'P' the potential can be expressed as in (1). The first term shows the effect of the space charge ρ_V, the second term describes the influence of the surface charges (ρ_A) situated on surface A_1, while the third term gives the potential value originated from the dipole moment v on surface A_2.

$$\varphi(P) = \frac{1}{4\pi\varepsilon} \int_V \frac{\rho_V}{r} dV + \frac{1}{4\pi\varepsilon} \int_{A_1} \frac{\rho_A}{r} dA + \frac{1}{4\pi\varepsilon} \int_{A_2} v \, \mathrm{grad} \frac{1}{r} dA \tag{1}$$

After the discretisation the integral equation is converted into a matrix equation (2), where the column vector q is an unknown and contains the surface charges appearing on the corona electrodes and the plates, q_s consists of the space charge values situated inside the discrete volume elements, while φ represents the potential values of a discrete surface element.

$$\underline{A}q + \underline{B}q_s = \varphi \tag{2}$$

Knowing the geometry, the elements of matrix $\underline{A}$ and $\underline{B}$ (which shows the effect of a small surface and volume element at the point where the potential is known) can be calculated. Regarding, that q_s and φ are input parameters, q can be calculated. When q_s and q are known, the electrical field strength and the potential can be easily calculated at the gridpoints of a predefined grid, and the field strength and potential distribution can be transferred to the flow field calculating module.

3.2 Flow field

The gas flow within the ESP channel can be described by means of continuity equation and momentum conservation law. The continuity equation for an incompressible medium is:

$$\frac{\partial v_x}{\partial x} + \frac{\partial v_y}{\partial y} = 0 \tag{3}$$

Regarding the assumptions mentioned above by neglecting the electrostatic forces acting on the gas phase and the interaction with the dust-phase a simple boundary layer equation can be applied for determining the 2D velocity field:

$$v_x \frac{\partial v_x}{\partial x} + v_y \frac{\partial v_x}{\partial y} = V \frac{dV}{dx} + \frac{\partial}{\partial y}\left(v_t \frac{\partial v_x}{\partial y}\right). \tag{4}$$

Turbulent viscosity v_t is evaluated on the bases of the mixing length model [11].

3.3 Concentration field.

Disregarding the streamwise diffusion of the dust-phase the concentration c can be computed with the help of a parabolic type transport equation:

$$v_x \frac{\partial c}{\partial x} + v_y \frac{\partial c}{\partial y} = \frac{\partial}{\partial y}\left(\frac{v_t}{Sc_t} \frac{\partial c}{\partial y}\right) - \frac{\partial}{\partial y}\left(c \cdot \overline{Wth}_y\right) \tag{5}$$

where in the last term $\overline{Wth}_y$ refers to the theoretical migration velocity of the particle due to the electric field strength. It can be derived from the equation of motion for the dust particle and can be expressed as:

$$\overline{Wth}_y = \frac{Q_p^\infty \overline{E}_y}{3\pi\mu d_p} Cu \tag{6}$$

where Q_p^∞ is the saturation charge of a spherical dust particle with a diameter d_p. The Cochet's charging equation is used in this study. Cu is the Cunningham correction factor, μ the fluid viscosity, Sc_t the turbulent Schmidt-number and E_y the y component of the electric field strength.

Knowing the flow field in the ESP channel, the concentration distribution of the suspended particles in the flow can be calculated. Equation (3)-(5) can be solved numerically, so obtaining a layer by layer solution for v_x, v_y and c marching along the axis x (see e.g. [1]). This computational model also allows an iterative method for determining the additional space charge due to the charged dust particles.

3.4 Particle transport

Since the suspended dust particles in a gas are regarded as a continuum, its flow can be characterized by the stream function Ψ. Equation (5) can be written in the following alternative form:

$$\text{div } \vec{j}_{tot} = 0 \tag{7}$$

where $\vec{j}_{tot}$ is the total flux vector of the dust phase:

$$\vec{j}_{tot} = \vec{j}_C + \vec{j}_D + \vec{j}_E \tag{8}$$

where $\vec{j}_C$ means the convective, $\vec{j}_D$ the diffusive and $\vec{j}_E$ the "electrostatical" flux vector-componente of $\vec{j}_{tot}$ total flux vector. Equation (5) can be expressed as:

$$\vec{j}_{tot} = c \cdot \underline{v} - \left(\frac{v_t}{Sc_t}\right) \cdot \text{grad} c + c \cdot \overline{Wth}_y \tag{9}$$

Based on equation (7) we can define the stream function Ψ as:

$$\vec{j}_{tot}\Big|_x = \frac{\partial \Psi}{\partial y} \quad , \quad \vec{j}_{tot}\Big|_y = -\frac{\partial \Psi}{\partial x} \tag{10}$$

Knowing the distribution of dust concentration c and v(x,y) velocity field, the flux vector-components and thereby the stream function $\Psi=\Psi(x,y)$ distribution can be calculated from equation (10). Then using the interpolation method, the streamlines of the dust-phase, $\Psi=$ constant, can be determined. Tangents of the streamlines of the dust-phase were defined to be parallel with the $\vec{j}_{tot}$ total flux vector, that is $\Psi=$constant lines "visualize" the dust-phase motion presented below.

4. Results

Fig.2. shows the equipotential lines in a horizontal cross-section. The upper part shows the potential distribution in the dust-free case, while the lower represents the potential distribution in the presence of dust. The supply voltage is 20 kV, the potential difference between any two neighbouring equipotential line is 1kV.

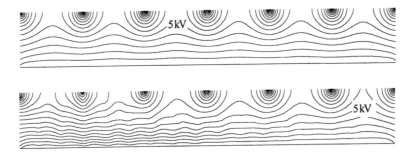

Fig. 2. Potential distribution in the half of an ESP channel

A FORTRAN computer code has been developed for the numerical solution of the two-dimensional transport problem. Fig.3. shows the contour plot of the stream function Ψ. These contour lines are the streamlines of the dust phase moving in the ESP channel from the inlet cross-section toward the outlet cross-section. Particles, in those streamtubes that reach the wall, were precipitated from the gas flow. Dust particles, indicated by the streamtubes passing through the outlet, are not deposited in the ESP. By neglecting the re-entrainment of particles from the wall, the collection efficiency can be calculated from the dust concentration distribution in each cross-section along the precipitator length (Fig.4).

Dependence of the precipitation process on various parameters was investigated with the help of the present numerical simulation. Numerical simulations have been carried out by changing the inlet values of the main parameters. The results are shown in the Figs 5-6. The inlet dust concentration profile was constant $c_{inlet}=10$ [g/m³]. Effect of the near-wall boundary layer on the transport of solid phase is clearly seen from the curvature of the streamlines. Furthermore, the turbulent rediffusion of the suspended dust toward the symmetry plane of the channel can also be seen from the contour plots of Figs 5-6.

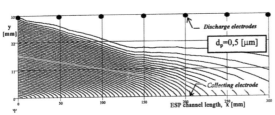

Fig.3. Dust streamlines, contour plot of Ψ at $V_x=2$ m/s

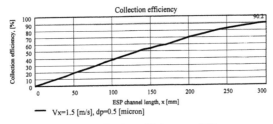

Fig.4: Collection efficiency, η [%]

304

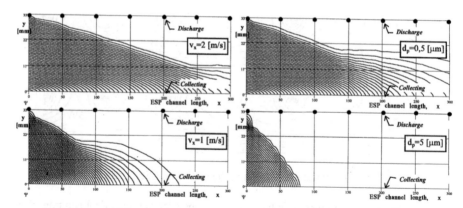

Fig.5 Influence of inlet gas velocity   Fig.6. Influence of dust particle diameter

5. Conclusion

As it is shown in Figs 5-6. the presented various parameters of inlet gas velocity and the dust particle diameter have a well pronounced effect on the collection efficiency. The precipitation of the particles strongly depends on the development of the boundary layer. In the case of the highest velocity flow (v_{inlet}=2 [m/s]) the dust particles pass through the channel rapidly due to the increased convective transport. As it can be seen in Fig.6. the collection efficiency increases with increasing particle size. Particles of diameter d_p=5 [μm] are collected already within the first third of the precipitator channel because of their relative high saturation charge i.e. for the larger particles their migration velocity and electrostatic flux is the predominant effect in the transport process.

References

[1] Fletcher, ·C.A.J. (1991), "Computational Techniques For Fluid Dynamics 2nd Edition, Volumes I-II" Springer-Verlag

[2] Gallimberti, I. : Recent advancements in the physical modelling of electrostatic precipitators. Proc. of 8th International Conference on Electrostatics, 1997, pp 1-30

[3].Houlgreave, J.A., Bromley, K.S.. Fothergill, J.C.: A Finite Element Method for Modeling 3D Field and Current Distributions in Electrostatic Precipitators with Electrodes of Any Shape Proc. of VI. ICESP, Budapest, 1996, pp 154-159

[4] Lu, C. and Huang, H. (1998), „A Sectional Model to Predict Performance of a Plate-Wire Electrostatic Precipitator for Collecting Polydisperse Particles" in *J. Aerosol Sci.* Vol. 29, No. 3, pp. 295-308.

[5] Medlin, A.J. et al (1996), "An Efficient Pseudo-transient Solution Method for Monopolar Corona with Charge Advection and Diffusion" in *Proc. 6th International Conference on Electrostatic Precipitation* (Budapest, Hungary) pp. 107-112.

[6] Riehle, C. (1992), "Bewegung und Abscheidung von Partikeln im Elektrofilter" Dr.-Ing. Thesis, University of Karlsruhe (TH)

[7] Riehle, C. (1996), "Precipitation Modelling by Calculating Particle Tracks in Simulated Flow Fields" in *Proc. 6th ICESP* (Budapest, Hungary) pp. 113-123.

[8] Szedenik, N., Kiss : Particle Charging in Industrial Electrostatics Proc. of 8th International Conference on Electrostatics, 1997, pp 88-92

Inst. Phys. Conf. Ser. No 163
Paper presented at the 10th Int. Conf., Cambridge, 28–31 March 1999

Electric field and space charge density analysis in an AC charger for small particles

K Adamiak* and X Deng[#]

*Dept. of Electrical and Computer Engineering, The University of Western Ontario, London, Ontario, Canada N6A 5B9
[#]Supra Products, Salem, Oregon, USA

Abstract. Numerical simulation of two different models of an AC charger is presented in this paper. The numerical algorithm is based on three techniques: the Boundary Element Method, the Finite Element Method and the Method of Characteristics. The results of simulation show that both chargers produce rather uniform electric field and space charge density distributions in the charging channel. However, one of them produces much higher electric field and charge density, so it can charge particles more intensively.

1. Introduction

As the collecting efficiency of conventional precipitators is poor when particles are small, electrostatic scrubbers are often used [1]; charged dust particles are passed between oppositely charged water droplets. In this case, particle charging and their collecting are separated. When particles are charged by a corona discharge, their trajectories are deflected towards the corona counter-electrode where they can be precipitated or give up their charge. In order to avoid unwanted precipitation of particles in the charger, they are simultaneously subjected to a unipolar ionic bombardment and an alternating electric field [2,3]. As a result, the particles flowing through the device are charged but they are not attracted to the electrodes.

2. Models of AC charger

Two different models of an AC charger are shown in Fig.1. Both have the same configurations of electrodes, but the corona wires and grid rods are excited in different ways. The chargers consist of two parallel grids, C and D, and two sets of corona electrodes, A and B.

In the first version (Charger A), the corona electrodes are excited by two high voltage AC power supplies connected in anti-phase, through the circuit made of diodes and resistors. Charger B has the same configuration of electrodes, but the corona wires are excited with DC power supplies of the same polarity, and the grids C and D are connected to a separate AC power supply.

The main advantage of the AC charger is that there is a negative space charge in the central zone of the charger all the time, but the electric field changes its direction, correspondingly to changes in polarity of the AC voltage source.

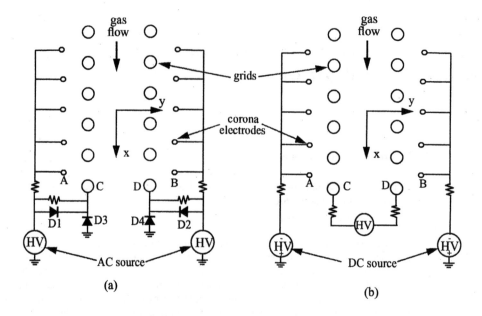

Fig.1 Schematic diagrams of AC chargers

3. Numerical algorithm

The space charge density produced by the corona discharge depends on the magnitude of the electric field. On the other hand, the electric field is function of the space charge density. Therefore, both problems are mutually coupled.

A very non-uniform distribution of the electric field in the area close to the corona wire is a source of difficulties in modelling the corona discharge. When the FEM is applied, a fine discretization of the domain is needed in order to have sufficient accuracy and smoothness of the solution [4]. The BEM usually yields much smoother solution, but it is very time-consuming for Poissonian fields, as the contribution of the non-uniform space charge must be integrated [5].

In this paper, a hybrid technique is used to simulate the corona discharge to overcome these difficulties. The BEM is employed to calculate the Laplacian field of the electrodes. It produces smooth and accurate solution and, as this component of the electric field dominates, the overall field smoothness is sufficient. The FEM is used to calculate the Poissonian component due to space charge only. Space charge is treated naturally in the FEM, so costly numerical integration needed in the BEM can be avoided. After the total electric field is known, the space charge density can be updated by means of the Method of Characteristics [6].

4. Results of simulation

In order to predict the particle charging it is important to know the electric field distribution.

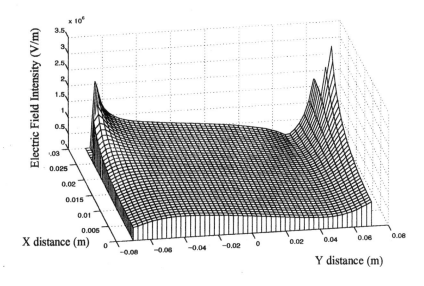

Fig.2 Electric field distribution in the charging zone of Charger A

The stronger the field, the more intensive charging. However, the electric field in the channel is usually much lower than that around the corona wire. Fig. 2 and Fig. 3 show the electric field distribution in the charging zone for both models. The distribution of the electric field in the charging zone of the Charger A is rather uniform with a quite large amplitude.

The Charger B has also a uniform electric field distribution. However, the electric filed intensity is pretty low even when the grid rod and the corona wire are supplied with relatively high voltages of 10 kV and 50 kV, respectively. In practical applications the grid voltage could not be higher than 10 kV, because the air gap between the rod and the wire is small and breakdown may occur, if this voltage is increased.

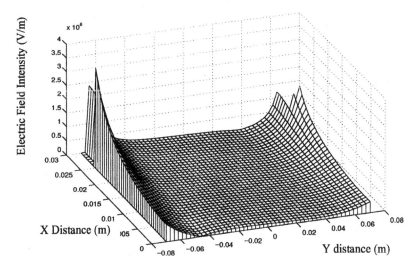

Fig.3 Electric field distribution in the charging zone of Charger B

308

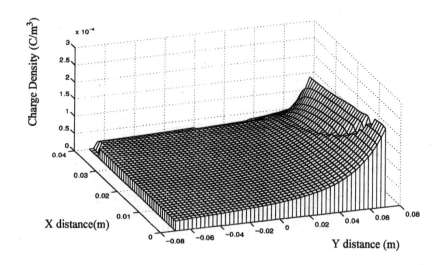

Fig.4 Space charge density in the charging zone in the Charger A

The performance of the charger also depends on the magnitude of space charge in the charge zone. The maximum value of the electric charge collected by a particle depends on the electric field magnitude only, but a particle needs some time to be fully charged. The higher the space charge density, the faster charging. Fig.4 shows the charge distribution in the Charger A. The charge density is rather uniform and again it is much higher in the Charger A than in Charger B.

5. Conclusions

The electric field and ionic space charge density in the charging zone were studied for two models of an AC charger. Both produce rather uniform electric field and space density distributions, independently of the voltage applied to the corona wires. However, their amplitudes are influenced by the voltages on the corona wire and the grid rod.

In Charger A, higher voltage on the corona wire leads to more intense current emitted by the corona wire, a stronger electric field and higher charge density. In Charger B, the electric field depends mainly on the voltage on the grid rod: the higher the voltage, the stronger the filed. The charge density is affected by both the current emitted from the corona wire and the grid voltage. The simulation results show that a higher grid voltage results in more charge drifting into the charging zone.

References

[1] Jaworek A, Krupa A and Adamiak K, 1998 IEEE Trans. on Ind. Appl., 34, 985-991
[2] Masuda S, Washizu M and Mizuno A 1978 Proc. of IEEE/IAS Annual Meeting, Toronto, 16-22
[3] Adamiak K, Krupa A and Jaworek A 1995 Inst.Phys. Conf. Ser. 143, 275-278
[4] Adamiak K 1994 IEEE Trans. on Industry Appl., 30, 387-393
[5] Adamiak K 1991 In C.A. Brebbia and G.S. Gipson (eds.) "Boundary Elements XIII", Computational Mechanics Publications, Southampton 407-418
[6] Butler A J, Cendes Z J and Hoburg J F 1989 IEEE Trans. Ind. Appl. 25 533-538

309

DISCUSSION - Section E - Applications and Processing

Title: Marking with Electrostatics
Speaker: D Hays

Question: M K Mazumnder
Could you give us some examples of photoconductive powders? Are these commercially available? What are the practical problems?

Reply: Your question relates to a discussion concerning the class of electrostatic marking physics in which an image dependent electrostatic force is obtained through a variation in the particle charge under constant electric field conditions. One method for obtaining an image dependent variation in particle charge is to use powder that is photoconducting. To my knowledge, there are no commercial sources of such materials since there are no products based on this imaging physics. However, about 25 years ago the Japanese ship building industry used this imaging physics to mark large steel plates for cutting. The process was to coat the plates with a powder (large size) containing zinc oxide, charge the powder in the dark, expose the powder to an image from a projector, and then remove the light exposed (discharged) powder with an air jet. Although this marking method was suitable for large scale marking, it is not practical for document printing that requires a particle size of approximately 10 micrometers.

Title: Enhancement of Transfer Efficiency and Appearance in Automotive Powder Coating
Speaker: R A Sims

Question: Trevor Willliams
Do you think you have anything to learn from the electrostatic crop spraying research which advocates alternating polarity to prevent charge build-up? This could help with back-corona problems.

Reply: Yes, we have discussed this as a way of reducing back corona. However, there are many problems involved with the switching of the high corona voltages. We have discussed methods to surmount these problems; including multiple guns (+ - + - etc) along a conveyor line.

Title: Electric Field Interaction with Debris Generated by Laser Ablation of Polyimide Films
Speaker: C Hayden

Question: D K Davies
Positive charging of polymers can occur under UV irradiation. Could this be used for debris removal with appropriate fields?

310

Reply: This is an interesting idea and could have possibilities. However, the principle of this work was to have an independently charged object to move debris from any substrate onto a replaceable object. Because the exact mechanism of charging is not known, we cannot be sure if ultraviolet charging of a polymer would be constructive. Also, the use of ultraviolet light would require an additional light source as the excimer laser produces a high intensity beam that ablates the material and does not expose the surface.

Section F
Materials Evaluation
and Applications

Inst. Phys. Conf. Ser. No 163
Paper presented at the 10th Int. Conf., Cambridge, 28–31 March 1999
© *1999 IOP Publishing Ltd*

Inherently-conductive polymers as antistatic agents for textiles

Timothy Derrick, Poopathy Kathirgamanathan, and Michael J. Toohey

Centre for Electronic Materials Engineering, School of Electrical, Electronic and Information Engineering, South Bank University, 103 Borough Road, London SE1 0AA, UK

John N. Chubb

John Chubb Instrumentation, Unit 30 Landsdown Industrial Estate, Gloucester Road, Cheltenham, Glos., GL51 8PL, UK

Abstract. Textile materials based on cotton and aramid have been coated with the inherently-conductive polymer, poly(pyrrole) by means of the *in situ* polymerisation technique. Surface resistances in the range of 10^7 to 10^{11} Ω can be obtained, with good adherence and uniformity of coating. Longer reaction times typically lead to lower resistance. Surface voltage decay time measurements after corona charging indicate that poly(pyrrole) coatings are effective in conducting static charges to earth, with decay times of the order of 25 ms for the more conductive samples. Poly(pyrrole) coating is a cheap and simple strategy for improving the static-dissipative properties of textile materials and finished garments.

1. Introduction

Conventional textiles used in clothing and upholstery have surface resistances from 10^{12} Ω (e.g. cotton) to over 10^{15} Ω (e.g. aramids). Triboelectrification of such textiles can cause serious problems: the high potentials involved can damage semiconductors, and sparks from charged textiles are able to cause ignition of flammable gases [1].

Methods for making static-dissipative textiles include weaving in conductive fibres, or applying chemical 'antistats'. Although effective, they have drawbacks: when garments are made from textiles containing conductive threads, special attention must be paid to ensure that adequate conductive pathways exist across seams (BS 2782 requires a cuff-to-cuff resistance measurement); chemical antistats working by adsorption of ambient moisture may have conductance which depends on relative humidity (RH).

Oxidation of aniline, pyrrole, etc. in aqueous solution yields 'inherently-conductive polymers' (ICPs), which have volume conductivities of up to 10^5 S cm^{-1} [2,3]. Reports of ICPs coated onto textiles by placing the substrate in the polymerising monomer solution stated that surface resistances (ρ_s) $10^1 - 10^7$ Ω were obtained [3-6]. Suggested applications included electrically-heated clothing [7].

We considered that such ICP-coated textiles would also have good static-dissipative properties, and therefore we prepared a range of materials of this type, to assess the accessible range of resistances, the quality of the coating, and their ability to dissipate static charges. We chose two materials with disparate chemical structure, moisture-absorbing characteristics and initial surface resistance levels to demonstrate the principle. Poly(pyrrole) was chosen for its ease of synthesis.

2. Experimental account

2.1 Preparation and characterisation of poly(pyrrole)-coated samples

We used an adaptation of the method in [4], for 40x40 cm pieces of cotton (thickness 1 mm) and aramid (thickness 0.8 mm) textile. Surface resistances before treatment were measured at 25% RH by the method specified in EN 1149-1.

Each was washed in detergent and suspended in a 10 litre reaction vessel filled with an aqueous solution of ferric chloride (technical grade, as supplied), to which pyrrole (redistilled) was added. In some cases, a cationic surfactant ('Glokil', Rhône - Poulenc) was also added. Reaction mixtures were stirred continuously at ambient temperature (ca. 15 °C). After 3-24 hr. reaction time, the sample was removed, and rinsed vigorously in tap water. It was then dried under vacuum at 60 °C for 3 hr. Table 1 lists the preparative conditions used.

Representative coated and uncoated samples were examined by scanning electron microscopy, and by thermogravimetric analysis under flowing N_2 (25-900 °C).

2.2 Electrical characterisation

Surface resistance of the coated samples was measured by the 2-probe method, using brass ring-disc electrodes. The upper limit for measurement was ~3 x 10^{12} Ω. Samples were conditioned for 24 hr in a constant RH (30 ± 5%) environment at room temperature before measurement. Table 1 shows the surface resistances of the samples 1 day after preparation and approximately six weeks later.

2.3 Charge decay and tribocharging measurements

Static-dissipative characteristics of the textiles were measured with a JCI 155 charge-decay meter. This instrument corona-charges the sample for a fixed period, then rapidly (within 10 ms) moves a field-meter over the charged region, to monitor the initial surface voltage, and its decay. Conditions were: corona voltage -9063 V for 20 ms, or -9688 V for 100 ms; samples resting on a metal ground plane; 19 °C and 59% RH.

We used a simple manual tribocharging method slightly adapted from that used by the

		Aramid			
Pyrrole concentration / mM	Ferric chloride concentration / mM	Surfactant concentration / % v/v	Reaction time / hr	Initial surface resistance^a, ρ_s / Ω	ρ_s after *ca.* 6 weeks^a / Ω
7.5	30	0.01	4.33	7.3×10^6	4.5×10^7
8.9	30	0.01	6.3	6.5×10^7	1.1×10^8
7.5	30	0.01	12.5	2.2×10^8	2.9×10^8
3.7	23	0	5.5	3.7×10^8	1.6×10^9
3.7	23	0	5	9.6×10^8	7.2×10^9
3.3	23	0	3.5	2.5×10^{10}	4.7×10^{10}
3.5	23	0	4.8	8.8×10^{10}	4.7×10^{11}
2.2	14	0.01	19.3	$>3 \times 10^{12}$	N/A
		Cotton			
10	30	0.1	14	3.1×10^6	7.0×10^7
4.5	30	0.1	14	3.9×10^6	4.0×10^8
4.5	30	0.05	12.5	1.7×10^7	1.9×10^8
4.5	30	0.03	8	1.9×10^8	5.5×10^8
4.5	30	0.03	10	1.6×10^9	4.8×10^{10}
0	30	0.05	5	2.9×10^9	7.9×10^9
4.5	30	0.03	3.3	2.5×10^{10}	$\sim 3 \times 10^{12}$
6	30	0.05	3	$>3 \times 10^{12}$	N/A
1	30	0.1	22	$>3 \times 10^{12}$	N/A

Table 1. Preparative conditions and surface resistance for poly(pyrrole)-coated textiles

a Mean of values at six positions on sample surface; standard deviations ~60%.

British Textile Technology Group [8]. Textile samples were held in a circular Perspex frame (electrically isolated), and manually rubbed with a wool pad. Sample charge was measured using a Faraday pail. Typically, we repeated the charging process 20 times, on a range of textiles with and without any antistatic component, and noted the *maximum* value of Q_{surf}, rather than the mean, to indicate the 'worst-case' behaviour of the textiles.

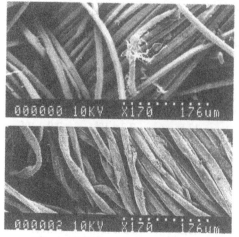

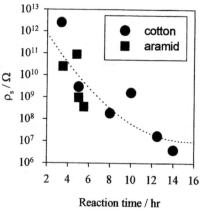

Figure 2. Tuning surface resistivity
by reaction time. Initial pyrrole concentration:
4.5 mM (cotton); 3.3 - 3.7 mM (aramid)

Figure 1. SEMs of poly(pyrrole)-

coated aramid (top) and cotton

As Figure 1 indicates, the treated fibres of both aramid and cotton have smooth, uniform coatings of poly(pyrrole), consistent with previous reports [4,5]. The poly(pyrrole) grows onto the fibres, rather than simply existing as separate particles trapped in the fabric weave. We estimate that the coatings are less than 1μ thick.

We performed TGA to assess the amount of poly(pyrrole) present, however, it was not possible to resolve a separate mass-loss region due to the coating. The decomposition temperature is clearly determined primarily by the substrate textile, although it is slightly reduced by the poly(pyrrole), by about 11 °C (aramid) and 30 °C (cotton).The thermal stability of textiles has safety implications; TGA answers the question as to whether ICP coatings have deleterious effects on thermal stability of the coated materials.

3.2 *Electrical characterisation*

Table 1 indicates that ρ_s of the coated textiles could be decreased by increasing the pyrrole concentration in the reaction mixture. For initial pyrrole concentrations ≥ 2.5 mM, ρ_s could be controlled by reaction time (Figure 2). The surfactant was included to improve the uniformity of coverage: the less-conductive coatings could be patchy without it.

We found that reproducibility of resistance values was in general rather poor, although this could be improved by effective mixing of the reaction mixture. This may be due to imperfect sample preparation, or to percolation threshold effects for coatings growing out from distinct nucleation sites.

Resistances increased immediately after preparation, but then remained fairly stable for about six weeks. Changes were about an order of magnitude, probably acceptable for antistatic applications. The increase in resistance is due to the degradation of the conjugated polymer backbone by oxygen and moisture, leading to oxidised forms of the pyrrole moiety.

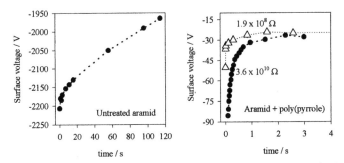

Figure 3. Decay of surface voltage following corona charging of aramid textile untreated, and with poly(pyrrole) coatings of different surface resistivity. Corona voltage: -9063 V, charging time: 20 ms

Larger changes were sometimes observed after this period, which merits further study, as we have not observed it for other ICP-coated materials.

3.3 *Charge decay and tribocharging measurements*

Figure 3 shows typical decay curves for corona-charged samples. Most of the decay curves did not fit a simple exponential decay, having instead a fast initial decay followed by a slower 'tail'. Also, very conductive samples dissipated almost all deposited charge in the interval between the end of the corona pulse and the start of measurement (about 25 ms). Here, the voltage at start of measurement, V_{pk}, is a better measure of dissipation than 'decay times' which only relate to insignificant residual voltages.

The untreated cotton had mean V_{pk} of -130 V. The decay curve was adequately represented by a single exponential, with a time constant of ~75 ms. All the poly(pyrrole)-coated cotton samples exhibited lower V_{pk} (-27±4 V). Thus, the cotton-based textiles - aided here by high RH - should not cause any static-related problems under these conditions. However, the untreated aramid displays high V_{pk} (mean -1124 V) and charge decay time constant (> 10^3 s). V_{pk} can be reduced to approximately 50 V by a poly(pyrrole) coating with ρ_s of 10^9 Ω or less.

The tribocharging method used, though simple, reproduces charging processes (ie the rubbing rates and pressures) expected for clothing, etc. Results were for isolated samples, thus show tribochargeability rather than charge dissipation. In practice, textiles will be earthed, either adventitiously through the wearer's body, or deliberately by means of a ground connection; but we considered that attempting to simulate this poorly-specified resistance to ground would complicate the measurement unduly. Figure 4 indicates that highly resistive (>10^{12} Ω), untreated samples charge highly, whereas samples with

318

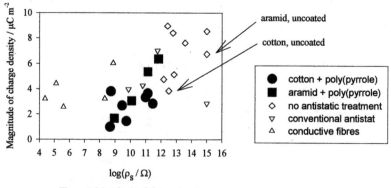

Figure 4. Magnitude of charge density
obtained by tribocharging various textiles

conductive threads and low overall ρ_s (10^4 to 10^8 Ω) are less chargeable. The poly(pyrrole)-coated materials show two characteristics:

(i) at least for the aramid-based samples, chargeability is strongly dependent on surface resistance (for some higher-ρ_s aramid materials, the substrate textile may also be involved)

(ii) at moderate resistances (10^8 - $10^9 \Omega$) we observe lower chargeabilities for these samples than for any of the other antistatic textiles examined, including the steel-containing ones.

This behaviour is likely to result from the homogeneity of the poly(pyrrole) coating. During the tribocharging process, charge on the textile is more able to flow back to the rubbing material because its coating has appreciable conductivity at all points of contact. Heterogeneous materials, when isolated, have insulating islands between the conductive threads on which charge can become trapped. When correctly grounded, the high potential difference between the 'islands' and the earthed conductive grid drives the charge to be rapidly dissipated, but the homogenous, ICP-coated materials may be useful where grounding is imperfect.

4. Conclusions

We have demonstrated that coating textiles with poly(pyrrole) by means of *in situ* polymerisation can produce uniform coatings with ohmic surface resistances in the range 10^7 - 10^{10} Ω. Tuning of resistance can be achieved by varying the reaction time or monomer concentration. Such coated textiles can have excellent charge dissipation characteristics. They also appear to be more difficult to tribocharge than many existing static-dissipative textiles which have inhomogeneous structures of comparable surface resistance. This may be advantageous in situations where grounding of the textile is imperfect.

5. Acknowledgements

The assistance of P. Holdstock (BTTG, UK) and J. Haase (STFI, Germany) is gratefully acknowledged. We thank South Bank University for support.

6. References

[1] Wilson N 1985 *J. Electrostatics* **16** 231

[2] Monkman A 1995 in *Introduction to Molecular Electronics* (Petty M C, Bryce M R, and Bloor D, eds.) (London: Edward Arnold) ISBN 0-340-58009-7

[3] Kathirgamanathan P 1991 in *High Value Polymers, Special Publication No: 87* (London: Royal Society of Chemistry) ISBN 0-85186-867-3

[4] Gregory R V, Kimbrell W C, and Kuhn H H 1989 *Synth. Met.* **28** C823-835

[5] Gregory R V, Kimbrell W C, and Kuhn H H 1991 *J. Coated Fabrics* **20** 167-175

[6] Kathirgamanathan P 1993 *Adv. Mater.* **5** 281-283

[7] Boutrois J P, Jolly R, and Pétrescu C 1997 *Synth. Met.* **85** 1405-1406

[8] P. Holdstock, British Textile Technology Group, personal communication

Inst. Phys. Conf. Ser. No 163
Paper presented at the 10th Int. Conf., Cambridge, 28–31 March 1999
© *1999 IOP Publishing Ltd*

W. Keith Fisher
Solutia Inc., 3000 Old Chemstand Road, Cantonment, Florida 32533 USA

Abstract. The purpose of this work was to investigate mechanisms of static dissipation in carpets containing electrically conductive yarn. Results indicate that ions produced by corona discharge around conductive fibers in carpet were an important mechanism of static dissipation. Cut pile and level loop carpets made from BCF (bulk continuous filament) yarn with and without conductive latex backing were investigated. Also individual carpet yarns were tested for their corona discharge characteristics.

1. Introduction

It is well known that conductive flooring (<~106 ohms to ground) and dissipative flooring (10^6 to 10^9 ohms to ground) are effective in limiting personnel voltage [1,2]. However, incorporating conductive fibers in carpet, even as isolated conductors, results in significant reduction, via corona ionization, of the amount of charge and therefore body voltage developed by a person walking on a carpet. [3,4]

2. Experimental

Figure 1 shows a cross-section of the conductive filaments used in this study. They are bicomponent filaments where the conductive (black) stripe was a carbon-nylon 6 concentrate and the remainder was nylon 6,6 [5]. These type of conductive yarns, produced by Solutia Inc., are widely used in carpet for static dissipation.

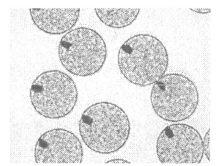

Figure 1. Cross Sections of Bicomponent Conductive Yarn

322

To make the anti-static carpets used in these experiments, one end of conductive yarn was inserted into a 3790 denier[1], 70 filament nylon 6,6 thread-line. Typically the conductive yarn is 1% to 2% by weight of the total yarn. This yarn combination was drawn, crimped, cabled with another identical end and then heat set. These yarns were tested as made or tufted into carpet.

3. Conductive Yarn Corona Test

This test was designed to measure the corona current as a function of applied voltage of carpet yarn containing conductive filaments. A 0.038 meter (1.5 inches) length of carpet yarn was lowered into a Faraday cup until the tip of the yarn was 0.0032 meter (1/8 inch) from the bottom of the cup. The Faraday cup was grounded through a sensitive current meter (Keithly Model 610 C). The other end of the yarn was connected to a dc power supply capable of providing up to 5000 volts. The end of the yarn which was connected to the power supply was coated with colloidal silver paint to ensure good electrical connection to the conductive filaments within the carpet yarn bundle.

Figure 2 shows typical results of this test for carpet containing No-Shock® yarn. The dc linear resistance of this yarn was about 10^6 ohm/cm The corona inception voltage was difficult to determine but was around 2.1 kV. The corona current increased approximately linearly with applied voltage for voltages greater than the corona inception voltage.

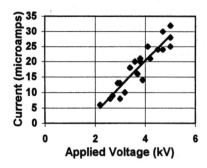
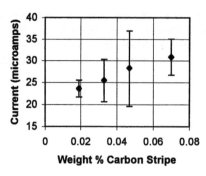

Figure 2. Corona current vs. applied voltage for No-Shock®conductive yarn

Figure 3. Corona current measured at 5 kV vs. Weight Percent Conductive Stripe

4. Relation Between Body Voltage And Corona Current

Figure 3 shows that the corona current tended to increase as the weight percent conductive carbon stripe content in the carpet yarn was increased. The data in Figure 3 was from finished carpet yarn just before tufting. The yarns from Figure 3 were tufted into a set of carpets of identical construction. The body voltage accumulated by a person walking on these carpets was measured, in accordance with standard

---

[1] Denier is a measure of yarn linear density. 1 denier equals 1 gram/9000 meters

AATCC method 134, "Walk Test," (20% rh, 21 C), and plotted in Figure 4 vs. the corona current results from Figure 3. The data show a trend of decreasing body voltage with increasing corona current.

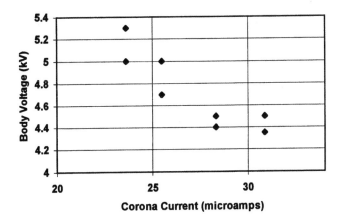

Figure 4.   Body voltage for cut pile carpets with standard latex vs. corona current at 5 kV

Figures 3 and 4 indicate that a point is reached where addition of more carbon does not give further decrease in body voltage, at least for the case of isolated conductors.

5. Carpet Corona Discharge Test

The purpose of these experiments was to develop a test method for evaluating the static dissipating capability of carpet. The test consisted of measuring the change in charge density on an initially charged Teflon™ sheet after the sheet had been passed over, but not contacting, a small carpet sample. The charged sheet was intended to simulate the charge distribution on an approaching shoe sole. The surface charge density on the Teflon™ sheet was measured using an electrostatic voltmeter, the sensing head of which was kept at a fixed distance below the sheet before and after it had been passed over the carpet. The voltage on the sensor was directly related to the surface charge density on the Teflon™ sheet. Details of this test apparatus are given in reference 3.

Figure 5 shows test results as plots of sensor voltage vs. time. The charge density on the Teflon™ sheet was reduced when passed over carpet containing conductive yarn. Since the carpet did not contact the charged sheet, the mechanism of static dissipation must be charge recombination with free ions produced by corona discharge around the conductive yarns. The presence of conductive latex resulted in a greater decrease in sensor voltage as well as lower body voltage than when the conductive filaments were isolated. The conductive latex provides a reservoir of electrons, sort of a pseudo-ground, to help maintain the potential difference between the charged sheet (or shoe sole) and the conductive filaments, thereby enhancing

324

corona discharge. Similar behavior, but to a lesser extent, was observed for level loop carpets compared to isolated conductors in cut pile carets.

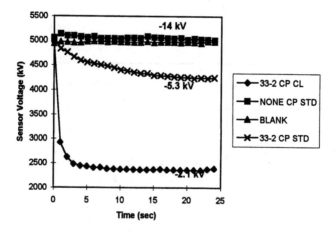

Figure 5. Sensor voltage vs. time for the Carpet Corona Test. Body voltage results are shown next to the curve for each carpet. CP is cut pile, STD is standard latex, CL is conductive latex, None indicates carpet with no conductive yarn and Blank means no carpet present.

6. Conclusions

Results of both of the tests described in this work show the importance of corona discharge in dissipating static electricity in carpet. The corona current measured in the Yarn Corona Test and the sensor head voltage from the Carpet Corona Test were correlated with observed body voltage. Both tests have promise as tools to predict in-use static dissipation performance of finished carpets.

References
1. Fowler, Stephen, "ESD Carpeting Comes of Age," Northern Telecom International Conference on ESD, May 19, 1994.
2. Rawlston, R.H., Why the Static, Textile Industries, p. 85, May, 1969.
3. Kessler, L. and W. K. Fisher, J. of Electrostatics, 39 (1997) 253-275.
4. MaClaga, B.and Fisher, W.K, in print.
5. Boe, Norm, US Patent 3,969,559 7/76
6. ESD Advisory 11.2, Tribocharging, ESD Association, Rome, NY., pp. 76, 119.
7. Von Engle, A., Electric Plasmas, Their Nature and Uses, Taylor and Francis, New York, p. 170 (1983).

Inst. Phys. Conf. Ser. No 163
Paper presented at the 10th Int. Conf., Cambridge, 28–31 March 1999
© 1999 IOP Publishing Ltd

Evaluation of charge retention properties of powder paints by thermally stimulated current spectroscopy

T.Ogiwara and K.Ikezaki

Department of Applied Physics and Physico-Informatics, Faculty of Science and Technology, Keio University, Yokohama, JAPAN

Thermally stimulated current (TSC) spectra of a powder paint for electrostatic powder coating were examined to clarify the storage temperature dependence of the charge retention capacity of the paint.   When the sample paint was stored at 50 °C for only 10 min, a drastic change in its TSC spectrum was induced and the charge retention capacity of the paint decreased.   TSC spectral change arising from the cure process of the paint was also examined for the sample paint pre-heated above 100 °C.

1. Introduction

For the electrostatic powder coating technology, charging characteristics of powder paints are one of the most important factors and the charging characteristics depend on both the generation and retention of charge.   It is generally known that charge retention characteristics are best examined by the TSC spectroscopy [1]. Recently, the science and technology of the electrostatic powder coating was reviewed [2]. However, reports of the electrical characteristics of powder paints themselves are very few [3], and as far as we know, TSCs of powder paints have not been reported yet.

Previously, we reported [4] charge decay characteristics of powder paints for electrostatic powder coating.   The charge decay at 40 °C obeyed the well-known charge decay empirical relation for powder systems [5, 6], while at 50 °C it progressed more rapidly and deviated from the empirical relation.   We found that this deviation and the observed decline in the charge retention capacity of the powder paint was due to change in the molecular aggregation state of the powder paint during the charge decay measurement at 50 °C.

Therefore, we will report in this paper pre-heating effect on the charge retention power of powder paints by the TSC technique in more detail.

2. Experimental

Sample powders of 0.12 g were pressed at a pressure of 55 kgf/cm² in an Al pan-shaped sample holder with a diameter of 20 mm and a depth of 1 mm in order to form thin disk-shaped sample compactions for observing the TSC spectra and the iso-thermal charge decay.   The Al sample holder was also used as an grounded electrode for TSC and charge decay measurements.

Effects of pre-heating at higher temperatures were also examined. For this purpose, sample powders were heated at different temperatures higher than 100 °C for 10 min for most of the treatment temperatures. For pre-heating at 150 °C, the sample powders were heated for different times up to 180 min.

For TSC observation, these pre-heated sample compactions were negatively corona charged for 5 min in air at room temperature. The initial surface voltage of these compactions was -500 V. TSC was observed at a constant heating rate of 3.5 °C in open circuit condition.

3. Results and Discussion

3.1 Low temperature heat treatment

Fig.1 presents the TSC spectra of the powder paint pre-heated at 50 °C for different times in air together with that of the untreated powder for comparison. As shown in Fig.1, the TSC spectra of the paint pre-heated at 50 °C drastically changed even for a short period of the heat treatment: TSC began to flow at 40 °C and the temperature of the TSC maxima shifted to a low temperature side by about 2 °C from that of the untreated one. With increasing heat treatment time, the intensity of the lower temperature parts of the TSC rapidly increased at the start and then saturated. It is also shown in Fig.1 that two TSC spectra for the samples which were pre-heated for 30 min and 9 h are almost the same. These TSC observations for the pre-heated samples clearly show that the charge retention capacity drastically decreased when the sample powder paint is stored at 50 °C for only 10 min.

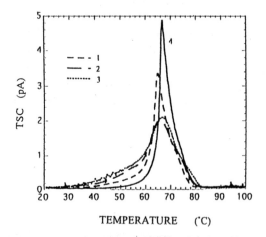

Fig.1 TSC spectra of powder paints pre-heated at 50 °C for (1) 10 min, (2) 30 min, (3) 9 h and (4) of the untreated paint.

To confirm the above-mentioned TSC result, we directly observed at 40 °C the iso-thermal surface voltage decay of corona-charged sample powders which were heat-treated at 50 °C for periods of 10 and 30 min. The results showed that the iso-thermal surface voltage decay was strongly affected by the heat treatment at 50 °C in agreement with the prospect from the TSC results.

3.2 High temperature heat treatment

The powder paint used in this study began to melt above 100 °C. Therefore, the sample paint changed in its form from powder to melt-cast film when the paint was heat-treated at high temperatures above 100 °C. For these melt-cast films prepared by heating at different temperatures for 10 min, we also observed their TSC spectra from room temperature to 120 °C. Observed TSC results are shown in Fig.2. As the treatment temperature increased, the small TSC band appearing around 50 °C faded. As for the main TSC band, the peak temperature shifted to the high temperature side with increasing treatment temperature and finally converged to 94°C when heated above 180 °C. The observed peak shift amounts to more than 25 °C unlike the case of low temperature treatment where the peak temperature of the main TSC band hardly shifted. The peak shift of the main TSC band is a remarkable difference between the low and the high temperature heat treatments.

Generally, in thermoset powder paints like the paint used in this study, heating the powder paint brings about a cross-linking reaction. Therefore, it is reasonable to attribute the observed peak shift of the main TSC band to change in the molecular aggregation state of the paint induced by the cross-linking reaction. From these TSC spectral change, we can explore not only the charge retention power but also the

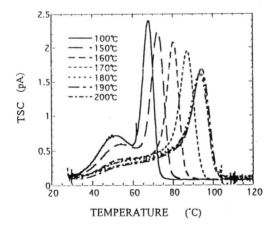

Fig.2 TSC spectra of melt-cast paint films pe-heated at different temperatures for 10 min. Heating temperature is presented in the figure.

degree of progression of the cross-linking reaction, that is, the curing process of the powder paint.

To know the degree of the progression of the cross-linking reaction, the peak temperature of the main TSC band was examined for the powder paint pre-heated at 150°C for different times. It was found that the peak temperature asymptotically approached the plateau temperature of 94 °C according to the following equation:

$$T_P(t) - T_P(\infty) = \{T_P(0) - T_P(\infty)\} \, exp\left(-\frac{t}{\tau}\right) \qquad (1),$$

where, $T_P(t)$, $T_P(\infty)$ and $T_P(0)$ are the TSC peak temperatures of the sample heat-treated for a period of t and a prolonged time, and the peak temperature of the untreated sample, respectively. The time constant τ in eq.(1) was found to be 45 min for the treatment temperature of 150 °C.

5. Conclusions

In this study, we examined the heat treatment effect on the TSC properties of the powder paint and some conclusions can be derived as follows :

(1) Charge retention power of the powder paint markedly decreased when heated at 50 °C for a short period of 10 min. On the contrary, when heated at high temperatures above 100 °C, it was prospected from the TSC spectra that the charge retention power greatly increased.

(2) The rate of cross-linking reaction proceeded in the heated powder paint can also be estimated from the peak shift of the TSC band. For the paint used in this study, the reaction proceeded in an exponential way with a time constant of 45 min at the cure temperature of 150 °C.

References

[1] Carr S H 1982 *Electrical properties of polymers* ed. Seanor D A (New York: Academic) p215-39
[2] Bailei A G 1998 *J. Electrostatics* **45** 85-120
[3] Cheever G D 1975 *J.Appl.Polym.Sci.* **19** 147- 63
[4] Ogiwara T, Nakayama F and Ikezaki K 1998 *ESA-IEJ Joint Symposium on Electrostatics 1998 Proceedings* (Morgan Hill, California: Laplacian) p252 - 61
[5] Takeuchi M, Nagasaka H 1982 *J.Electrostatics* **13** 175- 84
[6] Sugai M, Takeuchi M and Nagasaka H 1976 *Jpn.J.Appl.Phys.* **15** 1563-64

Inst. Phys. Conf. Ser. No 163
Paper presented at the 10th Int. Conf., Cambridge, 28–31 March 1999

The assessment of materials by tribo and corona charging and charge decay measurements

John Chubb

John Chubb Instrumentation, Unit 30, Lansdown Industrial Estate, Gloucester Road, Cheltenham, GL51 8PL, UK.
(Tel: +44 (0)1242 573347 Fax: +44 (0)1242 251388 email: jchubb@jci.co.uk)

Abstract: Many risks and problems from static electricity arise from charge retained on materials. Studies show that the nearby influence of tribo charge depends on the structure of the material. Simple fabrics show a low 'capacitance loading' effect that varies little with the quantity of charge. Complex fabrics, including conductive threads, may show much larger 'loading' effects. A high loading effect provides an alternative way to limit the influence of retained charge on materials with long charge decay times. It is shown that charge decay times (peak voltage to 1/e of this) need to be less than 0.2s to usefully limit initial peak surface voltages. Generally comparable results are obtained with corona charging. The opportunity is hence shown available for a fair and generally applicable way to assess materials by easy to use instrumentation.

1. Introduction

Assessment of the risks and problems from static electricity involve such basic questions as:

    a) whether significant potentials can arise and be retained on the surfaces of materials after they are rubbed

    b) whether incendive discharges can be drawn from charged materials (particularly if conductive components are included)

    c) whether materials charge other materials rubbing against them (e.g. walking on carpets or garments on people getting up from seating)

    d) whether shielding is provided against electrostatic field transients.

No one single test method will properly and fairly assess all these aspects of the electrostatic performance of materials for all applications.

Information on whether significant levels of static charge will arise on materials when they are rubbed and give significant surface potentials will, however, provide assessment for many of the major types of risks and problems. The present paper is concerned with new approaches to the assessment of materials in terms of the surface potentials likely to arise and their time duration.

2. Methods for assessment of retained charge

2.1 Method for tribocharge measurements

In practical situations electrostatic charge arises on materials by contact or rubbing actions with other materials. Methods to assess materials need to be based on triboelectric charging or be shown to relate to it.

330

A simple experimental approach has been developed based on measurement of the charge transferred by rubbing, the initial peak signal observed by a nearby fieldmeter and of the rate at which this signal decreases as the charge dissipates. A charge neutralised PTFE rod is used to 'scuff', wipe or impact the middle of an area of stretched sample material directly under a fieldmeter 100mm above the test surface. Observations by the fieldmeter show a brief initial excursion as the PTFE rod rises from the surface then a polarity reversal, as the rod is swung quickly away, up to a peak value from which the reading decays as charge migrates out over the surface of the rubbed material.

The quantity of charge transferred to the surface is measured by inserting the rubbed end of the PTFE rod into a Faraday Pail. Charge values are measured to about 10pC. Fieldmeter readings are measured with a resolution better than 1mV (corresponding to 1V of surface potential on a large plane target at 100mm). Readings are recorded either directly into a microcomputer with time steps of ¼s or using a Picoscope digital storage oscilloscope. It was shown that the fieldmeter signals relate to the quantity of charge available to couple to the fieldmeter sensing aperture and do not depend on the charge area if this is small. Isolated charged discs up to 26mm diameter gave a reading of 0.14mV/pC at 100mm.

2.2 Results from tribocharging studies

Three main results arose from the studies on a variety of 'simple' materials - a polyester fabric, a white lingerie fabric, a washed lingerie fabric, a handkerchief of cotton and other fibres and a polythene bag. First, the initial peak fieldmeter reading varied in proportion to the quantity of charge transferred. Second, the signals were less than applied for a freely supported charge. Third, the maximum initial fieldmeter signals reduced as the time for charge decay got shorter.

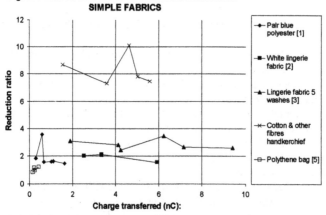

Figure 1: REDUCTION RATIO OF FIELDMETER SIGNALS - SIMPLE FABRICS

The above results suggest that fieldmeter signals are reduced by a 'capacitance loading' effect on the surface charge. This effect seems to be a simple, operator independent, way to assess the influence expected per unit of charge near a tribocharged material.

Rapid dissipation of charge on materials is a way commonly used to control static risks. Figure 2 shows the observed variation of fieldmeter response with decay time. The fieldmeter readings at very short decay times may be low partly, at least, because of the time to move the rubbing material away - but this is also the reality of practical situations.

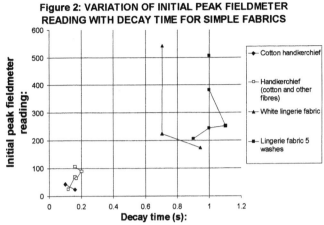

Figure 2: VARIATION OF INITIAL PEAK FIELDMETER
READING WITH DECAY TIME FOR SIMPLE FABRICS

Cleanroom garments are usually based on use of polyester filament fabric and a grid or stripe pattern of conductive threads into the fabric. The idea is to suppress surface voltages by proximity of charge to the conductive threads. The threads may have surface or core conductivity.

Tribocharging studies were made on a number of these complex structure fabrics having 5mm, 7.5mm and 10mm grid patterns and a 20mm stripe pattern. Results show the 'capacitance loading' effect can be much higher than for simple fabrics and may increase with the quantity of charge. The higher values are usually associated with a closer spacing of conductive threads and with surface, rather than core, conductivity. The high values of capacitance loading mean that initial peak values of surface voltage were low. Charge decay times were many seconds on most of these materials.

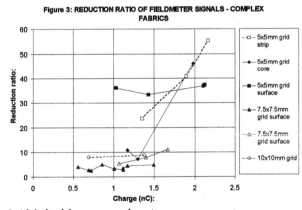

Figure 3: REDUCTION RATIO OF FIELDMETER SIGNALS - COMPLEX FABRICS

2.4 Method for corona charging measurements

The method for tribocharging measurements described above is simple, but only really suitable for experimental studies. Corona charging provides an easier to use, more consistent and less operator dependent way to make measurements of the charge decay characteristics of materials in industrial situations under selected conditions [1,2,3]. Studies were made with corona charging on many of the same materials as studied above to see whether comparable behaviour was observed.

A JCI 155 Charge Decay Test Unit was used as the basis for corona charging studies. The quantity of charge transferred to the sample was measured as a combination of the

332

charge conducted laterally outwards to the two plates mounting the sample and the charge induced on an electrode beneath the open backing region of the sample test area. The approach was an enhancement of that used previously [2]. The 'induction' electrode was an approximate geometric match for the sensing region of the JCI 155 above the sample - so about half the charge retained couples to the induction sensor. This proportion is established by corona charging, for example, a Melinex film sample and then carefully transferring this to a Faraday Pail for total charge measurement. The charges received by the induction sensor and by the mounting plates are measured by virtual earth charge sensing amplifiers. These measurements are recorded separately on a Picoscope digital oscilloscope and then appropriately combined to give the total charge received. The virtual earth charge amplifiers have built in relaxation time constants of 2s each, so they are effectively self-zeroing between tests.

The sensitivity to charge on a small isolated disc in the plane of the test aperture varied from about 0.83/pC for very small charges to 0.62/pC at 20mm diameter.

Measurements were made of the corona charge received by the sample (as above), the initial peak fieldmeter reading and the charge decay characteristics. 'Capacitance loading' values were calculated from the initial peak readings and the charge transferred in comparison to the readings expected for these charges in isolation.

2.5 Results from corona charging studies

With simple materials the 'capacitance loading' is fairly independent of the quantity of charge transferred. Values are comparable to those observed with tribocharging. With the complex fabrics that include conductive threads, the 'capacitance loading' usually increased with quantity of charge. At low quantities of charge, comparable to those used in the tribocharging studies, the capacitance loading effect is generally comparable. At high quantities of charge the loading tends to become proportional to quantity of charge - and this means that the readings are tending to plateau out at high levels of charge.

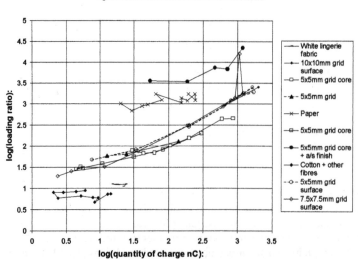

Figure 4: VARIATION OF LOADING FOR CORONA CHARGING

2.6 Interpretation in terms of surface voltages

Actual values of surface voltage in the above studies are very difficult to measure directly. Direct use of a voltage follower probe would require clever mechanics and be limited to slow charge decay situations. Values can be estimated assuming an initial diameter for the charge. Taking, say, 20mm diameter as the area charged in both situations the surface voltages in the tribocharging studies were about 11x the readings, and in the corona charging studies, with the JCI 155, about 1.6x. Hence in the tribocharge decay studies this means that local initial peak voltages were as high as about 1000V even for decay times as short as 0.2s.

3. Conclusions

Measurement of the quantity of charge transferred in tribo and corona charging studies at the same time as measurement of the reading on a nearby fieldmeter allows calculation of the 'capacitance loading' effect experienced by surface charge. It is noted that comparable values are observed with tribo and corona charging. The approach provides a way to assess how much the influence of surface charge is suppressed by the construction of the material.

Materials may be assessed as unlikely to generate any electrostatic risk from retained charge if the charge decay time is adequately short and/or if the capacitance loading effect is adequately large. It is proposed that, in general, the decay time needs to be below 0.2s and the capacitive loading ratio above about 50.

The similarity between results obtained by tribo and corona charging shows opportunity for easy to use instrumentation, based on use of corona charging, to assess materials by charge decay rates and the effective capacitance loading. Measurements need to cover charge levels comparable to those likely to be experienced with practical tribo charging.

It is hoped that notice will be taken, in discussions of formal Standards, of the ideas, results and conclusions of the present paper.

References

[1] J. N. Chubb *"Instrumentation and standards for testing static control materials"*
IEEE Trans Ind Appl 26 (6) Nov/Dec 1990 p1182.

[2] J. N. Chubb *"Corona charging of practical materials for charge decay measurements"*
J. Electrostatics 37 1996 p53

[3] J. N. Chubb *"Dependence of charge decay characteristics on charging parameters"*
'Electrostatics 1995', Inst Phys Confr, York April 3-5, 1995 p103

Inst. Phys. Conf. Ser. No 163
Paper presented at the 10th Int. Conf., Cambridge, 28–31 March 1999

Real-time Particle Size and Electrostatic Charge Distribution Analysis and its Applications to Electrostatic Processes

M. K. Mazumder, N. Grable, Y. Tang, S. O'Connor, and R. A. Sims

University of Arkansas at Little Rock, Department of Applied Science, College of Information Science and Systems Engineering, ETAS Building, 575, 2801 South University of Little Rock, AR 72204, E-mail: mazumder@eivax.ualr.edu

ABSTRACT Advancement of the Electrical Single Particle Aerodynamic Relaxation Time (E-SPART) analyzer for electrostatic charge and size distribution analyses on a single particle basis in real time with high levels of precision and accuracy for its application to powder electrostatics and particle electrodynamics is presented. The instrument uses an AC electric drive to oscillate the particles in air and measures the particle aerodynamic diameter (d_a) and electrostatic charge (q), as opposed to using an acoustic field superimposed on a DC electric field in its original configuration. We discuss here the relative merits with respect to the measurement of particle size and charge distributions of two excitation systems.

1. Introduction

There are a number of instruments that can be used to characterize the aerodynamic size distribution of particles. Likewise instruments are available to estimate the net average electrostatic charge of particles sampled. However, choice of instruments for real-time simultaneous measurements of both aerodynamic diameter (d_a) and electrostatic charge (q) distributions of particles on a single particle basis is limited. The Electrical Single Particle Aerodynamic Relaxation Time (E-SPART) analyzer [1-4] is used extensively for simultaneous characterization of particle size distribution (PSD) and charge distribution for toners. The analyzer can be used in the diameter range from 0.4 to 100μm and charges in the range from 0 to their saturation charge levels. The size range and charge levels for some typical industrial applications are shown in Table 1.

Table 1. Typical ranges of particle size and charge in selected electrostatic processes

Application	Size (d_a in μm) Range	Charge-to-mass ratio (μC/g)
Dry Powder Inhalers (DPI)	0.5 to 10	0.1 to 10
Electrophotography	2 to 14	8 to 15
Powder Coating	5 to 100	0.3 to 2
Coal and mineral beneficiation	0.1 to 700	0.05 to 10
Electrospray	0.5 to 100	0.01 to 10

2. Size (d_a) and Charge (q) Measurements in the E-SPART Analyzer

When an airborne particle is subjected to an oscillatory external force, such as an acoustic excitation, the resultant oscillatory motion of the particle lags behind the external driving field. This phase lag ϕ relates to the aerodynamic diameter (d_a) of the particle. To determine this phase lag, the analyzer uses a differential laser Doppler velocimeter (LDV) to measure the velocity $V_p(t)$ of individual particles subjected to either (1) a combination of an acoustic excitation and a superimposed DC electric field, or (2) an AC electric field. The force of resistance, F_d, for spherical particles $(1 \leq d_p \leq 100)$ moving through a viscous medium, with particle Reynolds number $Re < 1$, is given by Stokes's Law [5],

$$F_d = 3\pi\eta d_p (V_p - U_g) \tag{1}$$

where d_p = particle diameter, η = fluid viscosity, V_p = particle velocity, and U_g = fluid velocity. Using Stokes's law, a one-dimensional equation of motion can be written for a spherical particle,

$$\tau_p \frac{dV_p}{dt} (V_p - U_g) = \frac{F(t)}{3\pi\eta d_p} \tag{2}$$

where $\tau_p = \dfrac{m_p C_c}{3\pi\eta d_p}$, m_p = particle mass, t = time, and F(t) = any external (nonviscous) force.

The relaxation time (τ_p) and the aerodynamic diameter (d_a) of the particle are related to each other by:

$$\tau_p = \frac{\rho_o d_a^2}{18\eta} = \frac{\rho_p d_p^2}{18\eta}, \tag{3}$$

where $\rho_o = 1000 kg/m^3$ and ρ_p is the actual particle density. The E-SPART analyzers measure τ_p to determine d_a.

3. AC E-SPART Analyzer

In the AC E-SPART analyzer, particles are placed in an oscillating electric field ($E_o \sin \omega t$). The steady-state solution when $t > 3\tau_p$ can be written as:

$$V_p(t) = \frac{ZE_d}{\sqrt{1+\omega^2\tau_p^2}} \sin(\omega t - \phi) \tag{4}$$

where E_o = amplitude of the driving field, q = charge on the particle, $Z = \dfrac{q}{3\pi\eta d_p}$ = electrical mobility of particle, and $\phi = \tan^{-1} \omega\tau$. From the measured value of τ_p, d_a is calculated as

$$d_a = \left[\frac{18\eta \tan\phi}{\omega\rho_o} \right]^{\frac{1}{2}}. \tag{5}$$

The relaxation time is obtained from the phase lag in the same manner as for the acoustical E-SPART analyzer [1-4]. The amplitude of particle motion can be written as:

$$|V_p| = \frac{qE_o}{3\pi\eta d_a} (1+\omega^2\tau_p^2)^{-\frac{1}{2}}. \tag{6}$$

Since d_a and τ_p are determined from the phase lag measurement (equation 4), the charge q can be calculated by measuring V_p:

$$q = \frac{3\pi\eta d_a V_p}{E_o}(1+\omega^2\tau_p^2)^{\frac{1}{2}} . \qquad (7)$$

The direction of particle oscillation $V_p \sin(\omega t - \phi)$ may be in-phase or 180° out-of-phase with respect to the applied field $E_o \sin \omega t$, depending upon the polarity of charge. This direction provides polarity of the measured charge q. The advantages and disadvantages of the acoustic and AC E-SPART are shown in Table 2.

Table 2. Advantages and Disadvantages of Different Excitation Methods in the E-SPART Analyzer

Operational Features	Acoustic and DC Drive		AC Drive	
	Measurement of d_a	Measurement of q	Measurement of d_a	Measurement of q
Range of operation	from ϕ measurement in the range 0-70° [d_a can be measured from amplitude ratio (V_p/U_g)]	from V_{TE} measurement in the range 0 to $\pm q_{max}$	from ϕ measurement in the range 0-90°	from Vp/Eo measurement in the range 0 to $\pm q_{max}$
Need for Corrections	Stokes Law does not remain valid when $\phi > 70°$ (No corrections are needed if amplitude ratio measurements are used)	Particle Reynolds number (R_e) may exceed 1 for highly charged large particles	Stokes Law can be applied without significant error	Particle Reynolds number (R_e) does not exceed 1, even for the highly charged large particles
Counting efficiency and sampling error	Applicable to both charged and uncharged particles	Highly charged particles may be deflected away from the sensing volume	Applicable only to charged particles	No sampling loss caused by excitation
Change of size range/noise immunity	Change of acoustic drive frequency to change size range may need adjustments in electrode spacing	Flow turbulence and acoustically generated flow field affect q/m measurement	Range of operation can be changed continuously by changing frequency of the AC drive	No acoustically generated flow field noise. Measurement of q is insensitive to flow turbulence

4. Particle Motion and Drag Resistance

Stokes's law of drag resistance (equation 1) is valid when inertial forces are negligible compared to the viscous force, i.e. the Reynolds number (R_e), is less than 1. R_e is defined by:

$$R_e = \frac{d_p V_p \rho_g}{\eta},$$

(8)

where ρ_g is the density of the gas.

For $R_e < 1$, the coefficient of drag (C_D) is given by

$$C_D = \frac{24}{R_e} \qquad \text{(Stokes Regime)}$$

(9)

and for $1 \leq R_e \leq 800$, we can approximate the drag coefficient by [5],

$$C_D \cong \frac{24}{R_e}\left(1 + 0.15 R_e^{0.687}\right). \qquad \text{(Transition Regime)}$$

(10)

5. DC E-SPART Analyzer

We now examine the validity of Stokes's Law for operation of the Acoustic + DC E-SPART operation for measurements of particle size (d_a) and electrostatic charge (q) (Table 3). First, we consider the particle Reynolds number (R_e) for two cases: 1) gravitational settling velocity (V_{TS}); and 2) maximum electrical migration velocity when $q = q_s$, the saturation charge of a particle in air based on the maximum surface charge density $2.65 \times 10^{-5} C/m^2$, and $E = E_{max} = 5 \times 10^5 V/m$, a maximum electric field that can be applied without causing corona breakdown in air.

Table 3. Comparison between gravitational settling and maximum electrical migration velocities of particles in the range 1 to 100µm in diameter, calculated from Stokes's law.

Aerodynamic Diameter in µm	Gravitational Settling of Particles		Electrical Migration of Particles ($q=q_s$, $E_o = 10^5 V/m$)	
	V_{TS} (m/s)	$R_e(V_{TS})$	V_{TE} (m/s)	$R_e(V_{TE})$
1	3.5×10^{-5}	2.31×10^{-6}	2.5×10^{-1}	1.65×10^{-2}
10	3.1×10^{-3}	2.05×10^{-3}	2.5	1.65*
50	7.5×10^{-2}	2.48×10^{-1}	12.5	41.25*
100	2.48×10^{-1}	1.65*	25	165*

*R_e for the particles exceeds 1, therefore, correction is needed in calculating the drag force.

Table 3 clearly shows that while settling velocities of the particles under gravity in the aerodynamic size range ($1 \leq d_a \leq 100$µm) are fairly within the Stokes region, the maximum electrical migration velocity can be well outside the Stokes flow regime. In calculating the particle charge q, we measure the particle's electrical migration velocity V_p ($V_p = V_{TE}$) in a DC electric field. In order to measure q for particles that are charged only to a small fraction of the saturation charge q_s, the E-SPART's electrodes are often operated in a high-intensity electric field. However, under a high-intensity DC electric field, the particles that are charged close to their saturation charge may have V_{TE} outside the Stokes region, and a correction is needed for calculating the value of q. Since we desire to determine q from the measurement of the actual velocity V_{TE} using LDV, we can use Newton's equation for drag resistance,

$$qE_o = C_D \frac{\pi}{8} \rho_g d_a{}^2 V_{TE}{}^2 \quad or \quad q = \frac{\pi C_D \rho_g d_a{}^2 V_{TE}{}^2}{8E_o}. \tag{11}$$

To determine C_D, we first calculate R_e from the measured value of V_{TE} and d_a from the phase lag measurements. Second, from the value of R_e, we calculate C_D using the approximation given in equation 10 and calculate q. A software program is currently being developed for obtaining the correct value of q from the measured value of V_p when V_{TE} is measured under a DC electric field.

When an AC electric field is used, the amplitude particle velocity $V_p(max)$ in an AC field is reduced by a factor of $[1 + \omega^2 \tau^2]^{1/2}$, since

$$V_P(max) = \frac{V_{TE}}{\left[1 + \omega^2 \tau_p{}^2\right]^{\frac{1}{2}}}, \tag{12}$$

which reduces R_e to less than 1. Therefore, the AC E-SPART operation for both size and charge measurements is within the Stokes region.

6. Size and Charge Calibration

While it is relatively simple to generate monodisperse aerosol containing particles of known size, it is difficult to generate particles of known charge for calibration. A vibrating orifice generator (VOAG) [Berglund-Liu Aerosol Generator, commercially available from TSI Inc.] was used for size and charge calibrations. Using a 1% solution of alcohol and oil in the Berglund-Liu VOAG, we varied the droplet size by changing the orifice excitation frequency from 45 to 70 kHz in 5 kHz steps for a fixed voltage on the E-SPART drive electrodes. We repeated each data run for electrode drive voltages of 300, 500, 1000, 2000, and 3000 volts peak-to-peak, respectively, for a total of 55 data calibration runs. Since the separation between the electrodes is 2 cm, the maximum field was 1.5×10^5V/m. The CMD, MMD, and Q/M for oil droplets measured at each frequency were recorded. The flow rate through the Berglund-Liu was measured, and the bulk Q/M of the oil droplets was determined using a Faraday cup and a Keithley electrometer. Q/M measurements were preformed for droplets from the VOAG for each of the orifice excitation frequency.

In the VOAG, the particle diameter d_a can be calculated from the following expression:

$$d_p = \left[\frac{6C_L Q_L}{\pi f}\right]^{\frac{1}{3}} (\rho_p)^{\frac{1}{2}} \times 10^4 \mu m, \tag{13}$$

where C_L = concentration of oil in alcohol (= 0.01), d_p is expressed in μm, Q_L = flow rate of liquid through orifice (= .00295 cm³/sec), f = orifice excitation frequency = number of drops generated each second, and ρ_p = density of oil (g/cm³).

Tables 4 and 5 show the comparison between the calculated and measured count mean diameter and Q/M values measured by the Faraday Cage and E-SPART analyzer.

340

Table 4. E-SPART vs Calculated droplet diameters

Orifice Vibration Frequency	Calculated droplet dia. (μm)	E-SPART Measured Count Median Diameter (CMD) in [μm] AC Excitation Amplitude (V_{p-p})				
		300 V	500 V	1000 V	2000 V	3000 V
45 kHz	9.75	10.27	10.33	10.16	10.03	10.03
50 "	9.41	9.96	10.08	9.76	9.67	9.66
55 "	9.12	9.66	9.72	9.58	9.34	9.31
60 "	8.86	9.35	9.38	9.23	9.14	9.11
65 "	8.62	9.22	9.22	9.05	8.86	8.83
70 "	8.41	9.25	9.06	8.89	8.82	8.87

Table 5. E-SPART vs. Faraday Q/M Measurements

Orifice Vibration	Charge-to-Mass (Q/M) Measured by Faraday Cup	Q/M [μC/g] Measured by AC E-SPART AC Voltage (V_{p-p}) vs. Q/M (μC/g)				
frequency	μC/g	300 V	500 V	1000 V	2000 V	3000 V
45 kHz	7.76	9.35	9.24	9.31	9.35	9.45
50 "	8.51	9.54	9.52	9.52	9.54	9.54
55 "	8.73	9.85	9.71	9.63	9.73	9.73
60 "	9.42	10.03	9.76	9.70	9.82	9.92
65 "	9.08	10.10	9.79	9.88	10.04	10.04
70 "	9.42	10.23	9.71	9.96	9.82	9.82

7. Conclusion

An E-SPART analyzer with an AC electric drive can be used to measure aerodynamic diameter (d_a) and particle charge (q) distributions in real time and on a single particle basis, in a size range from submicron to 100μm in diameter. The particle dynamics are within the Stokes region. The calculated values of droplet diameter and Q/M determined using a Faraday Cage agreed fairly well with the E-SPART generated CMD and Q/M values, indicating resolution and precision of the E-SPART analyzer.

The analyzer can be used in several electrostatic process applications, including powder coating, electrophotography, coal beneficiation, electrosprays, and pharmaceutical aerosols.

8. References

[1] Baron P. A., Mazumder M. K., and Cheng Y. S., 1992 *Aerosol Measurements: Principles* (New York: Van Nostrand Reinhold) Chp. 17
[2] Mazumder M. K., Banerjee S., Ware R. E., Mu C., Kaya N., and Huang C. C. 1994 *IEEE/IAS Transactions* **30** 365-369
[3] Mazumder M. K., Banerjee S., and Mu C. 1994 *Dispersion and Aggregation* (New York: Engineering Foundation)
[4] Mazumder M. K. 1993 *KONA Powder and Particle* **11** 105-118
[5] Hinds W. C. 1999 *Aerosol Technology: Properties, Behavior, and Measurement of Airborne Particles* (New York: John Wiley & Sons Inc.)

Inst. Phys. Conf. Ser. No 163
Paper presented at the 10th Int. Conf., Cambridge, 28–31 March 1999

Visualization of distribution of electron traps on polymer surface by means of Scanning Micro Laser Probe

Yuji Murata and Keisuke Kubota
Science University of Tokyo, Yamazaki 2641, Noda, Chiba, 278- 8510
JAPAN

Abstract. An apparatus which visualize the surface distribution of electron traps was newly developed. Using this apparatus, various patterns of trap distributions were investigated. The surface of polyethylene after solvent-dissolved showed lower trap density than that before dissolve. The area rubbed with a needle showed high trap density. The trap density of spherulitic region of polypropylene showed lower density than outer region of that.

1. Introduction

Contact charging of high polymers with other material is considered to be due to electron transfer between the contacting surfaces[1]. In the case of polymers which tend to charge negatively by contact with other material, electron traps on the surface must play an important role.

As one of a useful apparatus for investigating electron traps, we developed an apparatus for measuring the surface distribution of electron traps on polymer surface. Using this apparatus we can obtain visually the density distribution of electron traps on the sample surface.

2. Apparatus

In order to obtain the pattern of the distribution of electron traps on a polymer surface, the measurement were carried out by the following procedure;
(1) the surface of the sample was exposed to an electron shower,
(2) then the sample surface is illuminated by a focused laser beam,
(3) the number of photoelectrons emitted from the sample surface was measured,
(4) the measurement was carried out on the previously determined area of the sample surface.

By exposing the sample surface to an electron shower, the surface traps are filled with electrons[2]. After this procedure, the photoelectron emission can be obtained in the wavelength region of visible light.

The apparatus used in the present study consists of a laser, a laser-beam focusing system, a laser-beam scanning system, a photoelectron counting system, a source of electron shower and a sample holder. The laser beam is focused onto the surface of a film sample. The diameter of the focused spot of the beam on the sample surface is minimum 8.5 μ m. The laser used is a He-Ne laser of 3 mW. Scanning of the laser beam is carried

342

out using a mechanical stage driven with two pulse motors in the direction of x and y axis. The minimum movement of the focused spot is 1 μm.

The sample to be investigated is maximum 10-mm square in size and fixed on a sample holder mounted perpendicularly to the laser beam. The electron counting system consists of an electron multiplier, a high voltage source, an amplifier, a discriminator and an electronic counter. The electron multiplier is provided in the vicinity of the sample. The out put charge pulses from the electron multiplier is amplified and after eliminating low-level noise pulses by means of a discriminator the number of the pulses are counted using an electronic counter.

The focused laser spot on the sample surface moves to the neighboring position after a short duration of measurement and it scans whole area previously charged. The minimum distance between neiboghring positions is 1 μm and the maximum scanning area is 10mm square. Both the counted results of the number of photoelectrons from the illuminated position and the information of the position on the sample surface are introduced into a computer and the distribution of the number of photoelectrons is converted to the difference in colors on a monitor screen. By this procedure, we can obtain a visualized image of the distribution of the number of electron traps on the sample surface.

The sample on the holder is previously exposed to an electron shower as described above. The source of an electron shower is produced by a tungsten filament for thermal electron source and a grid electrode. The charging is carried out in the vacuum chamber and it is kept in the dark before laser scanning.

The measurement is carried out in a vacuum of 10-3 Pa.

3. Results and discussion

3.1 Photoelectric emission from charged sample

Fig.1 shows the result of the measurement of photoelectric emission from polyethylene surface. The sample surface is previously exposed to an electron shower and is charged negatively. The acceleration potential of electrons is -300 V and the potential is increased very slowly during exposure in order to avoid surface damage of the sample by high energy electrons. Before this procedure, no electron emission is observed by illumination of such long-wavelength laser beam. The number of emitted electrons per unit time decreases with time duration after the sample surface is illuminated. This means that the photoemission of charge trapping centres are progressively emptied.

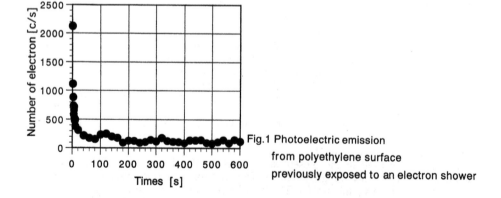

Fig.1 Photoelectric emission from polyethylene surface previously exposed to an electron shower

3.2 Trap distribution of a model sample

Fig.2 shows an example of visualized pattern of trap distribution of polyethylene sample covered with a sheet of aluminum foil with a small circular aperture at the center of it. The upper part of the figure corresponds to polyethylene surface exposed from the aperture of the foil. We can obtain a clear pattern of the trap distributed area which corresponds to naked polyethylene surface. We can not obtain photoelectrons from the foil surface.

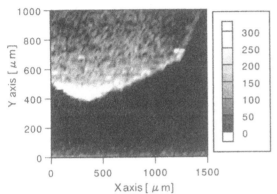

Fig.2 Distribution of electron traps
on polyethylene surface

3.3 Change in trap distribution after surface treatments

Measurement of trap distribution of polyethylene surface before and after treatment with organic solvent was examined. The surface of polyethylene sample received a droplet of acetone. The charging and the measurement was carried out after the acetone droplet evaporated. The dark area of Fig.3 about 1500μ m in diameter corresponds to the treated area. It is obvious from the figure that the trap density decreases after the treatment.

Fig.4 shows the pattern obtained after the surface of polyethylene was rubbed with a sewing needle. The bright line corresponds to the rubbed area.

The first case may show trap density of once solvent-dissolved surface is smaller than that of heat-casted surface. The second case shows the mechanical destruction of molecules generates high density traps.

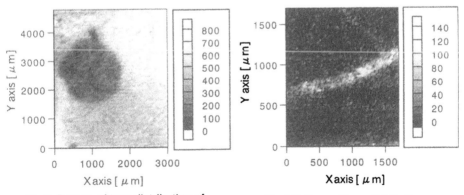

Fig.3 Pattern of trap distribution of polyethylene surface after treated acetone

Fig.4 Pattern of trap distribution of polyethylene surface after rubbed with a sewing needle

344

3.4 Trap distribution of spherulitic polypropylene
    Measurement of trap distribution of polypropylene films with spherulites was examined. Fig.5 shows the obtained pattern. From the figure, we can clearly recognize the pattern of a spherulite. This result coincides with that obtained by Ikezaki et al[3] where lower densities of charging are obtained in the region of a spherulite.

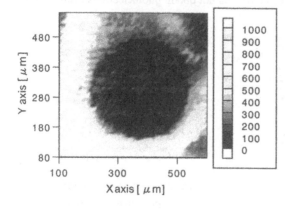

Fig.5 Trap distibution on polypropylene surface with spherulite

4. Conclusions

(1) Distribution of electron traps of polymer surface can be visualized using a focused laser-beam scanning apparatus. Using this apparatus various trap distributions are investigated.
(2) The traps of polyethylene after solvent-dissolved showed lower density in comparison with that before this treatment. The rubbed surface of polyethylene with a needle showed high density traps.
(3) The spherulitic region of polypropylene showed lower density of traps than outside region.

References

[1] Bauser H 1973 Dechema-Monoger. 72 11-28
[2] Murata Y 1979 Jpn.J.Appl.Phys. 18 1-8
[3] Ikezaki K, Yagishita A and Yamanouchi H 1994 Proc. 8th Int. Symp. on Electrets 428-433

DISCUSSION - Section F - Materials Evaluation and Applications

Title: Polymer Semiconductor Devices
Speaker: R H Friend

Question: B Makin
To obtain the pattern using inkjet (i) can the resolution be obtained , (ii) will the inkjet technology degrade the polymers by either (a) the aerosol evaporation process on electrostatic atomization and create particles ~20µm?

Reply: Our present work is based on piezo-inkjet heads, and we do not expect this process to cause degradation of the polymers.

Question: D K Davies
Electrostatics is used for ordering in L/C Technology, will this have a role in your device?

Reply: Electric-field ordering may be possible especially if the polymer is present (transiently during this process step) in a liquid crystalline phase.

Question: S Cunningham
You explained colourful different structures, however they show similar if not the same transport properties is this true?

Reply: Transport properties are generally governed by the role of disorder, and low mobilities result from hopping between localised states. Variations between materials are due to differences in order.

Title: Inherently-conductive Polymers as Antistatic Agents for Textiles
Speaker: P Kathirgamanathan

Question: Martin Glor
Can this agent also be applied successfully to PE bags?

Reply: Yes.

Question: D K Davies
Do your coatings exhibit good adhesion to the substrate and is there any pretreatment?

Reply: Yes, the adhesion is very good. Pretreatment usually includes an organic solvent or dilute acid etching.

Question: M K Mazumnder
(a) Could you apply conductive polymers by spray coating? (b) What surfactants can be used to increase wettability?

Reply: (a) Yes, we have developed solvent processable and water soluble conducting polymers which can be spray coated. (b) Depending on the polymer/ solvent system, cationic, anionic or non ionic surfactants can be used.

Title: The Evaluation of the Electrostatic Safety of Personal Protective Clothing for use in Flammable Atmospheres

Speaker: P Holdstock

Question: Martin Glor

It is important to mention that the need for such clothing depends on the sensitivity of the explosive atmosphere which may be present.

Reply: I quite agree - The standard from which this work developed, EN1149-1, is only intended for testing of clothing for use in very sensitive flammable atmospheres. The British Standard BS5958 gives recommendations for two situations: if MIE >0.2mJ there are no precautions required in relation to clothing; if MIE $\leq$ 0.2 mJ then surface resistivity of clothing should be less than 5.0×10^{10} Ω. The more relevant standard is draft Cenelec RO44-001, which states that clothing does not represent an ignition hazard except in very sensitive atmospheres, in which case surface resistivity shall be less than 5.0×10^{10} Ω as measured using EN1149-1.

Question: D K Davies

Was there a difference in waveform for incendive and non-incendive discharges?

Reply: No. With the measuring system that we used, there was no obvious difference between the electrical characteristics of incendive and non-incendive discharges. However, there were often multiple discharges generated. The measuring system could only record one discharge at a time. There is a possibility that a discharge recorded during an ignition event may not have been the actual incendive discharge.

Question: Norman Wilson

In your flow chart for testing fabrics you first measure the surface resistance. If this is < 10^9 Ω the fabric passes the test. If it is > 10^9 Ω you calculate the surface resistivity. Is this calculation valid for inhomogeneous materials, e.g fabrics with conducting yarns in the form of stripes or grids?

Reply: The purpose of carrying out the steps shown in the flowchart is to identify homogeneous and inhomogeneous fabrics. Although the concept of resistivity is meaningless for fabrics with small amounts of conducting fibre, it does allow us to distinguish between homogeneous fabrics and those with core conductive fibres, at least for the materials tested.

Title: Static Electricity Dissipation Mechanism in Carpets Containing Conductive Fibres

Speaker: W K Fisher

Question: Allel Bouziane
You use corona as a means for charge dissipation. Does that limit the thickness of your carpet?

Reply: No, to my knowledge it doesn't. The range of carpet thickness is small, between 1/8" to1/4" and in this range we have not seen differences in static dissipation caused by different carpet thicknesses.

Section G

ESD/EOS

Inst. Phys. Conf. Ser. No 163
Paper presented at the 10th Int. Conf., Cambridge, 28–31 March 1999
© *1999 IOP Publishing Ltd*

The effect of resistance to ground on human body ESD

D E Swenson

ESD Association, 7900 Turin Road, Bld 3 Suite 2, Rome, New York, USA 13440-2069

3M/Electronic Handling & Protection Division, 6801 River Place Blvd., Austin, Texas, USA 78726-9000

Abstract. The disk drive industry processes electronic parts that are extremely sensitive to electrostatic fields and direct discharges. These parts may in fact be the most susceptible parts to electrostatic forces ever mass-produced. Even though these parts are extremely sensitive, some of the manufacturers seem to think they can ignore Ohms Law and the connotations it brings to proper earthing techniques. This discussion will describe the fallacies of present thought in this important area of electronics manufacture.

1. Introduction

The computer industry most assuredly has been among the most important and prolific in terms of industrial innovation in the past decade. Faster, smaller, cheaper and more powerful are trademark terms for this industry. The computer industry has a subset that deals with the storage of all the information that is obtained, created, or otherwise manipulated. This is the hard disk drive industry. Modern storage devices are extremely powerful and are likely to gain in abilities, only limited by the available technologies themselves.

Consider that in 1981, the cost of information storage was on the order of $1000 (US) per megabyte (MB) for a 5 MB disk drive that was an option on the first IBM PC. Today, storage costs are about 10 to 15 US cents per MB in drives that have multiple gigabyte storage capacity. A principle reason for this dramatic change in storage ability at tremendously reduced cost is the development of the Magneto-Resistive read-write head technology in magnetic storage and retrieval systems. Recent developments have lead to the commercialization of the Giant MR head (GMR), which further increases the functional *areal density* of magnetic storage systems. While it appears that the limits of this technology are fast approaching, the industry believes that the present type of equipment will meet customer demands for the next 5 to 8 years. Next generation storage systems will probably not be magnetic, at least not in the mechanical configuration of "spinning disks" and "flying heads" built today.

From an electrostatics point of view, the read-write elements of the disk drive are extremely sensitive to changes in electrical potential across their structure. These parts are susceptible to direct discharges, rapidly changing electric fields in the vicinity of the parts, and possibly from electromagnetic interference (EMI/RFI) [1]. People are required to do the majority of the assembly of disk drives and thus are the source of the greatest electrostatic threat. Only limited success has been achieved with automated or robotic assembly due to the complexity of wire attachment to the head elements. Also, the production process changes very often due to part configuration modifications so the investment in robotics is not realistic at this time. The factory environment and materials within that environment represent possible electrostatic hazards as well. Grounding of people and everything else possible in the environment are extremely important aspects of electrostatic control for the disk drive manufacturer. A secondary but also important consideration is the reduction of radiated and transmitted electromagnetic emissions in the work place.

2. MR/GMR head sensitivity

Numerous investigators have presented dramatic findings related to the sensitivity of MR/GMR heads. The Proceedings of the EOS/ESD Symposia (US ESD Association) from 1995 through 1998 have many papers describing sophisticated test protocols and experiments that have helped to define the hazard levels and damage mechanisms for these sensitive parts.

Investigators agree that the typical MR head can be damaged by approximately 10 nanojoules of energy delivered by conductive contact in a human body model discharge (HBM) and most agree that the GMR head can be damaged by 1 nanojoule or less. Taking a simplistic approach results in the following calculation of electrical potential necessary to deliver these typical electrostatic discharge energies from a HBM equivalent circuit:

From the basic energy equation:

Eq. 1
$$E = \frac{CV^2}{2} \frac{R_h}{R_h + R_f}$$

Where:
E = Energy in Joules
C = Capacitance in Farads-HBM Circuit
V = Voltage
R_h = Resistance of MR head
R_f = Resistance of HBM Test Circuit

Rearranged to solve for voltage:

Eq. 2
$$V = \sqrt{2 \frac{E}{C} \frac{R_h + R_f}{R_h}}$$

Using the generally accepted value of 100 pF for the capacitance and 1500 Ω for the resistance in the HBM and 35 Ω as a typical MR/GMR head resistance, we can solve for the voltage applied through the simulator that is equivalent to a given amount of delivered energy:

For MR Heads at 10 nJ

$$V = \sqrt{2 \frac{1x10^{-8}}{1x10^{-10}} \frac{1535}{35}}$$

Eq. 3
$$V = \sqrt{8771}$$
$$V \cong 94v$$

For GMR Heads at 1 nJ

$$V = \sqrt{2 \frac{1x10^{-9}}{1x10^{-10}} \frac{1535}{35}}$$

$$V = \sqrt{877}$$

Eq. 4
$$V \cong 30v$$

It must be understood however, that the calculated values represent the expected delivered energy from a HBM electrical equivalent test model circuit. Damage to MR heads has not been observed from human contact by finger or hand at the HBM discharge level unless the hand happened to be holding a metallic tool, such as a tweezers or wire cutter. The waveform of a metallic contact has much higher peak energy transfer (fast rise time <5 nanoseconds for first pulse) and a "ringing" pattern whereas the HBM has a comparatively slow rise-time (5 to 25 nanoseconds) and exponential type decay. In addition, the HBM does not take into account the contact resistance between a person's finger and the receiving item. Investigators are currently attempting to describe and define a test model circuit that will properly emulate a person, holding a metal tool in resistive contact with the skin, or when the person is wearing dissipative or conductive gloves [2]. Since this is a complicated model and most certainly the industry will not readily agree across all segments, it may be some time before a so-called "Tweezers" discharge model will exist in practice. For now, most of the disk drive companies want to limit the potential on personnel to 10 to 20 volts or less.

3. Resistance to Ground (Earth)

As stated above, grounding of people is one of the most important measures that can be taken in establishing any sort of electrostatic control plan for protection of sensitive items. What is missed by most of those responsible for an electrostatic control program is the fact that people are not an exceptional electrical conductor since people have inherent skin resistance. Add shoes, clothing, and dry air and the effective resistance to ground (earth) increases substantially. Certainly, a grounded person is a good enough conductor to be electrocuted when contacted with high voltage but most people are not a good enough conductor to equalize *all* electrostatic charge when grounded. Given the fact that people also are mobile, their capacitance relative to earth varies with each motion. Human capacitance changes result in dramatic fluctuation in electrical potential with respect to earth. Since a human body cannot possibly attain zero resistance to earth, some electrical potential will exist, at least for a short time, with respect to earth. The relationship between charge, capacitance and voltage with respect to decay time is shown by:

$$V = V_0 e^{-t/RC}$$

$$V = \frac{Q}{C} e^{-t/RC}$$

Eq. 5

Where:
V = Potential on a body
V_0 = Potential on a body with resistance in the discharge path
R = Resistance
C = Capacitance

Therefore, V is the potential that results immediately after contact and separation in the absence of a discharge path. The potential of a body rises to V_0 in the time interval associated with the separation of the charging partners. When a leakage path is provided, the time constant of the discharge is given by:

$$\tau = RC$$

Eq. 6

Where:
τ = Time Constant
R = Resistance in Ohms
C = Capacitance in Farads

The balance between the charge generating and the charge decay processes determines the peak voltage attained in any practical system. Comparing Eq.1-4 with Eq. 5-6 shows that both capacitance and the resistance of the discharging circuit have great impact on the voltage and subsequent stored verses delivered energy.

In 1979 (which for all practical purposes was the beginning of the era where electrostatic control was accepted as a necessity in electronics manufacturing), a paper was presented that reported the electrical potential between people and the earth while bonded to the earth through various resistances [3]. This work, summarized in Figure 1, has been used by the electronics industry as a tool to assist in establishing electrostatic control programs. Several US and International standards have used this data as a baseline.

Resistance to Ground (Megohms)	Peak Voltage Created by Movement (Volts)
0.5	<1
1	<1
10	6
100	85
1000	240
∞ (no wrist strap)	440

Figure 1 - Wrist Strap Resistance to Ground Vs Peak Voltage

The study described by Figure 1 was repeated in 1993 by the International Technical Commission (IEC) Technical Committee (TC) 101- Electrostatics, in Working Group 5 (Electronics). This experimental work provided background information for their handling practices Technical Report, IEC TC101 61340 –5 Part 1 and Part 2 (User Guide).

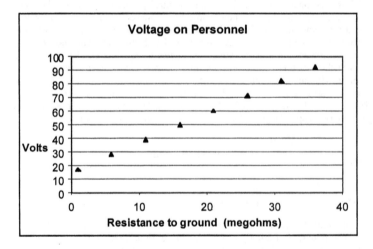

Figure 2 – Triboelectrically generated human body voltage vs. resistance to earth

It should be noted that the resistance values stated in Figure 2 should be increased by 1 megohm as the resistance shown on the x-axis is the "added" resistance above the customary 1 megohm resistor found in wrist strap grounding cords used in the electronics industry. In addition, the resistance of the person in the measurement circuit is not taken into account. Figure 2 reports the average of independent experiments conducted in four different countries.

While the numbers do not precisely agree between Figures 1 and 2, it is obvious that the trend is clear enough to make some relatively sound assumptions:
- As the resistance between a person and earth increases, the ability to develop an electrical potential increases proportionally.
- Electrostatic control levels can be based on resistance to earth values.
- Higher levels of sensitivity (lower threshold of electrostatic potential to cause damage) demand lower total resistance to earth.

In the first experiment, personnel charged themselves by scuffing their feet on an insulating plate while holding onto the lead of various resistors bonded to earth. Their potential was measured using an electrometer with a non-contact voltage probe. A fixture held a metal plate precise distances from the voltage probe. Direct charging of the metal plate with a known power supply allowed calibration of the measurement distance. An attached wire from the metal plate made electrical connection to a hand-held metal electrode. The peak potential verses resistance shown in Figure 1 is the result of averaging values recorded for 10 people.

The second set of experiments graphed in Figure 2 was done in much the same way as described above for Figure 1. However, the investigators used voltage probes fitted to oscilloscopes or other digital recording devices rather than electrometers to measure the voltage on personnel so the results are likely to be a bit more precise. In addition, each investigator wore a wrist strap electrically bonded to earth through a resistance substitution box. Thus, the resistance in the ground path could be changed readily without affecting the contact resistance between the skin and the wristband.

4 Electrostatic Control in the Disk Drive Industry

It is obvious that the disk drive industry knows there is a need to control electrostatics in the work environment and especially the need to control electrostatic charge where personnel are involved. The principal investigators generally understand the relationship between resistance to earth and the electrical potential that can be developed on people. Understanding seems to break down when this information has to be communicated to their manufacturing sites around the world. Also, there is a bit of disagreement amongst the knowledgeable investigators as to the level at which the electrical potential on a person, equipped with the proper tools, becomes *an actual hazard* to the sensitive parts. The method and procedure to measure the resistance of a person is also a matter of considerable discussion within the industry. It is reasonably clear to most of the electrostatic control material suppliers that the disk drive industry wants near "zero" electrostatic potential in their work environment. However, they do not seem to understand the need to reduce their resistance to ground specifications for personnel and their fixed factory installations such as flooring.

Clearly, Figures 1 and 2 show the need to maintain people at a resistance to earth approaching 1 megohm and certainly under 10 megohms for processing of MR heads. For GMR heads it may be necessary to reduce the upper limit to 1 megohm if 5 to 10 volts of electrostatic potential are *really* a hazard. Fortunately, the use of properly designed dissipative tools reduces the actual risk of high current discharge in the manufacturing process even at potentials (on personnel) up to 100 volts or more [4]. For extremely sensitive operations, personnel with inherently low skin resistance may be required to do the assembly.

Resistance measurement of personnel includes a relatively new concept in the industry called "wrist strap constant monitoring". While several different techniques are used, the most widely accepted variety makes a d.c. loop resistance (or continuity measurement) between the measuring apparatus, a person and the work environment. A dual conductor wristband and two-conductor earth bonding cord (ground cord in the US) are used with these systems. A generic example of this type of monitoring system is shown in Figure 3:

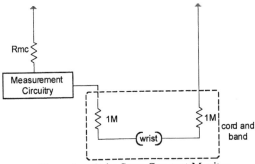

Figure 3 – Wrist Strap Constant Monitor

In this type of system, the main variation between manufacturers is the measurement voltage value and how it is applied through the ground cord to the person wearing the wristband. Some of the investigators of

MR/GMR head sensitivity worry about the applied potential of wrist strap monitors. Even the monitor with the highest measurement voltage (16 volts), can only apply that value when the wrist strap monitor alarm is sounding and only under unusual conditions (a major increase of resistance on only one side of the wristband). Figure 4 [5] shows the results of adding resistance evenly on both sides of the wristband for a monitor that applies 16 volts in a pulse. Note the point at which the alarm starts sounding, alerting personnel that something needs attention. Also, note that at 4 megohms added resistance the monitor contributes less than 4 volts to the person but the person can generate 11 volts simply by sliding a foot across a floor surface. If there is a major concern about very low electrical potentials, the resistance to earth values that are acceptable in the industry must be reduced accordingly.

Resistance Added Between the Wrist Strap and Ground and Between Wrist Strap and Monitor	Calculated Resistance to Ground	Average Maximum Potential Generated by Sliding Motion From Eq. 1: *** $$\frac{R_{tr}}{R_b + R_{tr}} = \frac{35}{330+35}$$		Contributed By Monitor	
		Potential	Energy	Potential	Energy
0 added resistance	1 megohm	3 volts	.043 nJ	1.5 volts	0.1 nJ
1 megohm	1.8 megohms	6 volts	.172 nJ	2.6 volts	0.3 nJ
2 megohms	2.6 megohms	9 volts	.39 nJ	3.1 volts	0.5 nJ
3 megohms	3.3 megohms	10 volts	.48 nJ	3.5 volts	0.6 nJ
4 megohms*	4 megohms	11 volts	.58 nJ	3.9 volts	0.8 nJ
6 megohms**	5.2 megohms	17 volts	1.3 nJ	4.2 volts	0.9 nJ
8 megohms**	6.2 megohms	20 volts	1.9 nJ	4.7 volts	1 nJ
10 megohms**	7.1 megohms	22 volts	2.3 nJ	4.8 volts	1 nJ

* Monitor intermittent alarm ** Monitor alarms continuously *** Resistance of MR head -35Ω, body-tool resistance 330 Ω derived from E=IR where: E = 4.2 volts and I= 12.8 mA, measured in reference #7, Fig. 10. Figure 4- Continuous monitor with 16-volt open circuit measurement voltage and 10-megohm loop resistance alarm limit.

4 Conclusion

Resistance to earth determines the electrical potential that can be developed on mobile systems such as personnel. Personnel have inherent resistance that affects their ability to make low resistance earth bonds. Added resistance in wrist strap assemblies or footwear impacts the attainable overall resistance to earth. The "household" version of Ohms Law, E=IR, must be considered in the design of an effective electrostatic controlled environment, including the role of people and their actual as well as practical resistance to earth (ground).

6 References

[1] Wallash A., Hughbanks T., and Voldman S., "ESD Failure Mechanisms of Inductive and Magnetoresistive Recording Heads", EOS/ESD Symposium, 1995, Proc., pp 322-330

[2] Lam C., "Characterization of ESD Tweezers for Use with Magnetoresistive Recording Heads", EOS/ESD Symposium, 1996, Proc., pp 14-21.

[3] D.M. Yenni, "Basic Electrical Considerations in the Design of a Static-Safe Work Environment", Proceedings, 1979 NEPCON/West, Anaheim, CA.

[4] Lam C., Salhi El-Amine, and Chim S., "Characterization of ESD Damaged Magnetoresistive Recording Heads", 1997 EOS/ESD Symposium, Proc. pp. 386-397

[5] Swenson D, and Wilson R., "Grounding of Personnel in ESD Sensitive Environments", Proceedings, DataStor Asia '98 Conference, Singapore, 1998

Inst. Phys. Conf. Ser. No 163
Paper presented at the 10th Int. Conf., Cambridge, 28–31 March 1999

357

Environmental tests of silicon microstrip detector devices for high energy space applications[1]

Philipp Azzarello and Davide Vitè[2]

Departement of Nuclear and Particle Physics, University of Geneva, Geneva, CH-1211, Switzerland

Abstract. In the framework of silicon detector development for high energy and space applications, we have performed a number of environmental tests on different devices. Most of the effects we have detected during the construction phases are possibly due to external contamination to the detector surface, as well as to electrostatic effects at surfaces and interfaces, with possible consequences on our future design, production and construction activities.

1. Silicon detectors in high energy space applications

Our group has been making use of silicon microstrip detectors for high energy physics applications since a few years; recently, we have been involved in the design and construction of a complex experimental apparatus to measure cosmic rays above the Earth surface, at an altitude of approximately 400 km. In particular, we participated, within an international collaboration, to the construction of a silicon tracker, composed of six planes of microstrip arrays. Our basic detector elements are 4 cm wide and 7 cm long, on a 300 μm thick crystal silicon substrate. A few thousand strips are integrated on both surfaces, along orthogonal directions, and daisy-chained to an external readout via kapton cables with gold-coated copper lines. Of the greatest importance is the electrical behaviour of such detectors, in terms of dark current and electrical noise, on top of long term resistance to the challenging environments typical to mass production and space applications.

1.1. Development and mass production of silicon microstrip detector modules

During the construction and assembly phases of some 70 modules, composed of up to 15 detector elements supported by light frames and complete with front-end readout electronics, we have encountered in our laboratory a few problems of performance degradation, both at the sensors and at the module levels, in a few cases critical to our expected overall tracking performance. In particular, we have been confronted to increases in the leakage current of the modules subsequent to a few construction and integration procedures, often involving contact to the electrically insulating surfaces of metal construction jigs. The leakage current is the dark current flowing through the detectors in absence of signal, and in our case was typically of the order of a few μA per sensor, measured at operating conditions.

    We have started a test program to investigate such problems and to understand how to avoid them in the future production phases. We suspect environmental conditions, coupled to electrostatic effects, to be responsible for such unwanted degradations. We have reproduced the different situations and environment as encountered by the sensitive elements throughout their life, from construction to operation, trying to separate different phenomena as to better understand the origin of the effects seen. As an example, we will focus in the following section on the leakage current as the driving parameter to this

---

[1] Work supported in part by the Swiss National Science Foundation
[2] E-mail: davide.vite@physics.unige.ch (corresponding author)

358

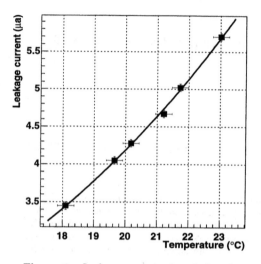

Figure 1: Leakage current of a single microstrip sensor as a function of temperature (100 V bias).

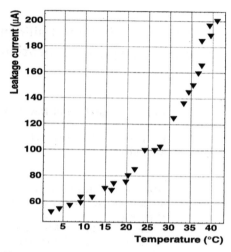

Figure 2: Leakage current of a complete 12-sensor module as a function of temperature (100 V bias).

purpose; alternative parameters would be noise level or number of noisy channels, defined appropriately. The Equivalent Noise Charge (ENC) contribution to the total electronic noise related to the leakage current I grows as $\sqrt{I}$.

2. Environmental tests

We have investigated possible effects on sensor electrical performance of external temperature, exposure to high relative humidity, high temperature annealing, mechanical and chemical cleaning, and mechanical stress. Though most of those effects are known and documented, at least qualitatively, a thorough program was felt necessary within our collaboration to better understand medium and long term behaviour of our devices. No systematic environmental studies were performed, to our knowledge, onto devices more or less similar to ours, i.e. silicon microstrip detectors. This test program will be continued and completed in the near future.

2.1. Temperature dependence

When in space, the overall temperature of the experimental apparatus is subject to variations because of the electronics dissipating heat and of exposure to or shading from the Sun; leakage current evolution and relative noise performance as a function of temperature impose time limits to possible configurations. Fig. 1 shows the leakage current of a silicon sensor as a function of external temperature, at an operating bias of 100 V; measures fit well with the behaviour theoretically expected, providing a reasonable value for the fit parameter c_1, the semiconductor gap energy. Knowledge of the leakage current behaviour of individual sensors is necessary to disentangle different effects in subsequent tests.

As an example of extension of these studies to a full-size detector, fig. 2 shows the leakage current of fully equipped 12-sensor module as a function of the external temperature, also at a bias of 100 V. This test was performed over a few days in a temperature controlled environment, at the University of Bern, Switzerland. Above 20 C, the leakage current increases with temperature at a higher rate, with the related noise performance degrading as a consequence.

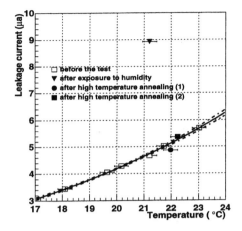

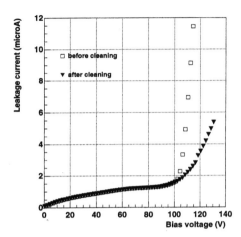

Figure 3: Leakage current of a single sensor after humidity exposure and annealing (100 V bias) – full and dashed lines are fits using central and extreme parameters, respectively.

Figure 4: Leakage current of a single sensor as a function of bias voltage, before and after mechanical and chemical cleaning.

2.2. Humidity tests

As a few sequences of module assembly require sensors to be positioned onto special jigs, with vacuum applied for a few hours, we aimed at investigating possible effects of such configurations. A few selected sensors underwent extended exposures to high relative humidity (i.e. >80%) under different conditions, namely on Teflon and Nylon jigs and in open storage boxes. High humidity modifies the distribution of charges accumulating at the oxide interface and compensating partially the positive charge trapped in the oxide, once bias is applied. After one exposure a breakdown, i.e. an abrupt increase in the leakage current due to a small increase of the bias voltage applied, was found only on one of the sensors under test. Further tests will be performed.

2.3. High temperature annealing

We exposed a few sensors to high temperature (150-180 C) annealing in a very clean and confined environment, to look for possible improvements in the sensor quality. Fig. 3 shows e.g. the leakage current of a single device as a function of temperature, and its corresponding value after long term exposure to high humidity and short term high temperature annealing. Empty markers are reference values before the test. The outcoming gases were analysed with a mass spectrometer, and water vapour was found, as expected, but no other substances were detected. The breakdown previously shown after humidity exposure disappeared.

2.4. Mechanical and chemical cleaning

Dust, particulate and chemical pollution may be removed from the unprotected surfaces of sensors by means of ultrasound cleaning in acetone-methanol 50%-50% and pure methanol mixtures. Fig. 4 shows the leakage current measurements of a sensor before and after chemical cleaning. The silicon sensor selected had been left for more than a year on the special latex support used for precise diamond cutting of the fully processed 4"-wafers[3]. The breakdown previously showing at about 105 V disappeared after cleaning.

[3] These supports are known not to be usable for long-term storage as they cause chemical contamination to devices

360

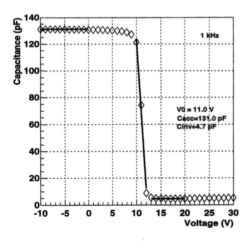

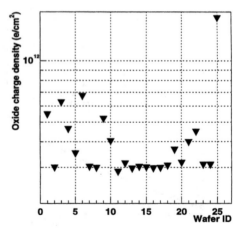

Figure 5: CV curve of an MOS capacitor, with fit to determine V_0 (post-production batch).

Figure 6: Oxide charge density of MOS capacitors for all wafers (post-production batch).

2.5. Mechanical stress

We have performed dedicated stress tests on single sensors. We have built a special jig with a system allowing an external and measurable force to be applied to the sensor middle point. Dark currents as a function of stress applied and deformation have been measured on a couple of microstrip sensors, and no detectable effect was found: the measured leakage currents remained stable at values of 12.3 and 5.0 μA for deformations of 1.25 and 1.46 mm of the middle point with respect to the edges (along the longer - 7 cm - side), respectively. This deformation has to be compared to the standard silicon thickness of 300 μm. Care has been taken in avoiding possible causes of detector damage, particularly at edges. A destructive test was also performed on a silicon mechanical prototype: under the same conditions, breakage occurred at a deformation of the middle point of 2.3 mm.

3. MOS capacitors and oxide charge densities

A few Metal-Oxide-Semiconductor (MOS) structures are integrated onto each processed silicon wafer, allowing a rather simple study of the positive charge unavoidably trapped in the surface oxide during wafer manufacturing and processing. This surface charge is also subject to variations during the life of the detector, particularly because of environmental conditions such as humidity. Such devices are also frequently used to study ionising radiation surface damage of silicon structures.

Typically, the MOS capacitors considered have a circular shape, with a total area of 3 mm^2, and an annular guard ring. As an example, fig. 5 represents a CV curve[4] of an MOS capacitor, with a fit to determine the flat-band voltage V_0, from which the oxide charge density n_q, in units of e^-/cm^2, is calculated with the formula $n_q = C_{acc} V_0/(e\,A)$, where C_{acc} is the measured MOS capacitance at accumulation, e the electron charge and A the MOS gate area. Fig. 6 summarises the MOS oxide charge densities for each of the 25 wafers processed during a test batch which followed the production.

3.1. Oxide charge densities and leakage currents

On each of the wafer processed and delivered, there are, among other, a full size silicon microstrip detector, a small size microstrip prototype ('baby'-detector) and quite a number of various test structures.

---

[4] Measurement of the MOS capacitance C as a function of the gate voltage applied V

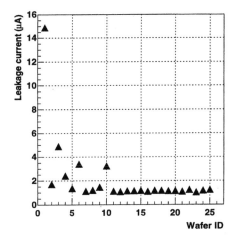

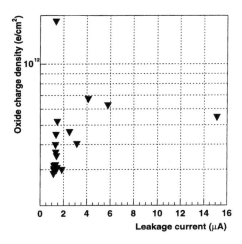

Figure 7: Leakage currents of large silicon microstrip sensors for all wafers (post-production batch).

Figure 8: Oxide charge densities of MOS capacitors vs. leakage currents of silicon sensors for all wafers (post-production batch).

Fig. 7 shows the measured leakage currents of the 25 full-size microstrip detectors of the post-production test batch considered, before diamond cutting (i.e. still on the original 4" wafer). Fig. 8 shows the relation between those values and the corresponding oxide charge densities of one of the MOS capacitors integrated onto the same wafer, consistently on the same test structure. In a few cases, higher values of sensor leakage currents correspond to higher oxide charge densities, except in the marginal isolated points.

The statistical correlation between leakage current I, rescaled at 20 C, and MOS oxide charge density n_q samples, calculated as $\rho(I, n_q) = \mathrm{cov}(I, n_q)/(\sigma_I \, \sigma_{n_q})$, has a value of 0.79 and thus shows a rather strong positive correlation, once two outliers[5] have been eliminated from the sample. A similar value (0.70) is found when the leakage current of the smaller size microstrip detector, also onto the same wafer, is considered against the MOS oxide charge, with one outlier left out. This leads to the conclusion that in this case the detector leakage currents and the oxide charge densities are correlated. Full-size and small-size detector dark current are strongly correlated (0.91, one outlier excluded).

4. Electrostatics considerations

During integration of the detector modules in our laboratories, which lasted a few months, we realised that leakage currents seemed to remain more stable if the construction jigs used, made of aluminium or glass, sometimes coated with insulating layers (teflon), were grounded. Typically, sensors are positioned on jigs, held by vacuum, whilst support frames or kaptons are glued onto the sensors. Waiting times, ranging from a few hours up to a day, are required by the operation sequence.

The reduction in the increase of the dark current or even its disappearance were interpreted to be a positive sign of electrostatic influence during operations, possibly coupled to environmental conditions such as temperature and humidity. Unfortunately, due to the unexpectedness of the phenomena observed, no systematic record was kept from the beginning of operations about the laboratory conditions, which are known to have varied within certain limits (temperature: 20 to 25 C, relative humidity: 40 to 70%). Our clean rooms have been substantially improved since then, offering now temperatures of 21±1 C and relative humidity lower than 50%.

---

[5] The origin of such values must therefore be found elsewhere (cfr. the two extreme points in fig. 8)

362

Electrostatic phenomena such as charge accumulation at the module interfaces, and particularly at edges, may be clearly responsible for the performance degradation found in a few cases. The MOS studies performed after the module production, onto a similar sensor batch, have shown correlations between surface charge trapped in the silicon oxide and the overall detector leakage current. After these unwanted phenomena appeared, we tried to control and minimise them, e.g. by grounding more carefully all contruction jigs and supports, as well as by building small nitrogen tents which protected the modules during construction from excessive humidity exposure and possible particulate, chemical or electrostatic contaminations, which would not be expected in a clean room environment anyway. These actions reduced, but not always, the increases in dark current seen.

Therefore, we also suspected an unknown and uncontrolled feature of the silicon sensors – as delivered – to be at least partially responsible for the effects seen during production. Nevertheless, we would not reasonably expect the substantial increases seen in about 5-10% of the modules processed in our laboratory to be due to modifications in the surface charge occurred during operations. We intend to perform in the near future more dedicated electrostatics studies on single sensors and complete modules.

5. Conclusions

The study of environmental influences on silicon sensors is a long-term project, particularly when trying to establish the possible qualitative – if not quantitative – electrostatics explanations of the effects seen. Tests performed to date show promising results. It has become even more clear than previously expected – within our community at least – that special care is needed throughout manipulation and storage of such sensitive devices. Our clean rooms facilities have been substantially improved to enable better control of the environmental condition such as temperature and humidity. A complete explanation of the dramatic increase of the module leakage current seen in a few cases remains to be found; further electrostatic and humidity tests will be performed to complete the present study.

Acknowledgments

The authors wish to thank the AMS group at the University of Geneva, the laboratory and workshop staff, among whom in particular Manfred Willenbrock, and Werner Kloeti of the spectrometry group. For the temperature test of the full modules we are indebted to Josef Fischer of the Physics Institute at the University of Bern[6].

Ms. Belinda Hopley of the IOP deserves special thanks for her work (and patience) before, during and after the Cambridge Electrostatics '99 Conference.

References

The AMS Collaboration 1994, AMS proposal
——1999 to be submitted to *Nucl. Instr. Meth.* **A**
The AMS Group 1996-1999 *Scientific Reports* (Geneva: University of Geneva)
Azzarello Ph 1998 *Diploma Thesis* (Geneva: University of Geneva)
——1998 *Internal Reports* (Geneva: University of Geneva)
——and Vitè D 1998 *Internal Report* (Geneva: University of Geneva)
Borchi E, Bruzzi M 1994 *Riv. Nuovo Cim.* **17** 1
Bosisio L 1997-1999 Private communication
Burger W 1997-1999 Private communication
Centre Suisse d'Electronique et de Microtechnique SA 1997-1999 (Neuchâtel)
Della Marina R, Weiss P 1997-1999 Private communication
Longoni A et al. 1990 *Nucl. Instr. Meth.* **A288** 35
Toker O et al. 1994 *Nucl. Instr. Meth.* **A340** 572
Vitè D 1996 *PhD Thesis* (London: Imperial College)

[6] Abteilung Massenspektrometrie und Raumforschung

Inst. Phys. Conf. Ser. No 163
Paper presented at the 10th Int. Conf., Cambridge, 28–31 March 1999

Simple passive transmission line probes for electrostatic discharge measurements

J M Smallwood

Electrostatic Solutions Ltd. 14 Courtland Gardens, Bassett, Southampton, Hants, UK.
email: jeremys@static-sol.com

Transmission lines feature wide (GHz) bandwidth characteristics and may be used as the basis of simple rugged probes for use in electrostatic discharge measurements with modern low cost fast sampling oscilloscopes. This paper describes some simple probes constructed using standard 50 Ω coaxial cable and carbon resistors. Measurement of ESD waveform characteristics such as peak current, charge transfer and discharge duration is demonstrated using discharges from the surface of charged PTFE, a small metal bolt, and the current generated by a capacitance discharge circuit.

1. Introduction

In electrostatic discharge (ESD) measurements it is of interest to determine characteristics such as the current waveform and charge transferred, and peak discharge current. These are of interest in electromagnetic compatibility, and investigations into the incendivity of electrostatic discharges, and characteristic of ESD arising from metal objects, insulator surface, textiles or other materials. Until recently, electrostatic discharges have mainly been investigated by charge transfer into a reference capacitor. This method, however, does not yield information on waveforms and discharge peak current, and cannot distinguish between single and multiple discharge events. This paper demonstrates that simple wide bandwidth probes can be fabricated using standard coaxial cable transmission lines and resistor networks.

2. A simple transmission line ESD probe

Transmission lines, such as the 50Ω coaxial cable, have wide bandwidth and low loss transmission properties. The coaxial cable can also be used as a very and rugged ESD probe for use with a fast digital storage oscilloscope.

The coaxial cable has a constant capacitance and inductance per unit length due to its well defined geometry and insulator dielectric properties (Jones et al 1993 pp2/23-2/26). The ratio of the voltage and current in the line is constant and defined as the characteristic impedance of the line (in ohms). In the simple case described here, a 50 Ω coaxial cable is used as a 50 Ω impedance ESD probe. At one end of the cable the inner core is bared for a short length and may, if desired be soldered to a small metallic bead or other convenient search electrode. An ESD wavefront injected into one end of the coaxial cable propagates along the cable as a voltage and current wave. If an impedance discontinuity is encountered, for example at a load resistor or oscilloscope input circuit, a reflection of the wave occurs, with part of the wave energy travelling back along the line, and part of the energy dissipated in the load. A correctly matched load is crucial in preventing reflections at the line termination. Many modern oscilloscopes have a 50 Ω input impedance option, or alternatively can be fitted with a 50 Ω terminator at the input socket.

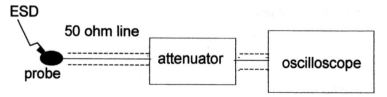

Figure 1. Simple transmission line ESD measurement arrangement

The oscilloscope must have a fast single-shot digitising capability to match the expected speed of the ESD waveform. Using standard coaxial cable and a modern low cost fast oscilloscope with a sampling speed of 2 giga samples/sec, a bandwidth of more than 400 MHz and resolution of waveforms in the nanosecond region is feasible. Modern oscilloscopes allow the digitised waveform data to be downloaded onto a disk for import into a spreadsheet. The waveform can be easily plotted and calculations made on the data.

A wide range of input waveforms may be investigated using this simple configuration, using the oscilloscope input attenuators to adjust the sensitivity. Care must be taken, however, not to exceed the oscilloscope input circuit ratings if larger energy discharges are investigated. Some additional attenuation and input circuit protection may be added by inserting attenuators between the probe and oscilloscope.

With correct matching, the ESD current pulse waveform is faithfully recorded as a digitised voltage waveform. The waveform rise time, fall time and duration can be measured from the oscilloscope display. Instantaneous or peak waveform voltage is converted into equivalent discharge current by a simple "Ohms law" calculation using the cable characteristic impedance. With an oscilloscope input sensitivity of 10V/division, a current sensitivity of 200 mA per division is achieved, and with a voltage sensitivity of 2mV/division, a current sensitivity of 40μA per division is possible.

If the complete discharge waveform has been digitised, then the charge transferred in the discharge may be obtained by summing the equivalent charge of all data points in the waveform.

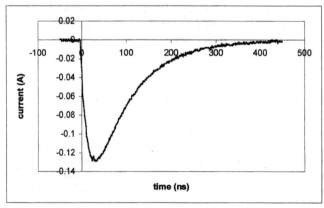

Figure 2. Discharge from PTFE surface tribocharged using paper towel

Figure 2 shows a typical discharge obtained from a charged PTFE surface. The probe consisted of a coaxial cable, which had the inner conductor connected to a small (approximately 6mm diameter) brass ferrule with a hemispherical tip. The relatively slow 23ns risetime and unidirectional form of the discharge can be clearly seen, with 16nC of charge transferred in a discharge of 100ns duration and peak current of 0.13A. In contrast, a discharge from a small metal bolt is extremely fast (Figure 3). In this case 0.7nC of charge was transferred in a discharge of about 1ns duration (measured at 50% pulse height), with a peak current of 0.82 A. The detailed form of the trace is interpolated by the oscilloscope based on a single shot digitisation with 0.5ns sampling interval, and accuracy of the measurements in this case must be considered doubtful.

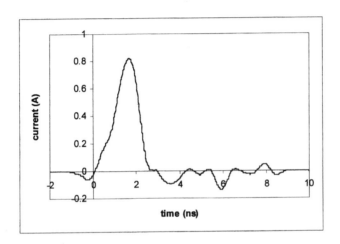

Figure 3. Discharge from M4x15mm bolt induction charged by proximity to charged PTFE sheet

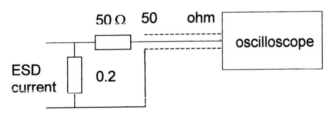

Figure 4. Current probe for monitoring capacitive discharge ESD current

3. An ESD probe using resistor elements to reduce input impedance and sensitivity

A resistor network can be added to reduce input impedance and allow high current discharge measurements. Figure 4 shows an example which was built into a capacitive ESD generator circuit and used in monitoring discharge currents. The ESD current passes through a 0.2 Ω shunt fabricated from five 1 Ω carbon resistors connected in parallel. Carbon resistors have a wide bandwidth (Mazda 1989, p 12/10) and an exceptional capability for withstanding high peak pulse current and power dissipation. Lead lengths were kept short to reduce inductance in the ESD current path. The voltage developed across the shunt resistor is monitored via an attenuator formed by a 50 Ω resistor in series with the 50 Ω transmission line input

366

impedance. The shunt resistor developed a voltage output of 0.2V/A, giving an overall probe sensitivity of 0.1V/A.

Figure 5 shows a discharge from a 188pF capacitor switched into a 2.2 Ω load using a vacuum relay. The 950nC charge is transferred with a peak current of about 200A and the waveform rings at 13 MHz. A considerable amount of VHF ringing is evident due to excitation of stray circuit resonances by the high initial rate of current rise.

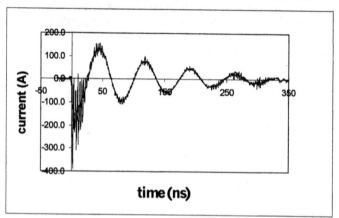

Figure 5. ESD from 188pF capacitor pre-charged to 4kV and switched into a 2.2 Ω 0.7μH load

A typical carbon resistor may have a parallel capacitance of about 0.5 pF. The performance of a circuit can be seriously affected at high frequencies when the impedance of this capacitance reduces, approaching the resistor value. Probe circuits may be calibrated using a well defined capacitive discharge waveform.

4. Conclusions

The transmission line probe is very simple, rugged, and high bandwidth measurements (>400 MHz) are easily achieved. The wave shape, peak current, rise time, and charge transferred are easily measured. The probe sensitivity is high and measurement of small discharges are easily made, but the impedance of the probe is rather high (50 Ω). Lower impedance probes for higher level ESD currents can be made using resistor networks with the transmission line. This type of circuit can be used to monitor ESD waveforms of hundreds of amps with bandwidths exceeding 100MHz.

References

Jones G R, Laughton M A, Say M G. 1993 *Electrical Engineer's Reference Book* 15th Edition. (Butterworth Heinemann) ISBN 7506 1202 9

Mazda F F (ed) 1989 Electronics Engineer's Reference Book 6th Edition (Butterworth) ISBN 0-408 05430-1

Acknowledgements

The author is grateful for the assistance of ERA Technology Ltd and Wolfson Electrostatics Ltd in this work.

Inst. Phys. Conf. Ser. No 163
Paper presented at the 10th Int. Conf., Cambridge, 28–31 March 1999
© *1999 IOP Publishing Ltd*

Multichannel measurement of ESD in silos

L Ptasiński[1], T Żegleń[1] and A Gajewski[2]

[1] Faculty of Electrical Engineering, Automatics, Informatics and Electronics, University of Mining and Metallurgy, al. Mickiewicza 30, 30-059 Kraków, Poland
[2] Physics Department, Academy of Economy, ul. Rakowicka 27, 31-510 Kraków, Poland

Abstract. A new multichannel system for recording ESD was used to simultaneous recording of discharge to two electrodes in the laboratory model of a silo within a pneumatic transportation system. One of the electrodes was flat, and it was attached to the wall of the silo made of perspex. The second electrode, called the discharging electrode, was made of a set of vertically-fixed wires forming the squirrel cage shape. The effect of certain factors on the ESD characteristics was examined using the described recording system..

1. Introduction

In order to reduce hazards caused by ESD in the pneumatic transport system, discharge wire electrodes are used. At the surface of the wires the strength of the electric field reaches high values even at a low volume density of the charge in the silo. Thus, it may be expected that the earthed wires force low-power partial discharges which themselves are nonincendive, and in this way they prevent accumulation of high charges and, consequently, occurrence of high-power discharges. Conductive fibres in the woven fabric of Flexible Intermediate Bulk Containers (FIBC) are expected to work in a similar way. However, a number of authors criticize this protection system against high-power discharges [1], [2] putting in doubt its effectiveness. The multichannel recording of ESD initiated to different conducting elements of the installation, including the discharging electrodes, should help to clear up these doubts.

2. Experiment

2.1. *Description of the experiment*

The investigations were carried out on a model of a pneumatic installation of the closed cycle presented in Fig. 1.

The model was made of perspex. The capacity of the silo is 0,15 m^3, with a diameter of 0,4 m. The product used in the experiment was polystyrene cylindrical granules with typical diameters of about 1,3 mm and a lengths of about 5 mm. The resistivity of the granules is $2 \cdot 10^{13}$ Ωm. In order to obtain the smoothness of the product stream, a feeding screw was used. The measurement results presented in this paper were obtained at the product stream of approx. 100 kg h^{-1} and the air stream of approx. 270 m^3 h^{-1}. The experiment was conducted at the air temperature of 24°C and humidity of 48%. Two electrodes were placed inside the silo: one electrode was flat (mounted on the silo wall) and the second electrode was made of wire in the shape of the squirrel cage and placed in half of the distance between the silo axis and its wall.

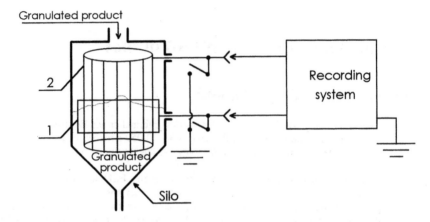

Figure 1. Schematic diagram of the experimental set-up, 1 – flat electrode, 2 – wire electrode

With the application of the new system presented at the conference [3] the charge-time characteristics were recorded. The system worked in the multichannel mode, triggered by each successive pulse. The system permitted a simultaneous recording of ESD for both electrodes and examining the influence of the earthing of the discharging electrode. Basing upon the recorded characteristics [4] and specially developed computer programme, histograms and those discharge parameters which reflect the ESD igniting power [4], [5] were determined. Histograms concern the two quantities: charge transferred in each ESD pulse (Q_t) and time intervals between the successive pulses (ΔT). The following ESD parameters were considered: maximum charge transferred in a single ESD pulse (Q_{tmax}), average value of Q_t (Q_{ta}), average value of ΔT (ΔT_a) and average value of electrical current (I_a). The ESD characteristics were recorded when electrodes were earthed through the recording system.

2.2. Results of the experiment

During the time of the experiment, the change of the shape of the heap of the granules was observed i.e. they were forming a concave heap as described in [6]. The speed of forming the concave heap was increased after removing the discharging electrode, but in its presence it depended upon the diameter of the wire and the earthing of the electrode. The sudden collapse of the granules on the surface of the concave heap was also observed. These effects depended clearly on the distribution of the electric field determined by the presence of the discharging electrode and the diameter of its wires, as well as by the chargeability of the granules and the space charge in the silo. For instance, during the tests carried out with the granules of lower chargeability, instead of a concave heap, a convex heap was formed, and the results of the measurements were characterized by greater repeatability. The observed effects had an influence on the recorded characteristics and were the cause of the unstationary process of ESD generating. Thus, drawing conclusions on the basis of the successively carried out measurements requires a great amount of caution. The complexity of the process requires a simultaneous multichannel observation of the tested site.

Exemplary measurement results are presented in Fig.2 and Fig.3. The measurements were carried out in the two-channel mode with both electrodes active. In the second experiment the squirrel cage type electrode was replaced by the similar one but made of wire of different diameter.

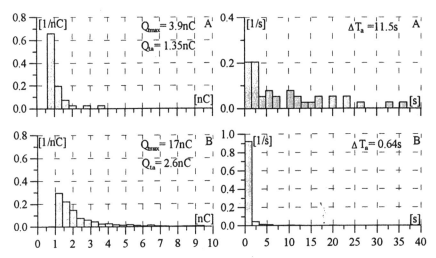

Figure 2. Histograms of ESD charges (Q_t) and time intervals between successive pulses (ΔT).
A – discharges to the flat electrode;
B – discharges to the discharging electrode with the wires of ϕ 1,15.

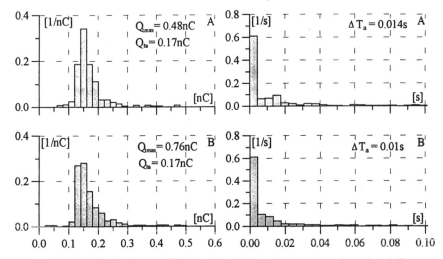

Figure 3. Histograms of ESD charges (Q_t) and time intervals between successive pulses (ΔT).
A – discharges to the flat electrode;
B – discharges to the discharging electrode with the wires of ϕ 0,2.

Table 1 shows the parameters of ESD initiated to the flat electrode in different conditions: after disconnecting the earthing of the discharging electrode and after removing it from the silo.
The reason to conduct these measurements was to check, whether incidentally broken connection between discharging electrode and earth results in increasing hazard, comparing with the situation, when the discharging electrode is not present at all.
An analysis of the obtained results enabled us to draw the following conclusions:
- Insertion the discharging wire electrode causes a decrease of the charges transferred in single ESD pulses to the flat electrode.

370

- Comparing the results of the two-channel recording, it can be noticed that ESD to the discharging electrode can have a higher igniting power than to the flat one.
- Decreasing the diameter of the wires of the discharging electrode causes a decrease of the ESD charges and shorts the time intervals between successive pulses. The problem of minimizing the hazards caused by ESD through the suitable geometry of the discharging electrode requires carrying out separate investigations.
- After the earthing of the discharging electrode is removed the effect of decreasing the ESD charge value can be still observed.

Table 1 ESD parameters to the flat electrode

Test No	Q_{tmax} [nC]	Q_{ta} [nC]	ΔT_a [s]	I_a [nA]
1	3,5	0,17	0,054	3,1
2	0,76	0,16	0,047	3,4
3	31	2,2	0,56	4,0

1 – with unearthed discharging electrode made of wires $\phi 1,15$;
2 – with unearthed discharging electrode made of wires $\phi 0,2$;
3 - without discharging electrode.

3. Summary

The results of the investigations have confirmed the purposefulness of applying the multichannel recording of ESD, above all on account of the observed complexity of the process of their generation and disturbances in its stationarity.

Insertion the discharging wire electrode causes a decrease of the ESD pulses magnitude. Disconnecting the earth from the discharging electrode does not eliminate the above effect.

The parameters of the discharges to the wire electrode prove that they may cause greater hazards than the discharges having place at the same time to the earthed construction.

Finding an answer to the question whether to install the discharging electrode in the silo requires carrying out investigations, similar to the described above, but conducted on larger plants.

In the industrial world silos are in most cases made from conductive materials. Therefore the electric field in such cases have different distribution than in the investigated silo model. The different electrical field distribution can influence the mutual relation between ESD initiated to the silo wall and to the discharging electrode.

References

[1] Glor M 1988 *Electrostatic Hazards in Powder Handling* (J. Willey&Sons Inc.)
[2] Davies D K 1997 *One Day Conference on Static and Textiles* (Inst.Phys. in London)
[3] Ptasiński L, Żegleń T, Gajda J 1999 *Electrostatics '99* (Cambridge Inst.Phys.Conf.)
[4] Ptasiński L, Żegleń T and Gajewski A 1996 *EOS/ESD Symposium Proc.*7.5.1-7.5.5
[5] Ptasiński L and Żegleń T 1997 *One Day Conference on Static and Textiles* (Inst.Phys. in London)
[6] Cheung W L and Bailey A G 1995 *Inst.Phys.Conf.*Ser.No.143 373-376

Acknowledgements

The work presented in this paper was sponsored by the State Committee for Scientific Research, Warsaw (contract No: 8T10C01012).

DISCUSSION - Section G - ESD/EOS

Title: The Effect of Resistance to Ground on Human Body ESD

Speaker: D Swenson

Question: T B Jones
Can you provide us with some quantitative data on the impact of ESD/EOS on the semiconductor industry?

Reply: Difficult to answer because company data is not readily available. Motorola does state that 5-7% of all parts produced are damaged - a 1992 report said 30.35% of all damage is from EOS/ESD.

Question: John Chubb
You have dealt very thoroughly with the question of controlling voltages by using wrist straps. The question I would like to ask is what limitations should apply to materials in proximity to sensitive devices - for example surfaces of magnifiers, lenses, and garments? In particular, I wonder about the role of 'charge device model' damage associated with the electric fields from any static charges retained on such nearby surface.

Reply: Most of the industry believes few devices today can be affected by electric fields/induction (except at very high values). The 1994 edition of RIA625 specified maintaining potentials in the environment to < 200V at 6 inches. The 1998 edition lists <1000V at 6 inches. Naturally, an insulator cannot be (properly) described by voltage - only by electric-field. The main reason for changing the specification is because it is unrealistic to attempt to maintain an environment at low potentials. Circuit board materials for instance are present and can have substantial electric-field values.

Title: Environmental Tests of Silicon Microstrip Detector Devices for High Energy Space Applications

Speaker: D F Vité

Question: Allel Bouziane
Why did you not investigate the effect of pressure?

Reply: We performed complete thermo-vacuum tests on most of the elements (detectors, electronics modules) prior to installation within the experimental apparatus. A couple of silicon microstrip modules underwent such test, as a sample, but not all of the 70 modules could be tested. Regarding the individual silicon sensors, we think humidity may play a major role, as it cannot be kept continuously under control, particularly during element transport, interactions between different production sites, etc. Both humidity and vacuum will be considered for our future production and integration phases.

Title: Simple, Passive Transmission Line Probes for Electrostatic Discharge

Measurements
Speaker: J M Smallwood

Question: D K Davies
Given the high level of activity in developing model circuits to simulate ESD, do you feel that more precise characterisation of "real" ESD is required?

Reply: Real ESD events are extremely variable and this may be very important for some types of situations. One example is the ignition of dust clouds, where there is evidence that the ESD characteristics affect the likelihood of ignition. Unidirectional discharges from capacitive discharge circuits have been shown to ignite dust clouds more easily than oscillatory discharges. Measurement of discharges from insulators and materials of intermediate conductivity show that unidirectional discharges may be obtained, in these circumstances. These observations have implications for MIE measurements and hazard assessment.

Question: D K Davies
Do you believe that the development of ESD simulation models are based on sound ESD waveform data for "real" events.

Reply: I believe there are notable gaps in our knowledge here, particularly in the simulation of discharges from insulators and surfaces of intermediate conductivity. These could be of great importance in ignition hazard assessment and sensitivity tests. They may also be of interest in the development of electrostatic damage prevention in electronics.

Question: Norman Wilson
You showed discharges from a positively charged insulating surface e.g Perspex®. These had durations of well below 1 μs. In tests with discharges from positively charged insulating fabrics we usually find that the duration of the spark is in the region of 100 to 200 μs. However occasionally we have produced discharges of duration below 1 μs. Can you comment on this?

Reply: I find it difficult to comment without examining in detail the experimental circumstances and measurement arrangement used. There are a variety of factors which could lead to such differences and I believe that adequate study of these discharges may not yet have been made.

Title: Multichannel Measurement of ESD in Silos
Speaker: L Ptasinski

Question: D K Davies
Why does your electrode system increase the ESD hazard?

Reply: It is difficult to say that this effect really occurs. We observed only that the charge transferred to the discharging electrode may be higher than that simultaneously recorded to the flat electrode (wall of the silo).

Section H

Scanning Probe Microscopy

Inst. Phys. Conf. Ser. No 163
Paper presented at the 10th Int. Conf., Cambridge, 28–31 March 1999

Measuring static charges by scanning probe microscopy

D. M. Taylor and P. W. C. Sayers

School of Electronic Engineering and Computer Systems, University of Wales,
Dean Street, Bangor, Gwynedd LL57 1UT, UK

Expressions are derived for the electrostatic force acting on the tip of an Atomic
Force microscope in the presence of (a) a point charge and (b) a uniform surface
charge on a polymer film. In a series of experiments in which a polypropylene
surface is 'uniformly' charged using a scorotron arrangement attempts at imaging
the charge in the 'lift-off' mode of the Nanoscope IIIA AFM appears to produce
highly resolved topographical images of the polymer surface. The role of charge-
tip interactions in producing such an image is discussed.

1. Introduction

Scanning probe microscopy (SPM) provides a means of spatially resolving surface charges to
submicron resolution. In conventional Atomic Force Microscopy (AFM) samples undergo
controlled translational motion under a sharp tip mounted on a flexible cantilever forming
part of an optical lever system for detecting its deflection when subjected to applied forces.
In the Electrical Force Microscope (EFM), the cantilever is deflected by electrostatic forces.
For greater sensitivity, modern instruments use modulation techniques which detect the
presence of force gradients [1]. Thus, it becomes possible to measure forces in the range
10^{-13} N enabling high-sensitivity (single electron) measurements to be made of charge
deposited onto microscopic areas of an insulator by contact and voltage-induced charging.
(For a recent review see [1,2]). A feature of previous measurements is the microscopic size
of the charged zone. There are occasions when more extensively charged regions need to be
examined e.g. corona-charged surfaces, biomembrane surfaces etc. Already, techniques are
being developed for measuring membrane charges in the presence of an electrolyte bathing
medium [3]. Here we concentrate on corona charged polymer surfaces.

2. Electrostatic Probe-Charge Interactions

In the following, the electrostatic interactions between the EFM probe and (a) a point charge
and (b) a uniform surface charge are considered to represent the two extreme experimental
cases that will be encountered. In the analysis, it is assumed throughout that the probe is an
earthed, conductive sphere of radius a. It is recognised that a probe is closer to a paraboloid
in shape but the simplification possible here offsets any errors introduced.

2.1 Point Charge
Figure 1 shows the EFM probe scanning horizontally over a point charge q located on the
surface of an insulating surface. It is assumed that the thickness of the insulator greatly

exceeds the height, z, of the centre of the probe above the surface. The interaction between the probe and the point charge is readily calculated from the interaction between the point charge and its image $q'=q.a/b$ which is located at A, a distance a^2/b from the centre of the probe on the line PC joining the real charge to the centre of the probe [4]. The vertical component, F_z, of the electrostatic force acting on the probe is then given from Coulomb's law as

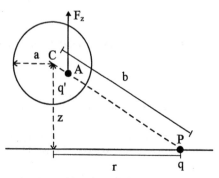

Figure 1 *Model used in model calculations. The point charge q induces an image charge q'=q.a/b at A where AC=a²/b.*

$$F_z = -\left(\frac{q^2}{4\pi\varepsilon_0}\right)\frac{az}{\left(z^2+r^2-a^2\right)^2} \qquad (1)$$

where ε_0 is the permittivity of free space and r the horizontal distance from the probe centre to the point charge. For $z>>a$, it is often assumed that the image charge is a point charge equal in magnitude to the real charge, e.g. [2]. This is clearly erroneous since, under this condition and assuming $r=0$, equation (1) reduces to $F_z = -(q^2/4\pi\varepsilon_0)(a/z^3)$. The force gradient, F'_z, on the probe is obtained by differentiating equation (1). Thus [5]

$$F'_z = \left(\frac{q^2}{4\pi\varepsilon_0}\right)\frac{a\left(3z^2-r^2+a^2\right)}{\left(z^2+r^2-a^2\right)^3} \qquad (2)$$

and shows a sign reversal when $r^2 > (3z^2+a^2)$ possibly leading to artefacts in the EFM image.

2.2 Uniform Surface Charge

The interaction between the probe and a charge sheet of density Q_S may be treated by assuming that the charge is confined to a circular patch of radius R with the probe located symmetrically above it. Developing the point charge case for this situation and assuming the probe radius is small, $a<<z$, it is readily shown that the total charge, Q_P, induced in the probe is given by $Q_P = -2\pi aQ_S[(z^2+R^2)^{1/2}-z]$ which reduces to $Q_P = -(2\pi aQ_SR)$ for $R>>z$ and $Q_P = -(\pi aQ_SR^2/z)$ for $R<<z$, the latter case being, of course, the point charge limit. The total force, F_z, acting on the probe is obtained by multiplying the induced charge, Q_P, by the electric field $(E_z = (Q_S/2\varepsilon_0)[1-(1+R^2/z^2)^{-1/2}])$ at the probe. Thus

$$F_z = \frac{\pi Q_S^2 az}{\varepsilon_0}\left[2-\left(1+\frac{R^2}{z^2}\right)^{\frac{1}{2}}-\left(1+\frac{R^2}{z^2}\right)^{-\frac{1}{2}}\right] \qquad (3)$$

which for extensively charged regions, $R>>z$, reduces to $F_z = (\pi Q_S^2/\varepsilon_0)aR$ in which case charge measurements based on monitoring the force gradient will fail since F_z' goes to zero.

Therefore, electrostatic charge sheets on polymers should give measurable phase signals only when the probe is close to the surface i.e. $z \sim a$.

3. Experimental

Polypropylene samples, 15 mm diameter and 57 μm thick were rinsed in isopropanol, left overnight to ensure solvent evaporation and mounted on earthed metal stubs on a rotating turntable where they were corona charged to predetermined potentials using a scorotron-type device [6]. The equilibrium surface charge density was measured with an induction probe [7]. Subsequently, samples were imaged in a DI Nanoscope IIIA STM/AFM with EFM extender accessory following a routine differencing technique (DI application note ????). Briefly, a tapping mode topographical image was obtained by maintaining constant cantilever oscillation amplitude (TMAFM) and collecting the phase image (TMPAFM) simultaneously during a linescan. Then the EFM image was collected by lowering the sample a pre-determined distance (typically 20 - 100 nm) and retracing the initial scan, following the stored surface contour. Residual forces acting on the probe should then be the long-range electrostatic forces which are recorded as a phase change in the cantilever oscillation.

4. Results and Discussion

Figure 2 shows a series of TMAFM, TMPAFM and EFM images of the polypropylene film (a) uncharged and (b) charged to a surface potential of 20 V. In both cases, the TMPAFM images show much greater detail than the TMAFM images owing to the edge enhancement present in phase imaging. EFM images for lift-heights of 10 and 100 nm for the uncharged sample are indistinct with signal strength decreasing almost to zero at 100 nm. For the charged sample, the EFM and TMPAFM images are almost identical for lift-heights up to 100 nm. At greater heights, the EFM image deteriorated such that at 300 nm the image was similar to that obtained for the uncharged film at 10 nm. Interestingly, resolution of features on the charged surface appeared greatest at a lift height of ~ 50 nm.

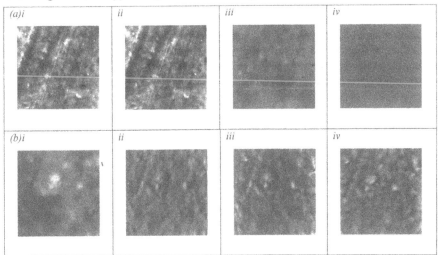

Figure 2 *(i) TMAFM, (ii) TMPAFM and EFM images for lift-heights of (iii) 10 and (iv) 100 nm for (a) an uncharged and (b) a uniformly charged polypropylene surface. All images 5 μm x 5 μm.*

When charged to 20 V the sample carries a surface charge of ~6 $\mu C/m^2$. If a = 50 nm and charges within 60 μm of the probe couple to it (reasonable since more distant charges will couple preferentially to the underlying earthed substrate), then F_z corresponding to a lift-height of say 500 nm ($z \gg a$) is estimated to be 3.83 x $10^{-11}N$. The spring constant of the cantilever used in this work was typically 50 N/m so that the expected static deflection is ~7.67 x 10^{-13} m and clearly too small to compensate for a lift-height of 300 nm! The effects on the phase and amplitude of the cantilever are also expected to be minute since the experimental conditions should yield an almost zero field gradient (equation (3) with $R \gg z$). Even if it is assumed that the surface charge acts over the whole cantilever as in a parallel plate capacitor this would lead at most to a factor 100-1000 increase in the total electrostatic force on the cantilever. Therefore, electrostatic forces *per* se acting on the freely vibrating probe can be discounted as the source of the EFM images in Figure 2. Rather, the effect probably arises from an electrostatically enhanced 'adhesion' to the surface preventing retraction of the probe. Thus, despite adjusting the z-piezo to the predetermined lift-height, when the probe is 'retraced' for the EFM image it remains in contact with the surface and produces the same phase image as obtained by TMPAFM. The presence of additional adhesive forces when retracting a *charged probe* from an earthed conductive surface has already been reported [8,9] and estimated to be ~ 1 μN [8] with -15 V applied to the probe. Assuming that the probe in the present work remained in contact with the surface even at a lift height of 100 nm, the force acting on the cantilever would be ~ 5 μN and close to that reported in [8]. When close to the surface, the probe may be assumed to act as a parallel plate capacitor so that the force acting is given by $F = Q_s^2 A/\varepsilon_o$ where A is the effective area of the probe and estimated to be ~ 2.5x10^{-7} m^2 which is orders of magnitude greater than the area of the whole cantilever let alone that of the probe. It appears, therefore, that electrostatic-induced adhesive forces of the magnitude observed here and by other workers cannot arise from simple coulombic interactions alone.

5. Conclusions

EFM images of uniformly charged polymer surfaces are identical to TMPAFM images of the same area even at a lift-off height of 100 nm. The effect is explained by an electrostatically-induced adhesive force which prevents the proper retraction of the probe. It is argued that the magnitude of this additional force is significantly greater than would be expected from a simple coulombic interaction between probe and charged surface.

References

1. R Weisendanger 1994 *Scanning Probe Microscopy and Spectroscopy: Methods and Applications* (Cambridge University Press).
2. GC Stevens 1998 in *Space-charge in Solid Dielectrics* 73-92, The Dielectrics Society.
3. WF Heinz and JH Hoh 1999 *Biophys J* 76 528-538.
4. WR Smythe 1968 *Static and Dynamic Electricity, 3rd Edition* (New York:McGraw-Hill Inc).
5. DM Taylor 1998 *Thin Solid Films* 331 1-7.
6. DM Taylor and PE Secker 1994 *Industrial Electrostatics-Fundamentals and Measurements* (Taunton: Research Press).
7. DK Davies 1970 *Advances in Static Electricity* 1 10-21.
8. R Schlaf, D Louder, MW Nelson and BA Parkinson 1997 *J Vac Sci Technol* 15 1466-1472.
9. JW Hong, ZG Khim, AS Hou and Sang-il Park 1996 *Appl Phys Lett* 69 2831-2833.

Inst. Phys. Conf. Ser. No 163
Paper presented at the 10th Int. Conf., Cambridge, 28–31 March 1999

Quantitative Methods for Non-contact Electrostatic Force Microscopy

Patrick J S Baird[1], John R Bowler[2] and Gary C Stevens[1*]

[1] Polymer Research Centre and [2] Department of Physics, University of Surrey, Guildford, Surrey GU2 5XH, UK

Abstract. We describe recent work on modelling the probe-sample interaction in electrostatic force microscopy (EFM) using a charge distribution method-of-moments calculation for a conical probe that can be generalised to a variety of probe-specimen geometries. This involves calculation of the force and force-gradient approach curves enabling comparison with a variety of experimental models from simple microdot electrodes to more complex multi-potential interdigitated electrode structures. We also demonstrate the use of non-contact heterodyne detection methods utilising cantilever amplitude and phase signals to quantify experimentally the probe-specimen interaction and to produce simultaneously surface potential and capacitance images.

1. Introduction

Electrostatic force microscopy (EFM) is one of an increasing number of methods used in scanning force microscopy that is gaining popularity as a technique for non-contact potentiometry and for charge and permittivity imaging at a microscopic level. It has already been applied to a wide number of problems from semiconductor devices to inorganic and polymer dielectrics [1]. Two important applications are integrated circuit failure analysis and imaging of charge effects on insulators such as polymers. The charge and spatial resolution is potentially much higher than that of existing macroscopic methods for potentiometry and surface charge measurement.

Despite a developing literature since 1988, non-contact methods have not been consolidated on commercial atomic force microscopes (AFMs) and little understanding has existed of the interaction between probe and sample and of optimisation of the technique, particularly in non-contact mode. The purpose of this paper is to present some of our recent work in modelling and experimentally validating the electrostatic interaction between the probe and a variety of substrates. The models have been used to generate force-distance interaction curves for various electrostatic situations and these have been compared with force curves

---

* Contact author, e-mail: g.stevens@surrey.ac.uk

obtained from experimental information based on a variety of signals obtained from the cantilever oscillation.

2. Theory of Forces involved in EFM

In true non-contact EFM, an alternating voltage is applied between a conducting cantilever and an electrically grounded plane beneath the sample under investigation. The resulting alternating capacitive force between the tip and ground plane produces an oscillation of the cantilever. The amplitude of the cantilever oscillation is detected optically and is kept constant using feedback control of the height of the probe above the surface. Any variation in the electrical characteristics of the sample will appear in the resulting height signal. Signals collected at different frequencies can be used to separate topographical and electrostatic information.

An approximate expression for the force in terms of capacitance between an oscillating cantilever tip and insulating sample with point charge q_s on its surface was formulated by Terris et al [2] for the case in which the sample thickness is much greater than the tip-sample separation. By considering image charges induced on the tip and substrate and for $V = V_0 \sin \omega t$ applied between tip and substrate we obtain the interaction force [3]:

$$F_z(z) \approx \frac{V_0^2}{4} \frac{\partial C}{\partial z} [1 - \cos(2\omega t)] - \frac{q_s}{4 \pi \varepsilon_0 z^2} [C(z) V_0 \sin(\omega t) + q_s] \qquad (1)$$

where C is the tip-substrate capacitance, z is the tip-sample separation, ε_0 is the permittivity of free space and $F_z(z)$ is the force in the z-direction. If the sample is a thin layer then additional interactions may need to be considered. In our case we measure the component of force at an angular frequency 2ω which provides the capacitance gradient image and the ω component which gives the charge or surface potential image.

In potentiometry, a null method is used in which the cantilever tip potential is made equal to the sample potential V_s. A dc voltage V_p is added to the tip such that $V_1 = V_p - V_s$ giving $V = V_1 + V_0 \sin \omega t$ and

$$F_z(z) \approx \frac{V_0^2}{4} \frac{\partial C}{\partial z} [1 - \cos(2\omega t)] + \frac{V_1}{2} \frac{\partial C}{\partial z} [V_1 - 2V_0 \sin(\omega t)] \qquad (2)$$

where V_p is servo-controlled until the signal $(V_p - V_s)$ collected at ω is zero. Van der Waals forces are also present but are negligible at separations greater than a few nm.

3. Modelling Tip-Sample Interactions

In the expressions above, the capacitance terms depend on the geometry of the cantilever tip and substrate, and are therefore not easily determined analytically. The well known Green's function integral equation for potential in terms of surface charge density was used (for a full description of this model see [3]):

$$V(r) = \frac{1}{\varepsilon_0} \int G(\mathbf{r}, \mathbf{r}') \sigma(\mathbf{r}') dS' \qquad (3)$$

where $\sigma(\mathbf{r}')$ is the charge density and the static free space Green's function is given by

$$G(\mathbf{r}, \mathbf{r}') = \frac{1}{4\pi |\mathbf{r} - \mathbf{r}'|} \qquad (4)$$

The method of moments was applied which involves dividing all charged surfaces (initially conductors for a cantilever tip and ground plane model) into discrete sub-areas of constant charge density. The potential on the conductors is known. By point matching to the known potential the integral equation to be approximated by a matrix equation. Solution of the matrix equation gives the charge densities on the conductor surface sub-areas. For a rotationally symmetric system of conductors whose surfaces are described using a cylindrical coordinate system the surface integral can be reduced to a line integral [4].

The important input parameters are the cone angle, cone height, radius of the spherical cap, cone potential, the number of sub-areas and the number/increment of the cone-plane separations. A typical tip radius is about 25nm. The equivalent of an infinite ground plane can be generated by using the method of images [5] in which the image cone has charge densities of equal magnitude and opposite polarity. The presence of a dielectric layer is accounted for by a thin cylindrical section of dielectric material, thus retaining the axial symmetry. The dielectric/air interface is also divided into sub-areas and an image dielectric used.

The forces are calculated using an expression for the force per unit area normal to a conductor surface:

$$\frac{dF}{dA} = \frac{\sigma^2}{2\varepsilon} \tag{5}$$

Summing over the surface sub-areas from the moment method [6],

$$\mathbf{F}(z) = \frac{1}{2\varepsilon_0} \sum_i [\sigma_i(z)]^2 \Delta S_i \mathbf{a}_i \tag{6}$$

where ΔS_i is the sub-area and $\mathbf{a}_i$ is a unit vector normal to the surface. Putting $\Delta S_i = 2\pi \rho_i(z)\Delta t_i$, we calculate the force on the cone surface in the z-direction as:

$$F_z(z) = \frac{\pi}{\varepsilon} \sin\alpha \sum_i [\sigma_i(z)]^2 \rho_i(z)\Delta t_i \tag{7}$$

The force on the spherical cap in the same direction is

$$F_z(z) = \frac{\pi a}{\varepsilon} \sum_i [\sigma_i(z)]^2 \Delta z_i \cos\beta_i \tag{8}$$

where α is the cone half-angle, a is the sphere radius and β_i is the angle of the arc between the vertical axis and the centre of sub-area ΔS_i. The force-distance curves vary with changes in the probe characteristics as shown in Figures 1, 2 and 3 where unless stated otherwise: a = 25nm, h = 12 μm, α = 17° and V = 5V in which the probe is interacting with an infinite ground plane.

4. Experimental Measurements

Our EFM measurements were performed on a commercially available Digital Instruments Nanoscope III AFM with minor modifications to the hardware to enable oscillation of the tip electrostatically. EFM on this instrument would normally be constrained to interleaved contact and non-contact scan lines, which is sufficient for potentiometry but not for surface charge imaging. Force-separation curves are not affected, as they require no feedback loop. We retain the interleaved scan method in order to separate topography and charge/potential information, but no contact is made except initially in order to determine the scanning parameters.

We set the drive frequency to half the resonant frequency of the cantilever to ensure that the amplitude signal picked up by the RMS detector is mainly due to oscillation at twice the drive frequency ω. This signal contains quantitative information on topography or permittivity and is useful for non-destructive evaluation. For information on charge or potential, the 2ω data is stored for each scan line and a second scan (interleaved) is taken moving the tip to the heights stored previously, thus minimising the topography and permittivity information in the charge signal (ω). If there is nor permittivity variation, the topography is effectively eliminated. The drive frequency for the interleaved line scan is set equal to the cantilever resonant frequency so that the phase signal picked up by the lock-in amplifier represents the ω term in equations (1) and (2) giving information on charge and on potential that no longer includes topography or permittivity information.

We used this arrangement to measure microdot electrode and interdigitated devices with externally applied voltages (Figure 4a) and charge regions deposited on thin polymer films through the probe (Figure 4b). Special procedures were required to set up the AFM for non-contact scanning and charge deposition and imaging. Following normal engagement of the tip and sample, parameters were set in force mode (vertical scan), the tip was moved to a remote position and the polarity of the vertical piezo scanner was reversed to take into account the attractive interaction. It was found that non-contact scans should begin at a tip-sample height of about 500nm or more. This separation could be reduced but not until the scan had stabilized (to avoid jumping into the contact region where the amplitude-distance curve is suddenly reversed).

The approximate height of the tip above the sample can be determined after the scan by moving to another region of the sample and using vertical scans ("force mode"), thus avoiding contact in the region being scanned over. Contact charging of the polymer samples through the tip was achieved by setting the scan size to zero and decreasing the setpoint until contact occurred. Problems arose in imaging the charge due to instability of the scanner and the subsequent difficulty in locating the deposited charge.

Model – experiment comparisons can be made with force curve data obtained from the AFM. For an oscillating cantilever this can be obtained from the phase signal θ (angle between driving a.c. signal and cantilever oscillation). At resonance should be $\theta = 90°$ and to obtain the actual phase angle for each point on the curve we apply the relationship between the phase angle and the shifted resonant frequency of the cantilever

$$\tan\theta = \frac{\omega\omega_0}{Q(\omega_0'^2 - \omega^2)} \qquad (9)$$

where ω_0' is the modified resonant frequency and the Q-factor of the cantilever-sample system is $\omega_0/\Delta\omega$ where $\Delta\omega$ is the width of the resonance peak. Then we obtain,

$$\frac{dF}{dz} = k[1 - (\omega_0'/\omega_0)^2] \qquad (10)$$

where k is the spring constant. The resulting force gradient data is then integrated with respect to the separation z to obtain the force curves. The optimum parameters for the cone model which best represented the tip of the cantilever used in Figure 5 were: cone angle = 17°; cone height = 12μm; sphere radius = 37nm. A tip-sample voltage of 5V was used throughout. A result of the model/data comparison is shown in Figure 5 which shows good agreement. In this plot the tip-sample separation in the static model can be assumed to represent the average position of the oscillating cantilever. Force curves over dielectric material can also be modelled and compared with experimental data.

385

5. Conclusions

We have shown that quantitative analysis of electrostatic data in EFM is possible on a commercially available AFM by comparing force curve data using appropriate tip-sample interaction models. These models also guide the choice of operating conditions and aid the interpretation of images.

The axisymmetric model is sufficient for force curve comparisons over simple sample structures, but breaking of symmetry in the model may be required to compare with situations where the sample structure is more complex. In addition, the model can be extended to include surface charge and sub-surface space charge simulation. In principle an entire EFM image (potential, charge and capacitance) could be simulated using model generated force curves.

The use of AFM for potentiometry and charge imaging has advantages over conventional macroscopic techniques. We can detect localised charge regions with spatial resolutions of the order of 50nm or better with different polarity that might otherwise be seen by macroscopic methods as a net charge distribution over a much larger region.

References
[1] Stevens, G.C. (1998), *Electrostatic Force Microscopy of Charges in Solid Dielectrics*, in Space Charge in Solid Dielectrics, Eds: J.C.Fothergill and L.A.Dissado, Dielectrics Society, ISBN 0 9533538 0 X, p73-92.
[2] Terris, B. D., Stern, J. E., Rugar, D. and Mamin, H. J. (1989), *Contact Electrification using Force Microscopy*, Phys. Rev. Lett., **63**, 2669-72.
[3] Baird, P.J.S., Bowler, J.S. and Stevens, G.C., to be published in J.Phys.D: Appl.Phys.
[4] Mautz, J. R. and Harrington, R. F. (1970*), Computation of Rotationally Symmetric Laplacian Potentials*, Proc. IEE **117**, 850-2.
[5] Smythe, W. R. (1968), Static and Dynamic Electricity, Third Edition, McGraw-Hill.
[6] Watanabe, S., Hane, K., Ohye, T., Ito, M. and Goto, T. (1993), *Electrostatic Force Microscope Imaging Analyzed by the Surface Charge Method*, J.Vac.Sci.Tech.B. **11**, 1774-81.

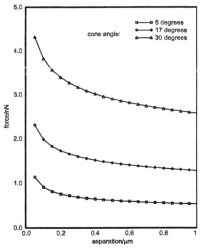

Fig.1. Effect of tip cone angle on tip-sample force for h = 12μm.

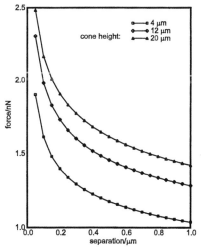

Fig.2. Effect of tip cone height on tip-sample force for 2α = 17°.

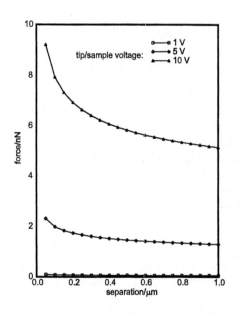

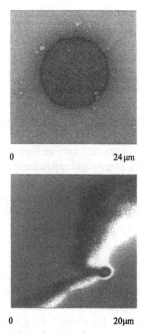

Fig.3. Effect of tip voltage on force curve for h=12μm, 2α = 17°.

Fig.4. (a) Phase-image of microdot electrode
(b) PMMA surface after multiple contact charging at the cusp point.

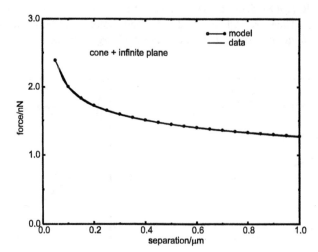

Fig. 5. Force-curve comparison of image cone-model calculation with experimental measurement for a r = 37nm, 2α = 17°, V = 5V.

Inst. Phys. Conf. Ser. No 163
Paper presented at the 10th Int. Conf., Cambridge, 28–31 March 1999

Electrostatic effects in mesoscopic quantum systems

Samia Cunningham, John G Williamson, Mahfuzur Rahman*, John M R Weaver

Nanoelectronic Research Centre, Department of Electronics and Electrical Engineering, University of Glasgow, Glasgow G12 8LT
*Physics Department

Abstract
A method for fabricating a gold microbeam has been developed so that it overhangs a planar quantum dot. The microbeam can be used as an out-of plane electrode to change the electrostatic environment of the dot, while being sufficiently far away to avoid the tunnelling effects. Preliminary experiments have been carried out on a single dot/ microbeam device, using conductance spectroscopy to determine the energy structure of the dot states. When no voltage is applied to the microbeam, the device displays normal quantum dot characteristics, with oscillations in conductance occurring in the Coulomb blockade region. When a bias voltage is applied to the microbeam, it is possible to induce a measurable broadening and shifting of the quantum dot conductance peaks. Thus the energy structure of the dot is influenced in a controllable way by its electrostatic environment.

1. Introduction

The study of mesoscopic systems offers the opportunity to provide answers on fundamental questions of physics that are important for the future generation of devices. One type of such devices is the quantum dot. Quantum dots are solid state structures of the order of 100 nm in size in which conduction band electrons are confined as standing waves in a 2-dimensional potential well [1].

They have been proposed as components of electron pumps and turnstiles [2], cellular automata [3], sensitive transducers [4], electrometers [5] and quantum logic systems [6]. The latter structures are complicated. To study movements and redistributions of charges due to the influence of electrostatic fields in these structures, multiple independent gates are more desirable and allow greater control.

In this paper we report a novel fabricated mesoscopic structure and explore its response to an applied electrostatic field. It consists of a quantum dot (150 nm in diameter) with a single in-plane gate $\approx$ 20 nm near it. This quantum dot is located beneath a 20 μm $\times$ 15 μm gold microbeam separated by a 3 μm air gap. The dot diameter is defined by dryetched trenches complete with entrance and exit ports. The combination of a small dot diameter and etched trenches permits greater device complexity (less gates) and a greater charging effect to be observed [6,7]. When a bias voltage is applied to the microbeam, it is possible to produce a uniform electrostatic field in the region of the dot without altering its geometry. This field induces changes in the energy of the quantum states of the electrons confined in the dot which can be measured by conductance spectroscopy.

The unique feature of the method lies in the use of the cantilevered microbeam as an independent additional gate. This allows the relative effect of the inplane gate and the microbeam on the device characteristics to be investigated.

Preliminary results shows that this device could be used to investigate the influence of an electrostatic field on the movement of charges on the dot surface.

2. Device description and fabrication

The quantum dot was fabricated to the design described in [8]. The Si-doped GaAs/AlGaAs heterostructure was grown by molecular beam epitaxy, with a 2-dimensional electron gas (2DEG) located approximately 30 nm below the surface. The 2DEG carrier concentration was 5.10^{11} cm^{-2} and the Hall mobility 3.10^5 cm / Vs at 4 K. Dry-etched trenches were used to define a 150 nm rectangular area connected by narrow entrance and exit junctions to the larger area of the 2DEG. To define the dot by etched boundaries rather than by the control electrodes that are often used in planar dot structures makes it possible to change the potential of the dot without significantly altering its geometry. Ohmic contacts were deposited to allow the dot conductance to be measured, and a single Ti/Au gate was located in close proximity ($\approx$20 nm) to the dot structure to allow the dot potential to be altered (nominal capacitance of 1.10^{-18} F).

Electron beam lithography was then used to fabricate a cantilevered gold microbeam of dimension 15x20μm which overhung the quantum dot with an air gap of 3 μm. The microbeam can be used as a control electrode to change the electrostatic environment within which the dot operates, but it is sufficiently distant from the dot to avoid tunnelling effects which can introduce complex switching and hysteresis behaviour in these systems [9]. The resulting device is illustrated schematically (not to scale) in Figure 1.

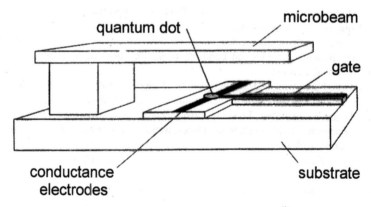

Figure 1: Schematic diagram of the fabricated structure.

3. Measurements and results

Measurements of the dot conductance were carried out in a cryostat at a temperature of 1.4 K using a standard low frequency lock-in technique with an excitation of 100 μV.

In the calibration procedure, the microbeam voltage (V_b) was held at zero while the in-plane gate voltage V_g was varied from 0 V to -0.3 V. The data acquisition sequence then consisted of the stepping of the in-plane gate potential in 0.1 mV steps from 0 V to -0.2 V, whilst the microbeam voltage was varied from 0 V to 7 V. The intention is to sweep V_b with successive increments of V_g in order to cover the whole range of operation of the device, see figure 2.

The dot has three regimes of behaviour: Firstly the regime between 0V and -0.10 V where the conductance oscillates slowly around a value of 1.5 e^2/h. This corresponds to an open system where the quantum dot is not fully formed and the conductance is characteristic of two quantum point contacts in very close proximity. Secondly there is a regime between -0.10 V and -0.13 V of rapidly decreasing device conductance corresponding to the pinch-off of the entrance and exit quantum point contacts. This region corresponds to the formation of the quantum dot. Thirdly the section below -0.13 V where the device is in the regime of Coulomb blockade of tunneling.

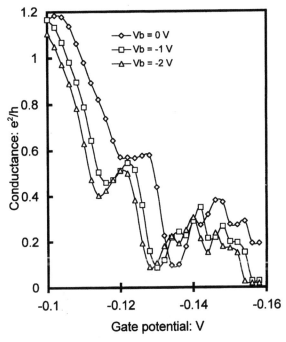

Figure 2: The device characteristics under different microbeam voltages.

Figure 2 also shows the conductance curves for potentials $V_b = 0$, V_g, $-2V_g$ of the dot/microbeam system. As the in-plane gate potential was lowered below -0.1 V, the conductance fell steeply due to depletion of the 2DEG carriers in the entrance and exit junctions. It is necessary here to distinguish between the physical effects of the in-plane gate voltage and the microbeam voltage. The spacing of the energy levels within the dot is a function of the dot geometry. However the microbeam voltage creates a uniform electric field over an area much larger than that occupied by the structure of the dot, and therefore does not change significantly the levels of the dot relative to its immediate surroundings. A well developed Coulomb blockade was achieved at around -0.13 V.

A comparison of the rate of change of conductance with respect to the gate voltage V_g to the rate of change of conductance with respect to the microbeam voltage V_b for different regimes of dot conductance reveals both qualitative and quantitative differences between potential changes induced by the two gates. To achieve the same change in the conductance with the microbeam as for the gate, 200 times the change in the voltage is required for $|V|<1V$. However the ratio is closer to 500 for $|V|>0.12$ V and $|V|<3.0$ V.

These preliminary results show that the structure described in section 2 allows a comparison of charge movement induced by the in- plane surface gate, which may change the heterostructure surface potential, with that due to an electrostatic field.

4. Conclusion

In this paper we report the use of a single electron electrometer (each peak corresponds to n. n+1 electron) as a sensor of the environment close to the Si-doped hetrojunction. The environment was controlled by a spatially separated microbeam produced by 3-dimensional lithography. The device is compact, non-intrusive and reliable. A comparison of charge movement induced by the in-plane surface gate to that of the microbeam is possible. A potential application of the device is as a micron-scale detector of local electrostatic fields.

References

[1] Meirav U and Foxman E B 1995 Semiconductor Science and Technology 10 255-284

[2] Kouwenhoven L P ,Johnson A T,Ven der Vaart N C, van der Enden A and Harmans C J M P 1991 Physical Review Letters 64 2691-2696

[3] Tougaw P D, Lent C S 1995 Quantum cellular automata Institute of Physics Conference Series 14 781-786

[4] Hauser A H, Smoliner J, Eder C, Ploner G, Strasser G, G Gornik E 1996 Superlattices and Microstructures 20 623-626

[5] Amlani I, Orlov A O, Snider G L, Bernstein G H 1997 Journal of Vacuum Science and Technology B 15 2832-2835

[6] Randall J N, Reed M A and Franzier G A 1989 Journal of Vacuum Science and Technology B 7 1398-1403

[7] Borsofoldi, Z., Rahman, M., Larkin I.A Long A.R., Davies J.H., Weaver J.M.R., Holland M.C.and Williamson J.G.1995 Applied Physics Letters 66 3666-3672

[8] Rahman M, Murad S K,Holland M C, Long A R, Wilkinson C D W and Williamson J G 1997 Microelectronic Engineering 35 91-94

[9] Duruoz C I, Clarke R M, Marcus C M,Harris J S Journal of Nanotechnology, 1996 7 372-375

Inst. Phys. Conf. Ser. No 163
Paper presented at the 10th Int. Conf., Cambridge, 28–31 March 1999

Vibrational response of cantilevered beams to piezoelectric drive

Nenad Nenadic and Thomas B. Jones, University of Rochester, Rochester NY, 14627, USA

Abstract – The dynamic responses of ideally- and non-ideally-clamped, cantilevered beams driven by surface-mounted piezoelectric transducers have been studied and compared. For ideal clamping, vibrations are confined to the beam only, whereas non-ideal clamping leads to coupling between the beam and the housing. An equivalent circuit model was obtained from free vibration experiments conducted first with ideally clamped beams. For a driven beam, the frequency response features well-isolated, second-order resonances which can be adequately modeled by a parallel arrangement of series RLC sections, one for each resonance. The shapes of the first three modes were also measured. The frequency response of a non-ideally-clamped beam was then investigated and compared to the results for ideal clamping. The non-ideal beam also exhibits well-isolated resonances, but the ratios amongst the resonant frequencies are changed.

1. Introduction

The conventional non-contacting electrostatic voltmeter (ESV) has a vibrating probe element -- such as a miniature tuning fork -- to sense the current induced in a small electrode. Manufacturing costs for tuning fork systems is high because each unit must be mechanically trimmed during assembly. Costs can be reduced substantially if the fork is replaced by a cantilevered beam. Our objective in this investigation has been to study the modal dynamics of such beams driven by surface-mounted piezoelectric transducers.

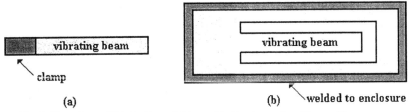

Fig. 1: Plan views of (a) ideally and (b) non-ideally-clamped cantilevered beams.

Because neither the mass nor stiffness of the clamp are sufficient to fully isolate vibration, "ideal" clamping can not be realized in practical probes based on cantilevered beams. The clamp always allows some coupling between the beam and the housing. Fig. 1 shows (a) an ideally-clamped cantilevered beam and (b) a practical, that is to say, a manufacturable realization of a non-ideal clamped beam.

2. Free vibration experiment

Though not realizable in a real probe, the ideally-clamped beam is well-understood theoretically and serves as an excellent vehicle to test the modal response of piezo-driven cantilevered beams. In the first experiment, an ideally-clamped beam, with a surface-bonded piezoelectric transducer mounted near its clamped end, was plucked and the damped vibrations recorded versus time. The beams, made of 0.38 mm brass sheet, were 31 mm long and 2.5 mm wide. The piezo-transducer output waveform was recorded with a digital oscilloscope. Fig. 2 shows (a) one cycle of the sinusoidal waveform, (b) several cycles of the damped sinusoid, and (c) an RLC equivalent circuit. The fundamental natural frequency f_{n1} and exponential decay rate, quantified in terms of the logarithmic decrement δ_n[1], were determined directly from the response.

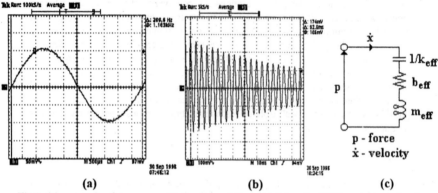

(a)   **(b)**   **(c)**

Fig.2: (a) one cycle of resonance, (b) the damped sinusoid, (c) series RLC equivalent circuit.

For the waveform of Fig. 2, the frequency and normalized damping are determined to be f_d= 209.6 Hz and $\zeta = 0.00764$. Because $\zeta << 1$, the quality factor can be approximated using $Q = 1/2\zeta \approx 65$. Furthermore, the natural frequency differs by less than 0.01% from the measured resonant frequency, that is,

$$f_{n1} = \frac{f_d}{\sqrt{1-\zeta^2}} = 209.606 \text{ Hz}. \tag{1}$$

Using ζ, f_{n1}, and the effective mass m_{eff}, one can extract the other two discrete mechanical system parameters, the spring constant k_{eff} and the damping b_{eff}.

$$k_{\text{eff}} = (2\pi f_{n1})^2 m_{\text{eff}} \tag{2}$$

$$b_{\text{eff}} = 2\zeta(2\pi f_{n1})m_{\text{eff}} \tag{3}$$

---

[1]...The logarithmic decrement measures the decay of a damped sinusoid for n cycles :

$$\delta_n = \frac{2n\pi\zeta}{\sqrt{1-\zeta^2}} = \ln\left(\frac{A_0}{A_n}\right),$$ where A_0 and A_n are the amplitudes of the two cycles separated by n-1

cycles. For example δ_1 is the natural logarithm of the ratio of the two consecutive cycles.

There is a degree of arbitrariness in choosing the effective mass; it is typically set equal to the beam mass [1]. Knowing the discrete mechanical system parameters, one then may invoke the direct electromechanical analogy to quantify the equivalent electrical circuit component values [3]. The equivalent circuit obtained this way is adequate from dc up to ~$5f_{n1}$, because the second resonance occurs at a frequency higher than $6f_{n1}$ and the beam has high Q.

3. Frequency response and mode shapes of an ideally-clamped beam

Plucking the beam only excites the fundamental mode. To study the higher modes, it is necessary to drive the beam with a piezo-transducer. Thus, for all frequency response and mode shape measurements, the input voltage to the piezoelectric drive transducer was fixed and the frequency varied from ~5 Hz up beyond the third resonance f_{n3}. A fiber optic displacement sensor mounted on a precision slide and connected to a lock- amplifier to eliminate the noise, was used to measure beam displacement. Amplitude and phase measurements near the first three resonances are shown in Fig. 3. As expected from the first experiment, the driven beam behaves like a set of simple, virtually decoupled, series RLC sections, with one section (Fig. 2c) for each of the resonances: f_{n1}, f_{n2} and, f_{n3}.

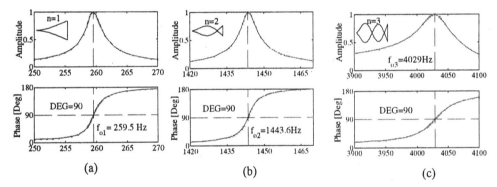

(a) (b) (c)

Fig. 3: Frequency response of cantilevered beam near the (a) first, (b) second and (c) third modal resonances.

The quality factor of the nth resonance, Q_n, can be computed from its well-known relationship to bandwidth ($BW_n = f_{high,n} - f_{low,n}$), where $f_{high,n}$ and $f_{low,n}$ are half-power points.

$$Q_n = \frac{f_{on}}{BW_n} = \frac{f_{on}}{f_{high,n} - f_{low,n}}. \tag{4}$$

The quality factor values for the first three resonances were found to be: $Q_1 = 66.7$, $Q_2 = 132.8$, and $Q_3 = 55.6$. Note that the value of Q_1 obtained in the free vibration experiment agrees with the result obtained here to within ~3 %.

For a general equivalent circuit model, the effective mass m_{eff} can be set equal to the total mass of the beam [2], the same for all modes. The quality factor Q_n, can be used to compute the damping factor for each mode: $\zeta_n = 1/(2Q_n)$. The damping

394

coefficient b_{effn} and the effective stiffness k_{effn} are then determined from the effective mass, the appropriate resonant frequency f_{on}, and the damping factor ζ_n, using Eq. (2) and (3). Because the resonances are well-isolated, the equivalent circuit is just a parallel connection of three series resonant circuits, with one section corresponding to each of the resonances

Using the displacement sensor, the first three mode shapes were measured and compared to the theory. One can see from Fig. 4(a), (b), (c) that experimental results are consistent with the theory.

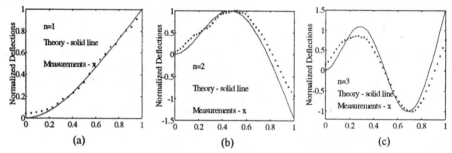

Fig. 4: Vibrational mode shapes for the (a) first, (b) second, and (c) third modes of the driven cantilevered beam. The solid curves are from classical beam theory [4].

4. Non-ideally clamped beam results

Measurement of frequency response was repeated for a non-ideally-clamped beam. Vibration amplitude and resonant frequencies in this case strongly depend upon the mounting of the housing, due to lack of vibration isolation between the beam and the clamp. Just like ideally-clamped beams, the frequency response exhibits well-isolated resonances, but the ratio amongst the resonant frequencies is altered.

Table 1: Ratios of resonant frequencies for ideally- and non-ideally clamped beams.

	f_{n2}/f_{n1}	f_{n3}/f_{n1}	f_{n3}/f_{n2}
Ideal Clamping	6.21	17.12	2.75
Non-Ideal Clamping	8.09	22.38	2.77

Table 1 shows the first three resonant frequencies and ratios amongst them, for an ideally- and a non-ideally-clamped beam. Note that the largest discrepancies occur among the ratios which involve the first resonance.

References

[1] Dimarogonas, A 1996 "Vibration for engineers", Prentice Hall, Second Edition, p 474
[2] Tilmans, H A C 1997 "Equivalent circuit representation of electromechanical transducers: II Distributed -parameter systems", J. Micromech. Microeng. 7, p 285-309.
[3] Neubert H K P 1975 "Instrument transducers", Oxford: Clarendon, Second Edition, p 7-11
[4] Craig R R Jr, 1981 "Structural dynamics: an introduction to computer methods", John Wiley & Sons, p 214-217.

Inst. Phys. Conf. Ser. No 163
Paper presented at the 10th Int. Conf., Cambridge, 28–31 March 1999
© 1999 IOP Publishing Ltd

Electrostatic Green's Function for Corrugated Interfaces

K. W. Yu[1], Hong Sun[2] and Jones T. K. Wan[1]

[1]Department of Physics, Chinese University of Hong Kong, Shatin, NT, Hong Kong, China

[2]Department of Physics, Shanghai Jiao Tong University, Shanghai, China

Abstract. The Bergman-Milton spectral theory is extended to calculate the field distribution for an interface of arbitrary shape, separating two media of different dielectric constants. The formalism is developed for an arbitrary interface and applied to a periodic interface. Results of this work will shed useful insight in obtaining valuable information from scanning electrostatic force microscopy data.

1. Introduction

Scanning electrostatic force microscopy (SEFM) has proven to be a powerful way of collecting information of surface structures [1]. Such instruments have been applied to image both conducting and insulating materials [2]. The distribution of field around the surface structures is required for a quantitative account of the electrostatic force between the tip and the sample. This is a formidable task for interfaces of complex geometry because of the complicated boundary conditions. In this work, we adopt the Bergman-Milton theory [3] to calculate the Green's function for an interface of arbitrary shape, separating two media of different dielectric constants. The solution also helps in the determination of image potentials in quantum dots of highly nonconventional shape [4].

2. Green's Function Formalism

For convenience, we consider an exterior problem in which a dielectric medium is embedded in a host medium with a unit point charge placed at $\mathbf{r}_0$ in the host medium. The electrostatic Green's function satisfies the differential equation:

$$\nabla \cdot [\epsilon(\mathbf{r})\nabla G(\mathbf{r}_0, \mathbf{r})] = -4\pi\delta(\mathbf{r}_0 - \mathbf{r}), \tag{1}$$

with standard boundary conditions on the interface, where $\epsilon(\mathbf{r})$ represents the dielectric constant (ϵ_1) in the host medium and that (ϵ_2) in the grain. Neglecting the dielectric contrast between the grain and medium, $\nabla^2 G_0(\mathbf{r}_0, \mathbf{r}) = -4\pi\delta(\mathbf{r}_0 - \mathbf{r})$, gives $G_0(\mathbf{r}_0, \mathbf{r}) = |\mathbf{r}_0 - \mathbf{r}|^{-1}$.

Following Bergman and Milton [3], we transform the original boundary-value problem [Eq.(1)] into an eigenvalue problem:

$$\Gamma\phi_n(\mathbf{r}) = s_n\phi_n(\mathbf{r}), \tag{2}$$

where $0 \leq s_n \leq 1$ and $\phi_n(\mathbf{r})$ are the eigenvalues and eigenfunctions of a hermitean operator:

$$\Gamma\phi(\mathbf{r}) = \frac{1}{4\pi} \int d\mathbf{r}' \theta(\mathbf{r}') \nabla' G_0(\mathbf{r}, \mathbf{r}') \cdot \nabla'\phi(\mathbf{r}'); \tag{3}$$

$\theta(\mathbf{r})$ equals unity when $\mathbf{r} \in$ medium 2 and zero otherwise. For an interface of simple geometry, e.g., planar interfaces or spheres, the eigenvalues and eigenfunctions of the Γ operator can be obtained analytically (see below), and form a complete set. The Green's function can be expanded in terms of the eigenfunctions. For $\mathbf{r} \in$ medium 2,

$$G(\mathbf{r}_0, \mathbf{r}) = \frac{1}{\epsilon_1} \sum_n \frac{s}{s - s_n} \langle \phi_n(\mathbf{r}) | G_0(\mathbf{r}_0, \mathbf{r}) \rangle \phi_n(\mathbf{r}), \tag{4}$$

where $s = \epsilon_1/(\epsilon_1 - \epsilon_2)$, while for $\mathbf{r} \in$ medium 1,

$$G(\mathbf{r}_0, \mathbf{r}) = \frac{1}{\epsilon_1} \left[G_0(\mathbf{r}_0, \mathbf{r}) + \sum_n \frac{s_n}{s - s_n} \langle \phi_n(\mathbf{r}) | G_0(\mathbf{r}_0, \mathbf{r}) \rangle \phi_n(\mathbf{r}) \right], \tag{5}$$

where the scalar product between two functions is defined by:

$$\langle \phi(\mathbf{r}) | \psi(\mathbf{r}) \rangle = \int d\mathbf{r} \theta(\mathbf{r}) \nabla \phi(\mathbf{r}) \cdot \nabla \psi(\mathbf{r}). \tag{6}$$

The Bergman-Milton theory thus allows a separation of the materials parameter s from the geometric information which is contained in Γ. With a given microgeometry, one can solve for the eigenfunctions. With the knowledge of the materials parameter, one can solve for the Green's function, thus providing a convenience way of representing the solution for interfaces of arbitrary shape.

To this end, we introduce a coordinate transformation $\mathbf{r} \to \tilde{\mathbf{r}}$. In space $\tilde{\mathbf{r}}$, the operator $\tilde{\Gamma}(\tilde{\mathbf{r}})$ can be written as $\tilde{\Gamma}(\tilde{\mathbf{r}}) = \Gamma_0(\tilde{\mathbf{r}}) + \Delta\Gamma(\tilde{\mathbf{r}})$, where $\Gamma_0(\tilde{\mathbf{r}})$ is the Γ operator for simple solvable geometry, and $\Delta\Gamma(\tilde{\mathbf{r}})$ represents the effect of nonconventional shape. The eigenfunctions of $\tilde{\Gamma}(\tilde{\mathbf{r}})$ in space $\tilde{\mathbf{r}}$ can be expanded with those of $\Gamma_0(\tilde{\mathbf{r}})$, which can be obtained analytically and form a complete set. As a result, the original eigenvalue problem can be reduced to a matrix eigenvalue problem which can be solved with standard algorithms.

3. Electrostatic Green's functions near a periodic interface

Let us first consider a planar interface. The Green's function formalism is readily applied to this simple case. The eigenfunction for the Γ operator reads:

$$\phi_{\mathbf{q}}(\mathbf{r}) = \frac{e^{-|\mathbf{q}|z} e^{i\mathbf{q}\cdot\rho}}{\sqrt{S_0|\mathbf{q}|}}, \tag{7}$$

where S_0 is the area of the interface (for normalisation), ρ is the position vector on the plane, z is distance from the interface and $s_{\mathbf{q}} = 1/2$ is the highly degenerate eigenvalue. By using Eqs.(4)–(6), we obtain the Green's function:

$$G(\mathbf{r}_0, \mathbf{r}) = \frac{2}{(\epsilon_1 + \epsilon_2)} \frac{1}{|\mathbf{r}_0 - \mathbf{r}|}, \quad \mathbf{r} \in \text{medium 2}, \tag{8}$$

$$G(\mathbf{r}_0, \mathbf{r}) = \frac{1}{\epsilon_1 |\mathbf{r}_0 - \mathbf{r}|} + \frac{\epsilon_1 - \epsilon_2}{\epsilon_1(\epsilon_1 + \epsilon_2)} \frac{1}{\sqrt{|\rho_0 - \rho|^2 + |z_0 + z|^2}}, \quad \mathbf{r} \in \text{medium 1}. \tag{9}$$

These results are standard [5]. Assume that the interface is described by a periodic profile of lateral period L_x: $z = f(x)$ with $f(x + L_x) = f(x)$. One can show that the eigenfunctions satisfy the Laplace equation: $\nabla^2 \phi_n(\mathbf{r}) = 0$ in medium 2. As a result, Eq.(3) reduces to

$$\frac{1}{4\pi} \int_S G_0(\mathbf{r}, \mathbf{r}') \nabla' \phi_n(\mathbf{r}') \cdot d\mathbf{S}'|_{z'=f(x')} = s_n \phi_n(\mathbf{r}), \tag{10}$$

a form being easier to handle in numerical calculations. The normal of the interface S points from medium 2 to medium 1. By a simple coordinate transformation:

$$\tilde{\mathbf{r}} = \mathbf{r} - f(x)\hat{\mathbf{z}}, \tag{11}$$

the interface S is transformed into a plane. In the transformed space $\tilde{\mathbf{r}}$, the left hand side of Eq.(10) becomes more complicated. However, it can be written formally as the Γ operator for a planar interface, plus a "perturbation" due to the corrugated interface. The eigenfunction of Eq.(10) in the transformed space $\tilde{\mathbf{r}}$ can be expanded with those of the planar interface:

$$\tilde{\phi}_n(\tilde{\mathbf{r}}) = {\sum_m}' A_{\mathbf{q}}^{(m)} \phi_{\mathbf{q}+\mathbf{K}_m}(\tilde{\mathbf{r}}) = {\sum_m}' A_{\mathbf{q}}^{(m)} \frac{e^{-|\mathbf{q}+\mathbf{K}_m||\tilde{z}|} e^{i(\mathbf{q}+\mathbf{K}_m)\cdot\tilde{\rho}}}{\sqrt{S_0|\mathbf{q}+\mathbf{K}_m|}}, \tag{12}$$

where $\phi_{\mathbf{q}+\mathbf{K}_m}(\tilde{\mathbf{r}})$ is the eigenfunction of the Γ operator for a planar interface, which forms a complete set. The in-plane wave vector $\mathbf{q}$ is restricted to the first Brillouin zone as determined by the lateral period L_x of the interface: $|q_x| \le K/2 = \pi/L_x$, with the reciprocal lattice wave vector being $\mathbf{K}_m = mK\hat{\mathbf{x}}$. The restricted summation ${\sum_m}'$ indicates that the function $\phi_{\mathbf{q}+\mathbf{K}_m}(\tilde{\mathbf{r}})$ with $\mathbf{q} + \mathbf{K}_m = 0$ is not included in the summation because constants do not contribute to the Green's function (determined by the derivatives of $\tilde{\phi}_n(\tilde{\mathbf{r}})$ or $\phi_n(\mathbf{r})$). Substituting Eq.(12) into Eq.(10) in space $\tilde{\mathbf{r}}$ yields a matrix eigenvalue equation which determines the expansion coefficients $A_{\mathbf{q}}^{(m)}$ and the eigenvalue $s_n = s_{\mathbf{q}}$, and can be solved with standard algorithm. In space $\tilde{\mathbf{r}}$, the electrostatic Green's function of the corrugated interface is given for $\tilde{\mathbf{r}}$ in medium 2 by:

$$G(\tilde{\mathbf{r}}_0, \tilde{\mathbf{r}}) = \frac{1}{\epsilon_1} \sum_{\mathbf{q}} {\sum_{m,m'}}' \frac{s}{s - s_{\mathbf{q}}} A_{\mathbf{q}}^{(m)*} A_{\mathbf{q}}^{(m')} F_{m'}(\mathbf{q}, \tilde{\mathbf{r}}_0) \phi_{\mathbf{q}+\mathbf{K}_m}(\tilde{\mathbf{r}}), \tag{13}$$

while for $\tilde{\mathbf{r}}$ in medium 1:

$$\begin{aligned} G(\tilde{\mathbf{r}}_0, \tilde{\mathbf{r}}) &= \frac{1}{\epsilon_1} G_0[\tilde{\mathbf{r}}_0 + f(\tilde{x}_0)\hat{\mathbf{z}}, \tilde{\mathbf{r}} + f(\tilde{x})\hat{\mathbf{z}}] \\ &+ \frac{1}{\epsilon_1} \sum_{\mathbf{q}} {\sum_{m,m'}}' \frac{s_{\mathbf{q}}}{s - s_{\mathbf{q}}} A_{\mathbf{q}}^{(m)*} A_{\mathbf{q}}^{(m')} F_{m'}(\mathbf{q}, \tilde{\mathbf{r}}_0) \phi_{\mathbf{q}+\mathbf{K}_m}(\tilde{\mathbf{r}}), \end{aligned} \tag{14}$$

where

$$F_m(\mathbf{q}, \tilde{\mathbf{r}}_0) = \int_S G_0[\tilde{\mathbf{r}}_0 + f(\tilde{x}_0)\hat{\mathbf{z}}, \mathbf{r}] \nabla \phi_{\mathbf{q}+\mathbf{K}_m}^*[\mathbf{r} - f(x)\hat{\mathbf{z}}] \cdot d\mathbf{S}|_{z=f(x)}. \tag{15}$$

398

Discussion and Conclusion

Having developed the Green's function formalism for two media, perhaps it is instructive to examine the more realistic case of three media. Consider a planar interface consisted of two media with ϵ_1 and ϵ_2. A cluster medium of volume V_3 with ϵ_3 is placed on the interface pointing into ϵ_1 to mimic typical SEFM geometry. The Green's function formalism is readily extended to this case.

$$G(\mathbf{r}_0, \mathbf{r}) = G_0(\mathbf{r}_0, \mathbf{r}) + \frac{1}{4\pi} \int_{V_3} d\mathbf{r}' \delta\epsilon(\mathbf{r}) \nabla' G_0(\mathbf{r}, \mathbf{r}') \cdot \nabla' G(\mathbf{r}_0, \mathbf{r}'), \tag{16}$$

where $\delta\epsilon(\mathbf{r}) = \epsilon_1 - \epsilon_3$ or $\epsilon_2 - \epsilon_3$ depending on whether $\mathbf{r} \in$ medium 1 or 2. It should be remarked that the unperturbed Green's $G_0(\mathbf{r}, \mathbf{r}')$ is given by Eqs. (8) and (9) rather than that of infinite space.

In summary, we have attempted to extend the spectral theory to the electrostatics of an interface of arbitrary shape. We developed Green's function formalism for an arbitrary interface and applied to a periodic interface. It is also shown that the formalism can be extended to the more realistic case of three media. We have presented analytic results in this work and will publish numerical results elsewhere [6].

Acknowledgments

This work was supported by the Research Grants Council of the Hong Kong SAR Government under Grant CUHK 4290/98P. H.S. wishes to acknowledge financial support of the Climbing (Pan-Deng) Program of the National Natural Science Foundation of China.

Reference

1. G. Binnig, H. Rohrer, Ch. Gerber and E. Weibel, Phys. Rev. Lett. **49**, 57 (1982).

2. J. Hu, X. D. Xiao, D. F. Ogeletree and M. Salmeron, Science **268**, 267 (1995).

3. D. J. Bergman and D. Stroud, in *Solid State Physics*, vol.**46**, edited by H. Ehrenreich and D. Turnbull, (Academic Press, Boston, 1992), p.147.

4. M. Henini, S. Sanguinetti, S. C. Fortina, E. Grilli, M. Guzzi, G. Panzarini, L. C. Andreani, M. D. Upward, P. Moriarty, P. H. Beton and L. Eaves, Phys. Rev. B **57**, R6815 (1998).

5. J. D. Jackson, *Classical Electrodynamics*, 2nd edition, (Wiley, New York, 1975).

6. K. W. Yu, Hong Sun and Jones T. K. Wan, unpublished (1999).

Inst. Phys. Conf. Ser. No 163
Paper presented at the 10th Int. Conf., Cambridge, 28–31 March 1999

Multiple degree of freedom model for driven cantilevered beam

Nenad Nenadic and Thomas B. Jones, University of Rochester, Rochester NY, 14627, USA

Abstract - Improved design capability for non-contacting electrostatic voltmeter probes will require accurate dynamic models for piezoelectrically-actuated cantilevered beams. In this paper, we describe a simple, multiple-degree-of-freedom model for a cantilevered beam driven by a surface-mounted piezo-transducer. The model allows one to determine the optimum position for maximized actuation of different beam vibration modes. Measurements performed with transducers mounted at different beam locations provide reasonable verification of the model.

1. Introduction

Some probes for non-contacting, electrostatic voltmeter (ESV) instruments employ a cantilevered beam as the vibrating element. The beam is driven by a surface-mounted piezoelectric transducer. To optimize performance of the vibrating element, it is desirable to have a dynamic model that accurately represents the effect of the transducer. Vibrating cantilevered beams with surface-mounted piezo-transducer drives have attracted growing attention in recent years. A variety of models for these systems, mostly based on finite element methods [1, 2, 3, 4], have been proposed. While very powerful, these models often obscure the physics of mechanical systems [5] and, furthermore, may ultimately prove unwieldy for design application, as in microelectromechanical systems (MEMS) technology. We describe here a multiple-degree-of-freedom (MDF) model, which offers the advantage of simplicity while yet retaining sufficient detail to describe all salient features of the system.

2. Discretization of the system - MDF formulation

The MDF model involves discretizing the beam, which is a continuum, into n parts of equal length. The coupled governing equations of motion for the discretized, undamped system may be expressed in matrix form [6].

$$\mathbf{M}_{n \times n} \ddot{\mathbf{u}}_{n \times 1} + \mathbf{K}_{n \times n} \mathbf{u}_{n \times 1} = \mathbf{F}_{n \times 1} \tag{1}$$

Here, $\mathbf{M}$ is the n×n mass matrix, $\mathbf{K}$ is the n×n stiffness matrix, $\mathbf{F}$ is the n×1 vector of applied normal driving force, and $\mathbf{u}$ is the n×1 vector of transverse deflections. $\mathbf{M}$ is a diagonal matrix with the elements roughly equal to the corresponding masses of the beam sections. $\mathbf{K}$ is the inverse of the flexibility matrix $\mathbf{A}$, whose elements are given by [6]:

$$a_{ij} = \frac{\left(\dfrac{iL}{n}\right)^2}{6EI}\left[3\left(\frac{jL}{n}\right) - \left(\frac{iL}{n}\right)\right], i = 1...n; j = i...n, \tag{2}$$

where L, E and I are the length, Young's modulus, and area moment of inertia of the beam, respectively.

The vector **F** represents the driving force exerted by the surface-mounted piezoelectric transducer, and determination of the elements of this vector is somewhat subtle. For simplicity, we concentrate the forces due to this transducer at the in-board and out-board edges of the transducer, as shown in Fig. 1a. These axially directed forces cannot be implemented in the MDF model directly; they are instead replaced by equivalent normal forces. The essential details of this development are depicted in Fig. 1: axial surface forces $\pm F$ shown in Fig. 1a are replaced by axial forces $\pm F_1$ acting along the neutral line plus equivalent moments $\pm M$ shown in Fig. 1b. The forces along the neutral line $\pm F_1$ may be ignored safely because any axial motion they excite will be very small compared to deflections of the beam.[1] The moments $\pm M$ are replaced by equivalent force couples, $\pm F_2$, as shown in Fig. 1c. The relationships between F, M and F_2 are given as follows:

$$M = \frac{h}{2} F \ , \tag{3}$$

$$F_2 = \frac{M}{dx} = \frac{h}{2dx} F \ , \tag{4}$$

where h is the thickness of the beam.

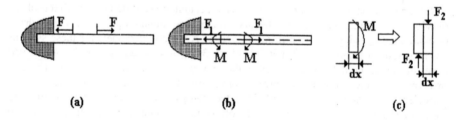

<table>
<tr><td>(a)</td><td>(b)</td><td>(c)</td></tr>
</table>

Fig. 1: Conversion of transducer forces by equivalent normal loading. (a) Forces imparted by surface-mounted transducer, (b) Resolution of moments M and axial forces F_1, (c) Resolution of moment into normal forces (F_2).

F has a total of four non-zero components, equal in magnitude to F_2 and, in pairs, acting in opposite directions upon the two adjacent elements of the discretized beam nearest the in-board and out-board edges of the actuating transducer. Fig. 1(c) shows the normal forces applied at the out-board edge of the piezo. With **M**, **K**, and **F** known, Eq. (1) can be solved for displacement **u**. The model, implemented using the matrix manipulation and solution capabilities of MATLAB®, may be used to determine the optimal location for the piezoelectric actuator.

---

[1] We may ignore F_1, because longitudinal vibrations of the beam will have a much higher natural frequency than the transverse vibration modes of interest here.

3. Optimization of piezo-actuator placement on cantilevered beam.

A surface-mounted piezoelectric transducer responds to AC electrical excitation by trying to change its length along the beam axis. This driven motion suggests that the optimal actuator placement is the region of maximum curvature of the beam for the desired deflection mode. For example, the best location for the first mode is very close to the clamped end, whereas the best location for the second (and all higher-order) modes is the modal peak.

To test the conjecture about the optimal location of the surface-mounted transducer for excitation of a given beam mode, we performed an extensive set of simulations for a 26 mm long by 2.5 mm wide by 0.38 mm thick beam, with two 6 mm long piezo elements (one serving as actuator, and the other as sensor). Simulations were performed for all permissible, non-overlapping locations of the piezos. To insure accurate representation of the continuum, the number of degrees of freedom was set to n = 100. Results of the simulations, performed for each of the first three modes, are shown, respectively, in Fig. 2a, b, and c.

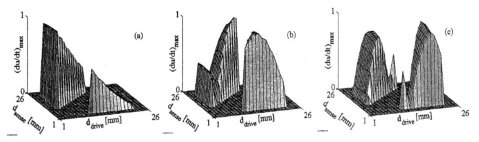

Fig. 2: Simulation results for maximum velocity of the free end of the beam as a function of the driving and sensing transducer locations: (a) first mode, (b) second mode, and (c) third mode

4. Experiment.

To test MDF modeling predictions, three beams, virtually identical in length, width and thickness, and differing only in the attachment points of their piezoelectric drive transducers, were fabricated. The three different transducer attachment points were chosen to optimize coupling to each of the first three cantilevered beam resonances: $f_{n1} \approx$ 240 Hz, $f_{n2} \approx$ 1450 Hz and $f_{n3} \approx$ 4000 Hz, as determined by MDF simulations. After fabrication, each beam was tested over the frequency range covering all three resonances.

Table 1

	Resonant frequencies			Displacements			Velocities		
	f_{n1} [Hz]	f_{n2} [Hz]	f_{n3} [Hz]	$u1$ [mm]	$u2$ [mm]	$u3$ [mm]	$v1$ [mm/s]	$v2$ [mm/s]	$v3$ [mm/s]
Beam1	271.3	1448.8	4032.0	31.18	2.67	2.97	8.46	3.87	12.0
Beam2	223.2	1408.3	4002.5	9.56	3.64	1.46	2.13	5.13	5.84
Beam3	220.7	1465.8	4041.6	2.03	2.68	1.89	0.45	3.93	7.64

Table 1 contains a summary of the experimental results. Indices of displacement u and velocity v signify the modal resonance (e.g., u_1 and v_1 correspond to resonance f_{n1}). Although all beams have virtually identical dimensions, their resonant frequencies differ

402

somewhat. Such differences are to be expected because the effective moment of inertia of the beam is influenced by the mounting location of the transducers. Though not taken into account in the present simulations, mass loading is easily accounted for in the MDF model.

Figure 3 a, b, and c show the measured velocity of a given mode for each of the three beams. In comparing Fig. 3a to 2a, Fig. 3b to 2b, and Fig. 3c to 2c, a general correspondence is apparent. Because we have tested only three attachment points, the experimental plots can not be expected to be as smooth as the plots generated from simulation, where 22 different placements for the driving transducer were studied. The discrepancy between simulation and experiment for the third mode is probably due to the more accurate placement required for the transducer.

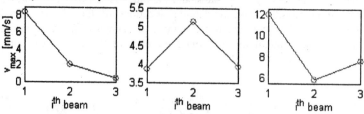

Fig. 3: Plots of measured beam displacement velocity for each beam, v_1, v_2, and v_3: (a) First mode; (b) Second mode, (c) Third mode.

5. Conclusion

A simple model for piezoelectrically driven cantilevered beams has been developed. Simulations performed using this model have been employed to find the optimum placement of a piezoelectric actuator to excite each of the first three vibrational modes of the beam. In general, the conjecture that optimum placement of a given mode corresponds to the location of maximum curvature is confirmed by the simulations. Experiments were performed upon three identical beams, where only the location of the drive transducer was varied. The experimental results clearly show that transducer placement has a strong influence on modal response which is generally quite consistent with the simulations.

References

[1] Crawley E F and Luis J 1987 "Use of piezoelectric actuators as elements of intelligent structures" AIAA J. 25 1373-85
[2] Shen H M-H "Analysis of beams containing piezoelectric sensors and actuators", Smart Mater. Struct. 3 (1994) 439-447
[3] Jaehwan K, Vasundara V V, Vijay K V and Xiao-Qi B "Finite-element modeling of a smart cantilever plate and comparison with experiments", Smart Mater. Struct. 5 (1996) 165-170
[4] Yang S M and Lee Y J "Interaction of structure vibration and piezoelectric actuation", Smart Mater. Struct. 3 (1994) 494-500
[5] Soderkvist J. "A phenomenological method of predicting the performance of piezoelectric beams", J. Micromech. Microeng. 1 (1991) 16-24
[6] Dimarogonas, A. 1996 "Vibration for engineers", Prentince Hall, Second Edition

Inst. Phys. Conf. Ser. No 163
Paper presented at the 10th Int. Conf., Cambridge, 28–31 March 1999
© 1999 IOP Publishing Ltd

The interpretation of tapping mode images of polymer surfaces in Atomic Force Microscopy.

J P Llewellyn and P Sayers

University of Wales Bangor, SEECS, Dean St, Bangor, Gwynedd LL57 1UT, UK

Abstract. The differential equation describing the motion of a vibrating AFM cantilever used in the tapping mode is solved numerically to find the phase of the vibrations. The effects of introducing a Coulomb interaction between the cantilever tip and the surface, and also an energy loss during the tapping process, are investigated.

1.Introduction

The recent proliferation of Atomic Force Microscopes (AFMs) has resulted in their widespread use for the investigation of surfaces. The most satisfactory method of using these instruments to study soft surfaces, such as polymers, is in the so-called tapping mode in which the sensing cantilever, while vibrating at its resonant frequency, is made to tap the surface under investigation. A picture is built up by monitoring either the amplitude or phase of the cantilever vibrations as it rasters over the surface. Images allegedly showing the surface topography, surface charge distribution or the surface potential distribution can be obtained in this way. We now report why caution should be exercised in interpreting these pictures in too simplistic a fashion; they may give a distorted representation of the feature they are nominally illustrating or they may even be illustrating a combination of features.

2. Theory

In order to investigate how the vibrations of the cantilever in the tapping mode are influenced by a range of different tip-surface interactions we have solved numerically the differential equation describing the motion of the tip,

$$m\frac{d^2z}{dt^2} + b\frac{dz}{dt} + kz = F_0\sin(\omega t) + F(z)$$

This is the standard equation for forced simple harmonic motion including a damping term: it also includes the term $F(z)$ describing a position dependent force to account for the tip-surface interaction. The value of the cantilever force constant k can be evaluated from the dimensions and Youngs Modulus of the cantilever [Stevens 1997]: if the values of k, and the resonant frequency, free vibration amplitude A_0 and Q of the cantilever vibrations are known then the effective mass m, the damping constant b and the force amplitude F_0 can be found.

Once the form of the function $F(z)$ has been chosen the differential equation can be solved numerically by the Bulirsch-Stoer method using the 'Bulstoer' programme on Mathcad 6. The phase of the resulting oscillations with respect to the driving force can then be determined by means of a simple curve fitting programme. The variable used in all these

404

calculations was α where $\alpha + A_0$ is the tip-surface distance when the cantilever is in its equilibrium position; thus tapping will only occur when $\alpha < 0$. It is also assumed that the probe tip is spherical and of radius 10 nm., that the cantilever is driven at its resonant frequency (300 kHz) and that the amplitude A_0 of free vibrations of the cantilever tip is 75 nm.

The precise form of the function F(z) depends on the particular tip-surface interaction that is being considered. We have investigated the cantilever response using expressions for F(z) that describe the following interactions

    (1) A Lennard-Jones interaction
    (2) A Coulomb interaction between trapped surface charge and an earthed tip
    (3) A Hertzian interaction, including a viscous energy loss.

3.Results and discussion

3.1 Lennard-Jones interaction

In this case the expression used for F(z) represents the integrated Lennard-Jones interaction force between a sphere (i.e. the probe tip) and a plane [Sarid 1994],

$$F_1(z) = L\left(\frac{x_0^2}{x^2} - \frac{1}{30}\frac{x_0^8}{x^8}\right)$$

where x is the plane-sphere separation $(x = A_0 + \alpha - z)$ and x_0 is the value of x at which the interaction potential is zero. L is a function of the tip radius R that can be evaluated.

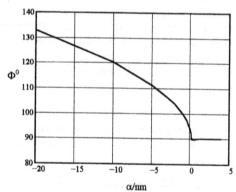

Figure 1. The variation of the phase Φ with the
set point α for a Lennard-Jones interaction

The phase Φ of the vibrations before tapping occurs is 90^0: the value of Φ then increases with the strength of the tapping, that is as α decreases below zero. In this, and subsequent results, the oscillation remains sinusoidal to a high degree of accuracy during the tapping process.

3.2 Coulomb plus Lennard-Jones interaction.

In addition to the Lennard-Jones interaction described above it is now assumed that there is also a Coulomb interaction between the earthed probe tip and charge q trapped on the surface immediately under the tip. The expression for this additional interaction force in terms of the tip radius R and x (defined above) is [Smythe 1950]

$$F_2(z) = \frac{q^2}{4\pi\varepsilon_0}\left(\frac{R}{R+x}\right)\left[(R+x)-\frac{R^2}{(R+x)}\right]^{-2}$$

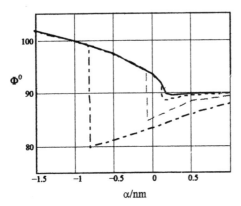

Figure 2. The variation of the phase Φ with the set point α for a Lennard-Jones plus Coulomb interaction. The full, dotted, dashed and dot-dash curves are respectively for $q = 0, 2, 5$ and 10 electron charges.

Surface charge is seen to have a marked effect on the response when the tapping strength is low. The phase then 'flips' at a critical value of α: comparable phase flips in the resonance curve have been reported by Kuhle et al [1997].

3.3 Lennard-Jones plus Hertzian interaction with energy loss

When a macroscopic sphere impacts on a surface the interaction forces can be regarded as arising from the elastic forces generated in the surface in the neighbourhood of the depression caused by the impact. We have modelled the tip-surface interaction in this way, using the expression [Chen et al 1994]

$$F_3(z) = gx^{3/2}$$

for the interaction force where $g = 8(2R)^{1/2}[3\pi(\kappa_1 + \kappa_2)]^{-1}$ and $\kappa_i = (1-v_i)^2(\pi E_i)^{-1}$. E_i and v_i are the Youngs modulus and Poissons ratio of the two surfaces and x is the distance of the bottom of the probe tip below the point at which the net Lennard-Jones interaction is zero; above this point it is assumed that the interaction is represented entirely by the Lennard-Jones force. In addition to the above elastic force a viscous force can be introduced [Tamayo and Garcia 1997] that is of the form

$$F_4(z) = \eta\sqrt{R\delta}\frac{dx}{dt}$$

where η is the effective viscosity and δ is the depth of the depression at time t.

The effect of using these elastic and viscous forces is shown in Figure 3. The full curve ($g = 10^6$ Pa [Chen et al 1994]) agrees well with the Lennard-Jones curve in Figure 1. The introduction of an energy loss via the viscosity term is seen to introduce a significant change, as has been reported by Cleveland et al [1998]. The value of η selected corresponds to an energy loss of the order of 10^{-16} J/cycle.

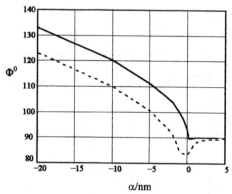

α/nm

Figure 3. The variation of the phase Φ with the set point α for a Hertzian interaction with viscous losses. Full curve; $g = 10^6$ Pa,. $\eta = 0$. Dotted curve; $g = 10^6$ Pa, $\eta = 10^3$ Pa s.

4. Conclusions.

The presence of a small electric charge on an insulating surface has been shown to have a significant effect on the phase response of an AFM cantilever used in the tapping mode: in particular there is an abrupt change in phase at a critical tapping strength which should result in a contrast reversal in the image obtained. It has also been shown that energy loss during the tapping process will influence the phase of the vibrating cantilever. It is therefore concluded that a tapping mode image obtained by monitoring the phase will depict a combination of topographical features,the surface charge and the surface stickiness.

5. Acknowledgements

The authors would like to thank Professor T.J.Lewis and Professor D.M.Taylor for their advice and encouragement and BICC Cables Ltd for financial support. One of us (PS) would also like to thank EPSRC for a CASE award.

6. References.

Chen J., Workman R.K., Sarid D. and Hoper R. Nanotechnology 5 199-204 1994
Cleveland J.P., Anczykowski B., Schmid A.E. and Elings V.B. Appl Phys Lett 72(20) 2613-2615 1998
Kuhle A., Sorensen A.H. and Bohr J. J. Appl Phys 81(10) 6562-6569 1997
Sarid D, Scanning Force Microscopy Publ O.U.P. 1994
Smythe W.R., Static and Dynamic Electricity Publ McGraw Hill 1950
Stevens G., Space Charge in Solid Dielectrics (Conf Rep) The Dielectrics Society 1998
Tamayo J. and Garcia G., Appl Phys Lett 71(16) 2394-2396 1997

DISCUSSION - Section H - Scanning Probe Microscopy

Title: Electrostatic Probe Microscopy
Speaker: H K Wickramasinghe

Question: Martin Taylor
              Were images obtained under UHV or air?

Reply: In air.

Question: Gary Stevens
              Has the Kelvin probe method been used to do true spectroscopy from far infrared to UV-VIS.

Reply: No, the Kelvin Probe Force Microscope as a temperature sensor as we described has not been used for spectroscopy beyond what I presented but in principle it is possible to do this.

Question: W Machowski
              The probes used in your experiments are made from sharpened tungsten wire bent at the 90^0. Could you please comment on the advantages of using such tips compared with a commercially available conical EFM tips.

Reply: The tungsten tips are conducting and therefore do not need to be coated like the silicon cantilever that are either coated or heavily doped. The resolution with the tungsten tips tends to be higher as we can make them sharper. However, with careful tip design, the silicon coated tips should come close in resolution.

Question: D K Davies
              The trap-hopping transfer of carriers through a dielectric is a fundamental process in the transfer of charge. Can these probes be used to characterise these traps and in particular, the effect of the ambient humidity?

Reply: Yes, we have observed the effect of humidity on charge but not characterized the transfer of charge through traps. It should be possible to do this however with commercially available EFM's.

Title: Measuring Static Charges by Scanning Probe Microscopy
Speaker: D M Taylor

Question: D K Davies
              Do these probes inject contact charge in the tapping mode and what is the magnitude in relation to your "sprayed" charge.

Reply: It is inevitable that some contact charging will occur because of the method used. The first scan 'taps' the surface to obtain the topography - while the rescan gives the electric field information in the non-contact mode. Scanning the same area repeatedly does not appear to change the electrostatic 'image'. So I assume we are not neutralising the surface charge.

Question: W Machowski
Is it possible that the charge deposited on the sample surface may induce a passive corona from the tip during scanning.

Reply: It is possible, but if it occurred then the electrostatic forces would weaken. The measured forces were greater than the image charge interaction analysis suggests.

Comment: A G Bailey
Macroscopic experiments on metallic ploughs pulled through the soil have shown that with the plough operated at a potential above ground the pulling force may be reduced considerably. The mechanism proposed to explain this is that water is attracted to the plough and then acts as a lubricant. Plough polarity is important. As Professor Taylor suggested, moisture may be responsible for the high attractive forces he has observed. The high field at the probe tip may be attracting moisture to the tip regime. Probe polarity may be important.

Section I

Atmospheric Electrostatics
and Environmental Aspects

Inst. Phys. Conf. Ser. No 163
Paper presented at the 10th Int. Conf., Cambridge, 28–31 March 1999

The interaction between air ions and aerosol particles in the atmosphere

K.L. Aplin and R.G. Harrison

Department of Meteorology, University of Reading, PO Box 243, Earley Gate, Reading, Berkshire, RG6 6BB, U.K.

Abstract

Charged particles are continually generated in atmospheric air, and the interaction between natural ionisation and atmospheric particles is complicated. It is of some climatic importance to establish if ions are implicated in particle formation. Atmospheric ion concentrations have been investigated here at high temporal resolution, using Gerdien ion analysers at a site where synchronous meteorological measurements were also made. The background ionisation rate was also monitored with a Geiger counter, enabling ion production from natural radioactivity to be distinguished from other effects. Measurements at 1Hz offer some promise in establishing the atmospheric electrical influences in ionic nucleation bursts, although combinations of other meteorological factors are also known to be significant. High time resolution meteorological and ion measurements are therefore clearly necessary in advancing basic understanding in the behaviour of atmospheric aerosol.

1. Introduction

Natural radioactivity and cosmic rays are a constant source of molecular ions in atmospheric air. Establishing the significance of these long-lived atmospheric ions is intimately linked with the importance of the atmospheric electrical system in atmospheric processes (Harrison, 1997), and the conventional view that the electrical properties of the atmosphere are completely insignificant is increasingly untenable. The ions produced are rarely single species, but form clusters of water molecules around a central ion. Atmospheric electrical fields lead to transport of small ions by vertical conduction processes. Typical atmospheric ion concentrations in unpolluted air and fair weather conditions are about 400-500 ions cm^{-3}.

It is of great interest to understand how ions could be implicated in particle production, and significant ion-induced nucleation of particles could be a relevant mechanism. Hõrrak *et al.* (1998) reported the spontaneous formation of intermediate size ions in atmospheric air, from a change in the ion mobility spectrum of urban air. Observations of direct aerosol formation in unpolluted marine air (O'Dowd *et al.*, 1996) and theoretical work on clustering reactions of ions (Castleman, 1982) suggest ions can be critical in gas-to-particle conversion processes.

If the volumetric ion production rate is q, the positive ion number concentration n_+, the negative ion concentration n_-, and if it is assumed that $n_+ = n_- = n$, then the variation of ion concentration with time is given by the equation

$$\frac{dn}{dt} = q - \alpha n^2 - \beta nZ - \gamma n \tag{1}$$

where α is the ionic self-recombination coefficient, β the ion-aerosol attachment coefficient (which varies with aerosol size and charge), Z the aerosol number concentration and γ the ion-induced nucleation coefficient. It has been usual to regard the aerosol and ions as distinct systems of particles, which interact purely by collisions. However the observations of Hõrrak *et al.* (1998) now suggest that the γ-term could be significant in new particle events, and this would couple the time dependence in the ion concentration equations to time dependence in aerosol concentration. The simultaneous measurement of ion and aerosol concentration is therefore of importance.

2. The Gerdien Condenser

Gerdien developed a method of measuring air conductivity from a cylindrical condenser in 1905. A voltage is applied between two cylindrical electrodes and the inner one is earthed via a sensitive ammeter (Harrison, 1997). In the tube, ions of the same sign as the polarising voltage are repelled by the outer electrode, and move in the electric field to meet the inner electrode where they cause a small current. In the Gerdien used for this work the currents are of order 10^{-15} A, and the conductivity is given by

$$\sigma = \frac{i\varepsilon_0}{CV} \tag{2}$$

where C is the capacitance of the Gerdien, and V is the polarising voltage across the electrodes. The Gerdien tube can be used to measure conductivity as long as the output current is proportional to the polarising voltage, indicating that a fixed fraction of the ions in the air are collected by the central electrode.

There exists a critical mobility μ_c such that only ions with $\mu > \mu_c$ contribute to the conductivity measurement. This is given by Wilkinson (1997)

$$\mu_c = \frac{(a^2 - b^2)\ln\frac{a}{b} \cdot u}{2VL} \tag{3}$$

where a is the radius of the tube, b is the radius of the central electrode, u is the flow speed through the tube and L is the length of the tube. Operating the Gerdien at different critical mobilities is easily executed by varying the voltage or flow rate, and enables it to act as a simple and effective mobility spectrometer.

3. Gerdien Control System

A Microcontroller-based system for remotely controlling and logging the Gerdien data has been developed. This is clearly necessary for measurement of mobility spectra, as the voltage across the Gerdien has to switch to different values. By this method, one Gerdien tube can also be used to measure quasi-synchronous positive and negative air conductivity by cycling between positive and negative polarising voltages.

The microcontroller used in this application[1] is programmable in BASIC and has eight I/O pins, which can be wired up for communication with other computers, or for digital inputs to external circuits. The microcontroller is cheap and compact, and can

---

[1] BASIC Stamp 1, Parallax Instruments Inc. 3805 Atherton Road, Suite 102, Rocklin, California 95765, USA

simultaneously control the Gerdien and send the data to the serial interface of another computer. Once the program is downloaded to the microcontroller it runs for as long it is powered, and the program resumes its operation if the power is cut and then restored.

The electronic control circuit employed essentially comprises two relays, driven by VN10KM MOSFET transistors controlled by digital outputs from the microcontroller, and a LTC1298 12-bit analogue to digital converter. One relay (a changeover device) was connected to a positive and negative bias voltage. The other relay was a high input impedance reed relay, connected across the feedback resistor of the picoammeter. When this relay is on it shorts the picoammeter feedback resistor and provides a measurement of the picoammeter input offset voltage, overriding the state of the other relay so that no current measurement is made. The analogue-to-digital conversion is carried out by a sub-routine in the microcontroller program (Parallax, 1998) and enables the voltage output from the picoammeter to be logged.

Name	Actions
Zero	Turns reed relay on to zero current amplifier
	Sets other relay to "0" to reduce power consumption
	Waits for 6s
	Takes 5 readings of zeroed state of current amplifier at 1Hz
	Sends readings to PC via serial protocol
+	Turns reed relay off to allow ion current measurements
	Sets other relay to positive bias voltage
	Waits 3 minutes
	Takes 10 readings of ion current at 1Hz
	Sends readings to PC via serial protocol
-	Sets relay to negative bias voltage
	As + above and repeats forever

Table 1: Outlines the actions performed by the program running on the microcontroller. The delay of three minutes after the change of bias voltage is required to allow complete recovery from the transient that occurs on switching.

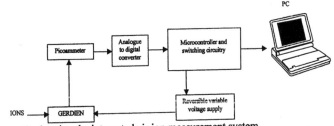

Figure 1: Schematic showing the integrated air ion measurement system

4. Results

Measurements of ion concentrations at different mobilities have been made in Reading by using the Microcontroller system to switch the Gerdien between two voltages of the same sign, –6V and –9V. Knowing the wind component into the Gerdien and the tube ventilation rate allows calculation of the critical mobility using (3). The ionic number density is approximated by

$$n = \frac{\sigma}{e\mu} \tag{4}$$

414

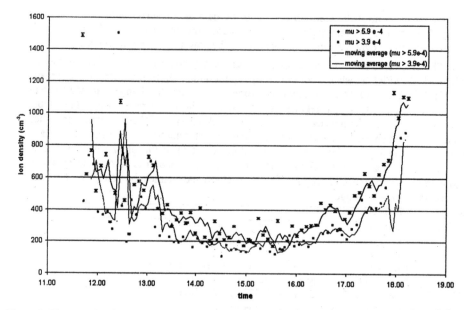

Figure 2: The negative ion number concentration at critical mobilities of 5.9 and 3.9 x 10^{-4} $Vm^{-2}s^{-1}$ for a typical day (23/2/99).

where e is the electronic charge, μ the critical mobility and n the concentration of ions with a mobility exceeding the critical mobility. The results are promising; for example the peak at about 12.30 shows a shift in the spectrum towards higher mass ions, perhaps implying a nucleation event.

5. Conclusions

These results indicate that the microcontroller-based ion counter system is ideal for measuring high time and mobility resolution ion spectra, and the microcontroller system can easily be expanded, such as for controlling the fan. Synchronous measurement of mobility spectra and meteorological parameters will assist in the understanding of ion-aerosol interactions.

References

Castleman A.G. (1982), In: Schryer D.R., *Heterogeneous atmospheric chemistry*, AGU, Washington

Harrison R.G. (1997), Climate change and the global atmospheric electrical system, *Atmospheric Environment*, **31**, 20, 3483-3484

Hõrrak U. *et al.* (1998), Bursts of intermediate ions in atmospheric air, *J. Geophys. Res*, **103**, 13909-13915

O'Dowd C.D. *et al.* (1996), New particle formation in the environment, *Proc. 14th Int. Conf. on Nucleation and Atmospheric Aerosols*, Helsinki, Eds: Kulmala M. and Wagner P.E., Pergamon Press, 925-929

Parallax Inc. (1998), *BASIC Stamp® Manual*, Version 1.9

Wilkinson S. (1997), *Determination of the characteristics of urban atmospheric aerosol*, MSc. Dissertation, Department of Meteorology, University of Reading

Inst. Phys. Conf. Ser. No 163
Paper presented at the 10th Int. Conf., Cambridge, 28–31 March 1999
© *1999 IOP Publishing Ltd*

On the effects of wind on the space charge formation in the atmospheric surface layer

Sven Israelsson

Meteorology, Department of Geosciences, Villavägen 16, S-752 36 Uppsala, Sweden

Abstract. The electrode effect near the ground surface is in most cases producing an excess of positive space charges and the space charge density decreases with increasing wind speed.

Negative space charge formation at ground level above a very dry ground surface has been measured at high wind speeds. This is opposite to the conditions of the electrode effect above a dry ground surface. A mechanism for this electrification is drifting particles from the ground surface.

Drifting snow carries a net positive space charge. The space charge density and the height of drifting snow can be a function of the wind structure. The positive space charge density is up to 100 times the normal, fair-weather value of the positive space charge density. The charge-producing mechansism is dependent on the temperature and is most effective at low temperatures.

The present study shows that in the wind speed interval 2-4.5 ms^{-1} at the height 10 m there is a fairly constant and low space charge density in the atmospheric surface layer. For lower and higher wind speeds there are, however, large values and signal fluctuations of the space charge density.

1. Introduction

The electrical processes in the lower atmosphere are complex, and vary within a large range of space and time scales. An overview of the electric phenomena in the lower atmosphere is given by Hoppel et al. (1986). A source of space charge at ground level is the electrode effect, see e.g. Israel (1971). The mathematical description of the electrode effect is given by a number of authors. Theoretical calculations and experiments have led to the conclusion that such an effect does exist, see Tuomi (1982) and Willett(1978).

Some studies of the electrification of sandstorms have been made. Latham's (1964) analysis shows that sandstorms exhibit similar electrical effects to those of snowstorms, and he suggested that the temperature gradient effect may provide an explanation for sandstorm electrification. Crozier (1964) reported a main negative space charge center in dust devils in New Mexico. Kamra (1972) noted that the most predominant trend in his electrical measurements during blowing dust with strong winds was for the potential gradient and space charge to become negative. Knudsen at al., (1989) showed that drifting dust was observed for wind speeds higher than 4.5-5 ms^{-1} at the height of 10 m.

The electrical effects associated with blowing snow were first studied by Simpson (1919), who found from several thousand measurements in the Antarctic that blowing snow was accompanied by a large increase in the normal positive potential gradient of the atmosphere close to the ground. Observations made by other workers have confirmed those of Simpson, see Israel (1973). Laboratory experiments in which the conditions obtaining in a snowstorm were simulated, and the charge was measured on the blowing snow particles gave similar results to the outdoor measurements.

Pierce and Currie (1949) and Itagaki (1977) have all observed considerable enhancement of the normal positive potential gradient near the earth's surface during drifting snow. The charge-producing mechanism is most effective at high wind speed and at low temperature. In general they attribute the charging to the friction and fracturing of blown ice particles. Laboratory experiments by Orikasa and Ohta (1983) also showed the occurrence of positive space charges in drifting snow, and a compensating negative charge residing partly on the heavier snowflakes and partly on the earth's surface. Knudsen at al., (1989) showed that drifting snow was observed for wind speeds higher than 4.5-5 ms^{-1} at the height of 10 m.

Attempts to explain these phenomena based on contact and friction electricity are given by Loeb (1953). Reynolds (1954) gave a certain explanation for the electrification of ice crystals on the basis of his results, according to which the electrical effects produced by ice-to-ice contact strictly depend on the temperature and degree of purity of both crystals. When rubbing ice on ice the piece having the higher conductivity becomes negatively charged, regardless of whether the different conductivities originate in different temperatures or different degrees of purity.

In the present paper we will study the fluctuations and values of the space charge density under different weather conditions. Especially the possibility of other mechanisms than the electrode effect for production of space charges at ground level during conditions with normal fair-weather electric fields and high wind speed will be investigated. The measurements should be carried out during different ground surface conditions; viz. in spring time with ploughed or harrowed fields when the occurrence of dry wind-driven dust ought to be more common and drifting snow conditions in winter.

2. Measurement site and instrumentation

The measurements were made at the Marsta Observatory (59°55'N, 17°35'E). The observatory is located in very flat farming area 10 km north of Uppsala, Sweden. The nearest forest is more than 1 km away from the observatory. The very flat surroundings and low temperatures in winter with snow cover lead to an occurrence of drifting snow. No industrial establishments by which condensation nuclei might be produced were in the surroundings of the observation place and no burning took place during the observations. A map of the Marsta Observatory is presented by Israelsson et al. (1973).

Measurements of the fine structure of the vertical space charge profile require an instrument which measures the space charge density in a thin layer. For this purpose the filter method was found to be suitable. Two identical instruments (Obolensky filters) were constructed which, apart from minor modifications, are copies of the Anderson apparatus (see Anderson, 1966). The instrument is shown schematically in a previous paper (Knudsen et al., 1989). In the present study average space charge density was recorded in a layer up to 1.2 m above the ground surface with a Faraday cage method.

In order to obtain an overview of the atmospheric electrical condition we measured the electric potential gradient with a radioactive collector at a of height 1 m and the electric polar conductivity (λ^+ and λ^-) with aspiration condensers according to the Kasemir-Dolezalek system (Israel, 1971 and 1973). The critical mobility of the conductivity meter was $2 \cdot 10^{-4}$ $m^2V^{-1}s^{-1}$. The conductivity and space charge sensors were not equalized to the ambient potential due to the high wind speed, see Knudsen et al. (1989). The micrometeorological parameters wind at 10 m height (u_{10}), temperature and relative humidity were continuously recorded with standard instruments.

3. Results

Results from measurements of space charge density and wind speed under different wind conditions are carried out during the years 1987-1993. A survey of the results are given in the Figure 1. Curve (a) represents the best fitted function of the u_{10}-data versus space charge density. The number of observations is 69 and they represent near neutral atmospheric stability conditions. The standard deviations of the observations are fairly large, see the vertical lines. Data from wind speeds lower than 1 ms^{-1} are not included due to that for conditions with very low wind speed a stratum of enhanced ionization can cause a reversed electrode effect with a stratum of negative space charges. The curve (a) in Figure 1 shows that space charge density decreases rapidly with increasing wind speed. Comparisons with the theoretical model for the electrode effect given by Tuomi (1982) show that the space charge caused by the electrode effect decreases rapidly with increasing wind speed. Further the ionization rate normally also decreases with increasing wind, resulting in a more pronounced decrease of the space charge caused by the electrode effect.

Wind-driven dust In order to test the space charge density in wind-driven dust we repeated during April 1993 the measurements of the total space charge density in conditions with positive fair-weather electric potential gradient and very dry ground surface. The surrounding fields were ploughed and some were harrowed. There was fair-weather, no clouds, very dry ground surface and high wind speed. In spite of a positive fair-weather electric potential gradient the total space charge density was negative during most of the measuring time.

The diagram, curve (b) in Figure 1, shows a close connection between wind speed (u_{10}) and a negative space charge formation. Further, due to the effect of wind on the formation of drifting particles, see Liljequist (1957), the curve represents the best fitted second order polynomial function, r the correlation coefficient and n the number of observations. The drifting dust was observed for wind speed u_{10} higher than 4.5-5 ms^{-1}.

Drifting snow In the present study we have also measured the space charge density in drifting snow using the Faraday cage method. Some comparisons with measurements with a method given by Itagaki (1977) show that the different methods of measurements agree fairly well. Simultaneous measurements of the electric field, polar conductivities and meteorological parameters were done during periods with drifting snow. Hourly mean values of space charge density, temperature, wind speed u_{10}, electrical conductivity and field strength were calculated from conditions with drifting snow, viz. 46 cases.

The vertical distribution of the drifting snow was estimated by visual observations on the flat country surrounding the observatory. For wind speeds from 4.5-5 ms^{-1} at the height of 10 m the observations and the measurements show that the drifting snow starts and the vertical distribution of snow particles for this wind speed is about 10 cm. The layer of drifting snow seems to increase with growing wind speed, and for winds 6 - 10 ms^{-1} the height seems to be around 1 m on an average.

418

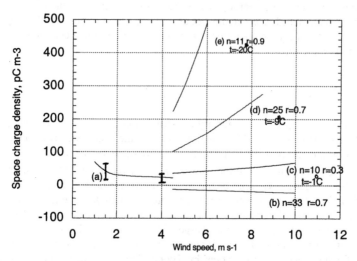

Figure 1. Space charge density versus wind speed u_{10}. Curve (a) represents the
electrode effect, curve (b) wind-driven dust, curves (c) - (e) drifting snow.
Correlation coefficients r and number of obsevations n are given.

All the measurements of the space charge density in the main drifting snow layer show
positive polarity but with very fluctuating values. The electric field strength shows very
similar variations. The space charge density increased suddenly when the wind caused the
drifting snow.

Due to the fact , that the electrical conductivity or dielectric constant of ice is temperature
dependent, (Hellwege and Hellwege (1959) and Shimizu et al., (1985) and that this can be
included in the charging process, the data are divided into temperature intervals, viz. T < -14,
-14 to -4, and T > -4°C. Further, due to the effect of wind on the formation of drifting snow,
see Liljequist (1957), space charge density is plotted versus the the wind speed at a height of
10 m according to a second order polynomial function. The correlation coeffients and the
number of obsevations are given by r and n. The curves (c)-(e) in Figure 1 show the calculated
curves. As can be seen in the diagrams the space charge density is always positive and
increases with increasing u_{10}. The charging effect in a snow drift increases with falling
temperature. From the diagrams we also observe that the value of the correlation coefficient
of the data increases with falling air temperature from +0.3 for temperatures around -0.8°C to
+0.9 for -19.9°C.

4. Discussion

By reducing the number of factors which influence the space charge density in the
atmospheric surface layer it has been possible to measure the effect of wind speed on the
space charge density for the wind speed interval 1-10 ms^{-1} in near neutral atmospheric
stability conditions. The results are in good agreement with the theoretical models of Tuomi
(1982) and Willett(1978). The electrode effect which increases in importance for lower

turbulent exchange with wind speeds lower than 2 ms^{-1} at the height 10 m.

The measurements show that negative space charge formation at ground level above a very dry ground surface is closely connected with growing wind speed. This is in contrast to the electrode effect which increases in importance for lower turbulent exchange.

It seems appropriate to consider the effect of wind in conditions with dry ground on the space charge formation at ground level to consist of two components; the turbulent component and the dust devil or human produced component.

The turbulent component is generally associated with the wind gusts that raise local dust into the lowest meters of the atmosphere. The fluctuations of the space charge density correspond to the turbulent structure of the atmospheric layer.

Latham's (1964) analysis showed that sandstorms exhibit similar electrical effects to those of snowstorms, and he suggested that the temperature gradient effect may provide an explanation for sandstorm electrification. The details of a mechanism by which this charge can be assembled are not clear, but Crozier (1964) pointed out an asymmetrical electrification of the dust particles at the ground surface, at the initial separation from the ground or upon subsequent impacts, or both, perhaps along the lines of the process suggested by Kunkel (1950).

In winter conditions visible drifting snow appears at wind speeds of 4.5-5 ms^{-1} (Knudsen et al., 1989). It is therefore possible that during our present measurements effects of drifting non-visible particles can be involved.

The electrical potential gradient in the surface layer and space charge density follow each other in most cases. It is difficult to make a distinction of large-scale electric field effects from the small scale. But the meteorological observations of cloudiness indicate that the large-scale electric field from clouds ought to be fairly limited.

From the observations of space charges in drifting snow it was ascertained that in the bulk of the drifting snow layer the space charge density is positive and is in most cases distributed in the layer up to 1 m with the highest values close to the ground surface. The height of the layer is dependent on the wind and surface conditions. This is in agreement with previous studies. The vertical distribution of space charges in the drifting snow seems to be a function of the size of the drifting snow pellets. Wendler (1987) reported that larger snow pellets are distributed close to the ground surface.

The observed charge distributions in the drifting snow seem to agree with observations published by Magono and Sakurai (1963).

According to Latham and Stow (1965) temperature gradients appear to be of prime importance in producing the charge transfer which accompanies blowing snow. The present study supports it. As has been proposed in the theoretical consideration and given in the present measurements, the temperature gradients at the snow surface in most conditions are negative, viz. directed into the snow cover with the coldest temperatures at the highest layer of the snow. The measurements show that the electrification of drifting snow is intimately related to the temperature gradients in the snow surface.

The present study may be important in the measurements of the global, regional and local electric currents within the coming Euorpean Network for studies of electrodynamic and charged particle effects on the stratosphere and troposphere. It is also of importance for the international program Global Atmospheric Electrical Measurement (GAEM, see Michnowski and Ruhnke, 1991), because the electrical processes connected to the micrometeorological phenomena in the lower layer can act as disturbances on the current measurements

Acknowledgements. The study was supported by grants from the Swedish Natural Science Foundation.

420

References

Andersson R A 1966 *J.Geophys. Res.* **71** 24 p 5809

Crozier W D 1964 *J. Geophys. Res.* **69** p 5427

Hellwege K H und Hellwege A M 1959 *Die elektrische Eigenschaften, Landolt-Börnstein, Zahlenwerte und Funktionen* (Springer Verlag, Berlin)

Hoppel W, Anderson R V and Willett J C 1986 *The Earth's Electrical Environment* (Studies in Geophysics, National Academy Press, USA) p 149

Israel H 1971 *Atmospheric Electricity* 1 (U.S. Dept. of Commerce Springfield)

Israel H 1973 *Atmospheric Electricity* II (U.S. Dept. of Commerce Springfield)

Israelsson S, Knudsen E and Ungethum E 1973 *J. Atm. Environment* **7** p 1127

Itagaki K 1977 *Electrical Processes in Atmospheres* (Steinkopff -Darmstadt) p 211

Kamra A K 1972 *J. Geophys. Res.* **69** p 5856

Knudsen E, Israelsson S and Hallberg B 1989 *J. Atm. Terr. Physics* **51** 6 p 521

Kunkel W B 1950 *J. Appl. Phys.* **21** p 820

Latham J 1964 *Quatr. J. Roy. Met. Soc.* **90** p 91

Latham J and Stow C D 1965 *J. Met. Soc. Japan* **43** 1 p 23

Liljequist G H 1957 *Energy exchange of an Antarctic Snow-Field* (Norsk Polarinstitutt Oslo)

Loeb L B 1953 *Thunderstorm electricity* (edited by H.R.Byers) p 150 Chicago

Magono C and Sakurai K 1963 *Journ. Met. Soc. Japan* **41** 4

Michnowski S. and Ruhnke L 1991 *Proc. on Int. Workshop on Global Atm. Electricity Measurements in Madralin Poland* (Institute of Geophysics, Academy of Sciences, D-35) p 238

Orikasa K and Ohta K 1983 *Proc. in Atmospheric Electricity* (A Deepak Publ. Hampton V-a) p 237

Pearce D C and Currie D A 1949 *Canadian J. Research* **27** Sec. A 1

Reynolds S E 1954 *Compendium of Atmospheric Electricity* (Rep. New Mexico Inst. Min. Techn. , Res. Dev. Div., Socorro, Dep. Ar Proj. 3-99-07-022. Sign. Corp. Proj.)172 B

Shimizu N, Kosdaki M and Horij K 1985 *IEEE* (CH 2115-4/85/0000 -539) p 539

Simpson G C 1919 *British Antarctic Exp. 1910-13 Meteorology* (Calcutta) **1** p 302

Tuomi T J 1982 *J. Atm. Terr. Phys.* **44** p 737

Wendler G 1987 *Antarctica Journ.* USA **22**(5) p 264

Willett J C 1978 *J. Geophys. Res.* **83** p 402

Inst. Phys. Conf. Ser. No 163
Paper presented at the 10th Int. Conf., Cambridge, 28–31 March 1999

421

STUDY OF INDOOR AEROSOL REMOVAL WITH ELECTRO-SPRAY

V. Smorodin[1,2], W. Balachandran[1] and C. Hudson[1]

[1]Dept. of Manufcaturing & Engineering Systems, Brunel University,
Uxbridge, MX, UB8 3PH, United Kingdom
[2]Dept. of Chemistry, Moscow State University, Moscow 119899, Russia

Abstract. Electro-diffusion coagulation of two systems of charged aerosol clouds of different size scales (sub-micron/micron dry particles and larger droplets of opposite charge) moving in a gravitational field has been studied. Numerical simulations were performed for different electro-aerosols and charged spray combinations. The characteristic time for particle collection is a function of initial concentrations, size and specific charge of aerosols. The results obtained clearly demonstrated that the scavenging of aerosol particles by charged sprays are several orders of magnitude greater than uncharged sprays.

1. Introduction.

A spray of fine water droplets is a well-known means of airborne dust removal. One method of improving the effectiveness of water sprays is by applying an electrical charge to the spray that is opposite in polarity to the charge on the dust to be suppressed. It has been found that most industrial pollutants and naturally-occurring fugitive dust acquire an electrostatic charge, well above the Boltzmann equilibrium, as they are dispersed in the air. If this charge of airborne material is exposed to oppositely charged water spray, contact between the particulate matter and the water droplets is enhanced. After contact is made, the wetted particulate matter agglomerates rapidly and falls out of the atmosphere. A dust removal technology using a charged spray was analyzed by Prem & Pilat (1978). Particle collection efficiency in the particle radius range of 0.05-10 μm by a single droplet was calculated numerically for the particle equation of motion. Among the electrostatic forces Coulombic force predominated. Effects of wet precipitation of charged aerosols were also studied by Wang et al. (1978), Peters et al. (1985) and many others. In reported works, authors made the assumption of constant charges during particle collisions. Balachandran et al. (1996) have stated that there are some limitations in achieving efficient precipitation due to low specific charge levels and lack of understanding of the optimum droplet size required for such applications. It is believed that the effectiveness of these charged sprays can be improved by controlling the spray drop size distribution and the specific charge. An idea of this research is to develop a mathematical model to be capable of predicting the efficiency of precipitation of an aerosol of known size and charge distribution in an enclosed environment.

2. Initial Model and Mechanism of Particle Removal.

Following experimental data showing a strong dependence of the electro-spray scavenging efficiency, we assume that:
1) A mechanism of smoke particle removal with electro-spray consists of the electro-coagulation oppositely charged droplets and smoke particles, followed by gravitational sedimentation. The main moving forces are due to (a) an electrical field of charged aerosols, (b) thermo-fluctuations and (c) gravitation;
2) Since typical sizes of water droplets (10-50 μm) are sufficiently larger than the particle those (0.01-1 μm), we may guess that under reasonable conditions, we can distinguish between (a) a fast particle collection (the electro-diffusion coagulation) on collecting droplets and (b) a relatively slow droplet sedimentation due to the gravitational field. Thus, the characteristic time of the particle collection is assumed to be much smaller than the droplet/particles gravitational sedimentation;
3) Every collision leads to association and not reflection, thus collecting droplets present slowly moving point centres "absorbing" fast particles around;

4) At an initial time, t = 0, both clouds of particles and droplets are homogeneously distributed ("spread") in 3D-space, and at t = 0 there is no movement;

5) Initial concentrations of droplets (N_{do}) and particles (N_{po}), and the droplet radii (R_d), satisfy conditions:

$$N_{po}/N_{do} >> 1 >> R^3_d N_{do} \qquad (1)$$

Therefore a "cell" model of the electro-diffusion coagulation and a gravitational precipitation is appropriate. The second inequality in (1) may be re-written as $(R_d/R)^3 << 1$, where $R = (3/4\pi N_d)^{1/3}$ is the cell radius. At given initial particle and droplet concentrations number of the particles in an elementary cell equals:

$$\varphi = 4\pi R^3 N_{po}/3 = N_{po}/N_{do} \qquad (2)$$

Experiments showed that a negatively charged spray was very effective for smoke precipitation. Therefore it may be concluded that smoke particles are mainly positively charged. For the smoke particles we assume Boltzmann's charge distribution. The charge of the water spray produced with a nozzle and an ionizer, is limited by Reilaygh's criterion. Polydispersity of both aerosol clouds will be described by a log-normal distributions.

3. Governing Equations.

If the inertial motion of the particle is not large compared to the motion induced by diffusion, temperature and concentration gradients, and external forces, then fairly good results are obtained by solving the continuity equation with initial and boundary conditions (Williams and Loyalka, 1991):

$$\frac{\partial N_i}{\partial t} = -\nabla J_i \qquad (3)$$

where $J_i = J_{el}^i + J_{dif}^i + J_{ext}^i + ...$ is the droplet (i=d) and particle (i=p) current density with a local concentration $N_i(t)$ caused by the various mechanisms, such as electrical, diffusion, external force etc. induced currents. In part:

$$J_{dif}^i = -D_i \nabla N_i; \qquad (4)$$

$$J_{el}^i = v_{el}^i N_i; v_{el}^i = B_i \vec{E} = -B_i \nabla \Phi^i \qquad (5)$$

$$J_{sed}^i = v_i N_i = \frac{2\rho^i g C_c^i R_i^2}{9\eta} N_i \qquad (6)$$

Analysing the electrical field (E) between the droplet and the particle interacting due to their electrical charges (treated as "points"), we may write:

$$E = -\nabla \Phi = Q_p Q_d(t)/\varepsilon r^2, \qquad (7)$$

where $\qquad Q_d(t) = Q_d^o + Q_p N_{p,col}(t); \quad and \quad N_{p,col}(t) = 4\pi R_d^2 D \nabla N_p(t), \qquad (8)$

r is the distance between centres of the droplet and the particle, v_i is the connective (hydrodynamic) velocity, D_i is the aerosol diffusion coefficient, Q_p is the particle charge, Q_d^o is the initial droplet charge, η and ε are the air viscosity and dielectric permittivity, respectively; B_i is the aerosol mobility, C_c^i is the Cunningham factor, Φ is the electro-potential provoking the particle motion; g is the gravitational constant, ρ is the droplet density.

Choosing an origin of the spherical coordinates in the centre of the collecting droplet, the initial and boundary conditions may be specify as:

$$N_i(t = 0) = N_{io} \qquad (9)$$

$$N_i \to N_{io}, \quad at \quad r \to \infty \qquad (10)$$

$$N_p(r = R_d + R_p) = 0 \qquad (11)$$

Since the number of particles onto the collecting droplet depends on a pre-history of this process (how many particles to have fallen before) a correct description of this phenomenon will transform the initial equations into non-linear ones which would be difficult to solve.

4. Electro-diffusion coagulation and precipitation in the averaged cell model frameworks.

To resolve the problem in the cell model frameworks we may first re-formulate the classical task arrangement and replace the external boundary condition (10) with:

$$N_p = N_{p\infty} \quad \text{at} \quad r = R \tag{12}$$

A detailed analysis of other appropriate boundary conditions and their comparison is a matter for separate discussion and beyond a scope of this paper. Using known formulas of coagulation theory (Williams and Loyalka, 1991) we get a result for a kernel of the electro-diffusion coagulation and the enhancement factor, p, due to electrical charging, in a standard form. Averaging the enhancement factor, p, over a number of collected particle assembly in a cell (or over the collection time, according to the principles of statistical physics), we obtain a mean value of the enhancement factor for monodisperse aerosols:

$$P_e = (1/\varphi) \int_0^\varphi p(N_{p,c}(t)) \, dN_{p,c} \tag{13}$$

A mean enhancement factor, P_o during collecting of all particles within the cell, for strongly charged droplets, when $-G = -\Phi/kT \gg 1$, can be presented as:

$$P_o \approx -G \approx \frac{1}{\varepsilon}\left(4\zeta\sqrt{\frac{2R_d R_p \sigma}{kT}} - \frac{1}{\pi}\frac{R_p}{R_d}\frac{N_{po}}{N_{do}}\right) \gg 1 \tag{14}$$

Here ε is the air static dielectric permittivity; $\zeta = Q_d/Q_{Re}$, $Q_{Re} = 4\sqrt{\pi\sigma R_d^3}$; σ is the surface tension of water droplets; k is Boltzmann's constant, T is temperature. After averaging over assemblies of both droplet and particles clouds, a mean value of the electro-diffusion coagulation kernel can be found. A formula for the characteristic time of electro-diffusion coagulation may then be written as:

$$\tau_c = \frac{3\eta\varepsilon}{kTC_p N_{po}}\left[4\zeta\sqrt{\frac{2\sigma R_d^3}{kTR_p}}Exp\left[\frac{1}{8}(9\ln^2\sigma_d + \ln^2\sigma_p)\right] - \frac{1}{\pi}\frac{N_{po}}{N_{do}}\right]^{-1} \tag{15}$$

where σ_d and σ_p, R_d and R_p are mean geometrical deviations and mean geometrical radii of the droplets and particles, respectively. This formula is only valid if

$$4\zeta\sqrt{\frac{2\sigma R_d^3}{kTR_p}}Exp\left[\frac{1}{8}(9\ln^2\sigma_d + \ln^2\sigma_p)\right] > \frac{1}{\pi}\frac{N_{po}}{N_{do}} \tag{16}$$

This criterion limits the parameters of effective scavenging of the charged spray for indoor aerosols. For highly charged droplets with parameters: $R_d > 10$ μm, $N_d < 10^5$ cm^{-3} and $\zeta < 0.1$ the characteristic time of decay $\tau_d > 10^2$ s. Therefore, it may be concluded that the space charge decay effect is slight, in a contrast to the case of ions of high mobilities. The total time of aerosol precipitation, T_p, from a height H can be estimated as:

$$T_p = \tau_s + \tau_c \approx \tau_s = 9\eta H/(2\rho g R_d^2), \quad \text{if} \quad \tau_c \ll \tau_s \tag{17}$$

Having evaluated the characteristic time, one may evaluate an efficacy of the smoke particle precipitation and the water spray yield needed for full indoor aerosol scavenging. Precipitation times of smoke particles using water spray ($R_d = 10$ μm) for charged ($\zeta = 0.01$, $T_p \approx 10^2$ s) and uncharged droplets ($T_p \approx 10^3$ s) from a height H = 2 m shown an efficacy of the charged spray.

Some results of numerical calculations of characteristic times of the smoke particle collection and the precipitation using the electro-spray with different regulating parameters are shown in Figures 1-2 below:

424

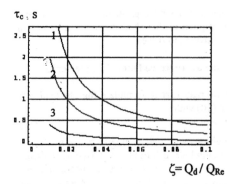

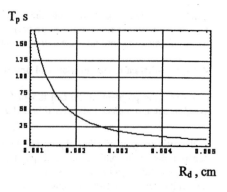

τ_c, s

$\zeta = Q_d / Q_{Rc}$

T_p s

R_d, cm

Fig.1. Dependence of the collection time, τ_c (s), on the droplet charge parameter, ζ : $N_{po} = 5*10^4$ cm^{-3} (1), 10^5 cm^{-3} (2) and $5*10^5$ cm^{-3} (while $R_d = 30$ μm, $\sigma_d = \sigma_p = 1.5$)

Fig.2. Dependence of the full precipitation time, T_p (s), on the droplet radius, Rd; H = 2 m.

5. Summary.

1. A "cell-averaging" approach has been proposed for a theoretical description of wet indoor aerosol particles using electro-spray technology. An initial model is valid for droplet radii $R_d > 10$ μm and for particle of minim radius 0.05 μm.

2. It was shown that the main mechanism of the effective indoor smoke precipitation with electro-spray consist of two type of processes: (a) a fast process of the electro-diffusion coagulation and (b) a relatively slow gravitational sedimentation.

3. It was found that the proposed initial model is realistic and adequate for describing and optimizing the advanced technology.

4. Using the initial model, for the given dust cloud, the critical parameters of the charged spray for efficient precipitation were calculated. The evaluated numerical results confirm that smoke can be effectively precipitated using electro-spray of appropriate specific charge and low volume of the spray.

References.

Balachandran W, Groemping M and Machowski W 1996 *Proc.12th Intl. Conf., Lund, Sweden*, p 2123

Peters M H, Jalan R K and Gupta D A 1985 *Chem. Eng. Sci.*, **40** (5) 723

Prem A and M J Pilat 1978 *Atmos. Environment* **12** 1981

Wang P K, Grover S N and Pruppacher H R 1978 *J. Atmos. Sci.*, **35** 1 735

Williams M M and Loyalka S K 1991 *Aerosol Science. Theory and Practice.* (Oxford, New York, Seul, Tokyo: Pergamon Press)

Inst. Phys. Conf. Ser. No 163
Paper presented at the 10th Int. Conf., Cambridge, 28–31 March 1999
© *1999 IOP Publishing Ltd*

Studies of aerosol deposition on a charged spherical collector

A. Jaworek[1], K. Adamiak[2], A. Krupa[1] and G.S.P. Castle[2]

[1]*Institute of Fluid Flow Machinery, Polish Academy of Science P.O. Box 621, PL-80952 Gdansk, Poland, ajaw@imppan.imp.pg.gda.pl*
[2]*Department of Electrical and Computer Engineering, The University of Western Ontario London, Ontario, Canada N6A 5B9, kaz@gauss.engga.uwo.ca*

Abstract. The experimental results of investigations of aerosol particle deposition on an electrically charged spherical collector are presented and compared qualitatively with numerical simulations. It is shown that charging both the collector and the particles substantially increases the particle deposition on the collector. A theoretical model is presented to describe the deposition process and is shown to be in good agreement with the experimental results.

1. Introduction

Investigations of deposition of aerosol particles on an object (collector) are of key importance in electrostatic painting or coating, crop spraying, electrostatic scrubbing and electrostatic probes designed for measurement of charge or mass flow rate in two-phase flows. The problem of deposition is usually solved for the case of both the collector and particle being charged. The trajectories of the particles are determined numerically from the Newton equation taking into consideration the air drag, electrical and gravitational forces. The experimental work is not widely represented in the literature, probably because the statistical nature of the charging and deposition processes makes quantitative comparison of theory with experiment difficult.

Some experimental results of investigation of aerosol particle deposition on a fixed spherical collector are presented in this paper. A qualitative comparison of these results with numerical simulations is also given. The experimental studies were carried out in a duct of rectangular cross section, in which a brass ball was mounted. Fine dust particles were charged and transported pneumatically through the duct. The traces of particle trajectories were recorded using a CCD camera. The numerical model for determining the trajectories of the particles was solved taking into considerations the electrical Coulomb and image forces, and mechanical Stokes, inertial and gravitational forces upon the particle. The Navier-Stokes equations were solved in order to determine the flow field around the collector.

2. Experimental investigations

The measurements were carried out in the experimental stand shown schematically in Fig.1. The experimental channel of rectangular cross section of 160x160 mm was made of clear acrylic plastic. A brass ball of diameter of 9.5 mm (3/8 inch) was used as the test collector. The ball was mounted on a thin stainless steel wire, 1 mm in diameter, in the vertical plane of symmetry of the channel and was connected electrically to a high voltage DC source, which voltage can be controlled up to 20 kV.

426

The particles of polymer powder of diameter ranging from 20 to 250 μm was used as test aerosol. The particles were fluidized by compressed air, and introduced into the channel through a thin pipe, directly to the AC field charger. The dust was charged by the AC charger developed by the authors of and described previously in the papers [1,2]. The charged dust was transported by the flowing air along the channel. The air flow through the channel was forced by a sucking blower placed at the outlet of the channel. The ball was illuminated by two collimated light sources: one light beam illuminated the ball from the upstream side, and the second light beam illuminated the rear side of the ball. The light beam was about 2 mm thick, and only the particles flowing close to the ball near the plane of symmetry of the channel were illuminated in the dark room.

The trajectories of the particles near the collector were recorded with a CCD camera coupled to a video tape recorder. A special sight-glass mounted in the channel side walls was used for these observations. Continuous light and light pulses of 3μs duration, generated by a stroboscopic lamp were used for illumination of the sphere. The electronic shutter of the CCD camera was set to 1/60s to record the particle trajectories. The recorded pictures were reviewed frame-by-frame with the aid of a frame grabber. From the recorded pictures the mechanisms of the particle deposition were reconstructed.

Sample photographs of the particle trajectories taken by the CCD camera are presented in Fig.2. A difficulty with recording the trajectories of the particles approaching the sphere in a selected plane is that the particles approach the sphere surface from all directions, and on the photographs there are some traces of different aspect angles which are not easy to distinguish from each other. One can see on the photographs that the particle trajectories are bent to the collector due to electrostatic forces.

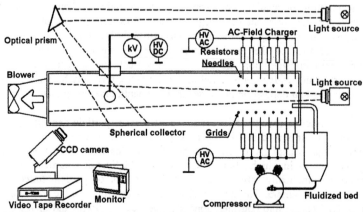

Fig.1. Scheme of the experimental arrangement for investigation particle deposition on a fixed spherical collector.

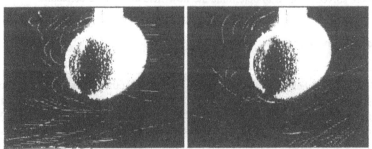

Fig.2 Photographs of the particle trajectories near a spherical collector for U_s=15kV.

3. Theory

The aerosol particle trajectories approaching a charged and fixed spherical collector can be determined numerically by solving the equations of motion of the particle in the system of coordinates placed at the collector center. It is assumed that the aerosol particles are spherical, and the flow field in the vicinity of the collector is not disturbed by them.

The trajectory of a particle of mass m_p is given by the vector differential equation:

$$m_p \frac{d\vec{w}}{dt} = \frac{C_d \, \text{Re}_p}{24} \vec{F}_s + \vec{F}_e + m_p \vec{g} \tag{1}$$

in which w is the particle velocity.

The Stokes drag force is:

$$F_s = 6\pi \eta_g R_p (\vec{u} - \vec{w}) / C_c \tag{2}$$

The electrostatic force on the particle is:

$$F_e = \frac{Q_p Q_c}{4\pi\varepsilon_0 r^2} + \frac{R_p Q_c^2}{4\pi\varepsilon_0 r^3}\left(\frac{r^4}{(r^2 - R_p^2)^2} - 1\right) + \frac{R_c Q_p^2}{4\pi\varepsilon_0 r^3}\left(1 - \frac{r^4}{(r^2 - R_c^2)^2}\right) \tag{3}$$

where Q_p and Q_c are the charges on the aerosol particle and the collector, respectively, R_p and R_c the radius of the particle and the collector, respectively, r is the distance between the particle and the collector centres, u the gas velocity, η_g gas viscosity, ε_0 is the permittivity of the free space. C_c is the Cunningham slip correction factor [3,4], and C_d the non-Stokesian drag coefficient, which for $\text{Re}_p < 2*10^5$ is given by Brauer and Sucker equation [5].

For the collector maintained at high potential U instead of being charged to Q_c, the substitution of

$$Q_c = 4\pi\varepsilon_0 R_c U \tag{4}$$

can be made, if all other objects are sufficiently far from it.

In the general case, the flow field around the spherical collector is governed by the vector Navier-Stokes equations, which form a set of non-linear partial differential equations with unknown components of the velocity vector and pressure. The problem is simplified by introduction of two scalar variables: the stream function Ψ defined as:

$$p = \frac{1}{r}\frac{\partial \Psi}{\partial r}, \qquad q = -\frac{1}{r}\frac{\partial \Psi}{\partial z} \tag{5}$$

and the vorticity:

$$\omega = -\frac{\partial p}{\partial r} + \frac{\partial q}{\partial z} \tag{6}$$

where p and q are components of the gas velocity vectors.

Using these two new variables, the Navier-Stokes equations can be converted to [6]:

$$\frac{\partial^2 \Psi}{\partial r^2} - \frac{1}{r}\frac{\partial \Psi}{\partial r} + \frac{\partial^2 \Psi}{\partial z^2} = -r\omega \tag{7}$$

$$\frac{\partial^2 \omega}{\partial r^2} + \frac{1}{r}\frac{\partial \omega}{\partial r} - \frac{\omega}{r^2} + \frac{\partial^2 \omega}{\partial z^2} = \text{Re}\left(q\frac{\partial \omega}{\partial r} - \frac{q}{r}\omega + p\frac{\partial \omega}{\partial z}\right) \tag{8}$$

After such an operation, the problem is governed by a set of two partial differential equations for the stream function and vorticity, one of them being nonlinear. These equations can be solved simultaneously, but iterative algorithms are also very effective. Assuming zero vorticity equation, equation (7) can be solved for ψ. Then, the stream function is substituted into equation (8), which is solved for ω. The process is continued until the convergence is reached. The rate of convergence depends on the value of Re; while only a few iterations are sufficient when Re is close to zero, a much higher number is required for large Re.

The boundary conditions result from the assumption that far from the collector the flow is undisturbed and uniform, and that both components of the velocity vector vanish on

428

the collector surface. The last condition leads directly to the homogeneous Dirichlet boundary condition for the stream function. The vorticity function satisfies the following condition [7]:

$$\omega = -\frac{\partial^2 \psi}{\partial n^2}$$

(9)

The equations were solved iteratively using the Finite Element Method with triangular discretization and linear interpolation of the solution. A monotone streamline upwinding was used to stabilize the computation process for large Reynolds numbers. Due to symmetry, only one half of the domain was discretized into about 2500 triangular elements. Inside of each triangle both functions: stream function and vorticity were interpolated linearly in terms of the nodal values. This procedure reduces the number of unknowns. Using the weighted residual technique an algebraic set of equations was created, which, after introduction of the boundary conditions, was solved for the nodal values of unknowns. The flow field was determined only for some values of the Reynolds number Re_c. The gas velocity vector for any other Reynolds number, was interpolated linearly from two adjacent sets of results.

The trajectories around the collector were obtained from the solution of equation (1). The Runge-Kutta method was used to obtain numerical results. Fig.3 shows a few examples of aerosol particle trajectories in the z=0 plane, near a fixed charged spherical collector for three values of particle size. It was assumed that the particles were charged up to 10^{-14}C.

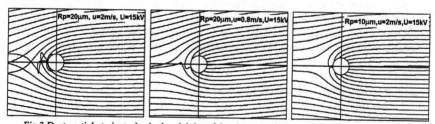

Fig.3 Dust particle trajectories in the vicinity of the charged spherical collector in the plane z=0.

4. Conclusions

The experimental investigations and numerical simulations of dust particle trajectories near a charged spherical collector are presented in the paper. The deposition of an aerosol particle on a collector is governed by the electrostatic forces which are proportional to the charge on the particle and the collector and inertial force which are determined by the particle mass and particle velocity. With the increase in the charges the deposition of the particle also increases, however with the increase in the inertial forces the electrostatic effect diminishes. The trajectories determined numerically are close to those obtained experimentally taking into consideration the statistical nature of the deposition process.

Acknowledgments

This paper has been supported by the Natural Sciences and Engineering Research Council of Canada and by the State Committee for Scientific Research of Poland (Grant No.940/T10/96/10). The scientific cooperation was supported by NATO Science Programme and Cooperation Partners Linkage Grant No.SA.12-2-02 LG.971755.

References
1. Jaworek A., Krupa A., J. Electrostatics **23** (1989), 361-70
2. Adamiak K., Krupa A., Jaworek A., Electrostatics '95, Inst. Phys. Conf. Ser. No.143, Bristol, 1995, 275-8
3. Cunningham E., Proc. Royal Soc. **83A** (1910), 357-65
4. Rader D.J., J. Aerosol Sci. **21** (1990) No. 2, 161-8
5. Brauer H., Sucker D., Int. Chem. Eng. **18** (1978), 367-74
6. Viviand H., Berger S.A., J. Fluid Mech., **23** (1965), 417-438
7. Ninomiya H., Onishi K., Flow Analysis Using a PC. Computational Mechanics Publ., Southampton 1991

DISCUSSION - Section I - Atmospheric Electrostatics and Environmental Aspects

Title: The Interaction between Air Ions and Aerosol Particles in the Atmosphere
Speaker: K Aplin

Question: S Gerard Jennings
(a) Could you tell me the basis for your assumption of an inverse relation between electrical mobility and aerosol mass? (b) How was aerosol number (or mass) concentration measured?

Reply: (a) As the mass of the aerosol particles increases their behaviour is controlled more and more by mechanical forces and less and less by electrical ones. Conversely small ions' paths are influenced almost completely by electrical forces; thus the relationship $(mass)^{-1/2}$ has been suggested. (b) Aerosol mass concentration was measured by an instrument called a TSI Dust Trak which has a nominal 15 μm upper cut off. This mass concentration was converted to a number concentration using an approximation for aerosol density.

Question: N L Allen
Does small ion concentration go down as the numbers of large neutrals and large ions go up?

Reply: Yes, because the small ions have a tendency to attach to larger particles. As the number of larger particles increases more and more small ions attach to them and their concentration (η) decreases. This is described by the equation $q = \alpha n^2 + \beta nZ$ where β is the attachment coefficient between ions and the aerosol.

Title: On the Effects of Wind on the Space Charge Formation in the Atmospheric Surface Layer
Speaker: S Israelsson

Question: R G Harrison
Is it possible that the radioactive potential probes used could produce additional local space charge?

Reply: The associated radioactive ionisation can cause errors in the space charge, especially at wind speeds less than about 0.5 m s^{-1}.

Title: Study of Indoor Aerosol Removal with Electrospray
Speaker: V Smorodin

Question: S Gerard Jennings
Previous scavenging and removal of aerosol particles by droplets show poor agreement between prediction and observation - measured scavenging rates have been much greater than predicted. Did you compare your predictions

with measurements?

Reply: No

Comment: R G Harrison
            The Boltzmann equilibrium charge is likely to underestimate charge levels in
            artificial environments in which there is usually substantial ion asymmetry.

Question: If your results were to be genuinely applicable to climatic aerosols as
            suggested, wouldn't you have to include inertial forces in scavenging
            calculations? (They are currently zero wind speed case.)

Reply: Yes, probably, I would like to do it in the future and thank you for your
            essential comment.

Question: Jen-Shih Chang
            The model should consider charge-to-mass ratio in a 3-dimensional
            distribution in electro-spray. Should be different with water spray.

Reply: Thank you for your question. The 3-D charge distribution was taken into
            account through the statistics of aerosol sizes (size distribution). Every
            droplet or particle size (radius) was attributed with a concrete value of
            charge. For droplets it was the Rayleigh limit and for particles it was
            Boltzmann's statistics. Then we have averaged all aerosol sizes in 3-D space.

Title: Studies of Aerosol Deposition on a Charged Spherical Collector
Speaker: A Jaworek

Question: Jen-Shih Chang
            When the Reynolds number is larger than 20, the effect of wake flow on the
            particle deposition near the back stagnation point must be considered as in
            the model of Chang et al (1976).

Reply: We observed the eddy effects from the rear side of the droplet, but only for
            small Coulomb numbers. In the simulations presented the coulomb number
            (i.e. the charges of the collecting droplet and the particle) was assumed
            sufficiently high that the effects you have mentioned were not observed.

Question: V Smorodin
            (a) Why have you talked about a spherical symmetry, when in a reality you
            are dealing with cylindrical symmetry (a gravitational field axis)? (b) Have
            you compared results on the collection efficiency by your new definition and
            a previous one used in a literature?

Reply: (a) On the cylindrical symmetry we can talk only in two particular cases: the fixed
collector and the collector freely falling down. In the first case only if the gravitational effect

is negligible and in the second because the gravitational field is symmetrical to the trajectory of to the collector trajectory. In any other cases the symmetry can not be assumed to be due to the gravity effects. (b) The definitions are quite different, also they can give similar results, within a few percent. The cross-section based definition requires the assumption of cylindrical symmetry that is not possible in the case of a droplet falling within the flowing gas. This is the reason the volume-based definition was proposed by us.

Section J

Instrumentation

Inst. Phys. Conf. Ser. No 163
Paper presented at the 10th Int. Conf., Cambridge, 28–31 March 1999
© *1999 IOP Publishing Ltd*

435

Using the Electrostatic Super Corona Gun and Electrostatic Powder Coating Diagnostic Instrument (EPCDI) to measure the thickness & adhesion of powder coated substrate

B. Makin and K. Dastoori

Department APEME, University of Dundee, Dundee, DD1 4HN

Abstract: The Super Corona Gun is a recent development that reduces back ionisation by cutting out much of the space charge that causes the build up of charge in the powder layer. The space charge is removed by placing grounded corona points behind the high voltage corona point on the barrel of the gun. The grounded needles react to the strong negative charges created by the gun to produce positive ions to cancel out the space charge. If the grounded points are well positioned the charge transferred to the powder particles is not be affected significantly. The Electrostatic Powder Coating Diagnostic Instrument (EPCDI) designed and manufacture in University of Dundee measures the uniformity of the powder coating by moving the powder sample and measuring the light transmission through it. The thickness of the powder layer is determined by measuring the transmission of an infra red light beam. It also tests the adhesion by blowing the powder off in a measured way. There are many variables that can affect the powder coat thickness and adhesion.

1. Introduction

The Super Corona Gun was made by adding a PTFE ring with grounded needles sticking out at regular intervals (see figure 1).

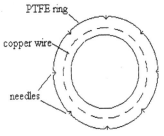

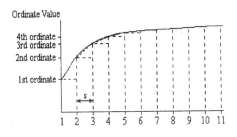

Figure 1- Ring with grounded corona added to corona gun

Figure 2 - Dividing the area under a curve into points sections

The needles were soldered to the copper wire (shielding taken from a coaxial cable) and pushed into the groove inside the PTFE ring. Around 30cm of the copper wire was left outside the ring and was sealed in shrink-wrap. The ring electrode was connected to ground.

436

2. Adhesion Value

The adhesion value is calculated for constant pulse adhesion tests. It is a measure of how difficult it is to remove the powder coat from the sample. A low adhesion value indicates the powder was not clinging strongly to the slide surface. The adhesion calculation is based on the Simpson rule [5]. This rule calculates areas under curves by dividing the region into sections and calculating the approximate areas of each of these sections (see figure 2). Each of the sections is split into a rectangle and a triangle. The area is given by

$$A = \frac{s}{3}\left[(F + L) + 4E + 2R\right]$$ Where s = width of each strip, F = first ordinate (labelled 1),

L = last ordinate (labelled 11), E = the sum of all the even ordinates (= (2+4+6+8+10)), R = the sum of the remaining odd ordinates (=(3+5+7+9))

In most cases the result would be an approximation of the area under the curve. In the case of the graphs from the Electrostatic Powder Coating Diagnostic Instrument (EPCDI) the value will be exact if the number of sections the curve is divided into is the same as the number of air pulses. The adhesion value uses the inverse of the area calculated by the Simpson rule (see figure 3). The inverse area is calculated using the equation:
Inverse Area   = Maximum Area - Simpson Area = 1000 - Simpson Area,
max. area = 10 (pulses) × 100 (%), Total Area = (Max Value- Min Value) × 10(pulses)
Adhesion Value = (Inverse Area/ Total Area) × 100 %

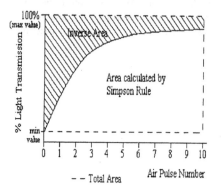

Figure 3 - Calculation of Areas for adhesion Value

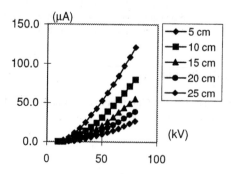

Graph 1 - Super corona ring current v. corona gun supply voltage with separation of corona Ring from the gun electrode.

3. Experimental Results

The ring with the grounded corona points slid over the gun barrel and the copper wire connected to the earth. The side of the barrel was marked off with a scale in centimetres of the distance from the corona tip. When the ring removes space charge from the area in front of the corona point a positive current flows from the ground connection to the needles. A multimeter was put into the circuit to measure current. The current was measured through the ring when the gun was operated i.e. three states no air supply and no powder, with air supply but no powder and with air supply and powder. The distance of the ring from the corona point was varied between 5 and 25 cm for each of the three states. In each state current reduced as the ring was slid further back. The results given in Graph 1 are obtained using the gun with an air supply and powder. The general shape of these results was the same for all three states, although they differed in magnitude. Graph 2 shows the results of the super corona ring current flowing through grounded ring when using no air, air and no powder and air and powder mix with the ring at 5cm from the corona point. With the ring set 5 cm from the corona point the amount of ions collected when the air is on is exactly the same as when the air is off. The presence of the powder has a little effect on the collected current. This result was continued at all distances from the corona point. Graph 3 shows the change of super corona ring current with distance of ring from corona point (50Kv for the three states tested). The currents collected when the air supply is on are very similar to the current when the supply is off. The rate of these drops off proportionately to the distance of the ring from the corona point. The current from the gun with powder drops off faster and is always less than the first two cases. These results agree with the theory [1]. The charge developed is directly related to the supply voltage of the corona gun. The grounded corona points on the ring will react according to the strength of the negative space charge around them, so as the ring is moved back the field is reduced and less current flows. The ions are able to move much faster than the air stream, so the current from the ring is not affected by the presence of an air stream. When a powder is blown in the air stream it collects many of the ions from the space charge and carries them toward the workpiece, away from the ring. Since these charges are attached to the powder they do not have the mobility of the free ions and cannot move quickly towards the grounded ring. When the ring is placed between 10 and 15cm from the corona point the current collected drops off (see graph 3). This may indicate that in this range it is the ideal place to put the ring, a position where it is able to greatly reduce the space charge but is unable to pull the charge off the powder. Tests were carried out on powder samples to determine adhesion and powder thickness while varying the distance of the ring from the corona point. An initial test varying the corona gun voltage with the ring set at 5cm from the corona point showed a decrease in powder adhesion and thickness. This indicated the ring was effective in reducing the space charge around the corona point. The reduction in powder thickness and adhesion also suggested the ring was too close to the corona point and was neutralising some of the charge on the powder particles. Tests for the variation in powder thickness and adhesion were carried out using the corona gun voltage set at 30kV, 50kV and 80kV. The test results for powder thickness was quite consistent. The plots of current flowing through the ring against distance from corona point indicated a significant change around 15cm so readings were taken around this distance. Graph 4 shows the average powder thickness results when using 30kV gun voltage.

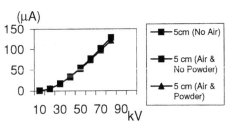

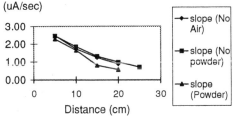

Graph 2- Results of the super corona ring current at 5cm from the corona point.

Graph 3 - Shows when the ring is placed Between 10 and 15cm from the corona point the current collected drops off

The graph has quite an unusual shape. It reaches a peak thickness with the super corona ring around 12cm from the corona point of the gun. Either side of this point the powder layer is much thinner. As the ring distance increases the gun behaves more like a normal corona gun. Graph 5 shows the average powder thickness plots using a 50kV gun voltage. The graph has a clear peak in thickness around the 14 - 15cm region. This time the powder layer continues to become thinner as the ring moves away from the 15cm region.

438

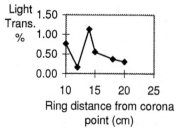

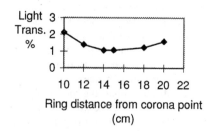

Graph 4- Powder thickness with varying ring distance with the gun at 30kV

Graph 5- Powder thickness with varying super corona super corona ring distance with the gun at 50kV

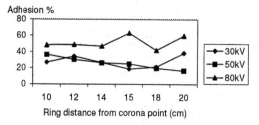

Graph 6 - Powder adhesion with varying super corona ring distance

Since the powder thickness and adhesion are both related to the amount of charge on the powder particles it was expected that the adhesion curves would follow the same patterns as the powder thickness plots. Graph 6 shows the adhesion plot for a 30kV, 50kV and 80kV corona gun voltage. The adhesion graphs closely followed the shapes of the powder thickness curves and confirmed the 12-15cm optimum distance of the super corona ring. At the peak the ring is removing the excess space charge while not affecting the charges on the powder particles.

4. Conclusions

The following conclusions were suggested by the experimental results conducted. The tribo gun produced evenly coated samples. The thickness and adhesion of these was related to the length of time the gun was used. In the case of the corona gun, care had to be taken to produce an even powder thickness. The powder layers were greatly influenced by the shape and positioning of the workpiece. Corona powder samples were at a maximum thickness around 50kV, however this dropped off after 50kV and picked up again as the gun voltage approached 80kV.The optimum positioning for the super corona gun ring was related to the voltage source used. The best distance of the ring from the corona point was in the range 12 - 15cm (for the 30 - 80kV gun voltage range). This was critical as having the ring in the slightly wrong position greatly reduced the powder thickness and adhesion.

References

1. Makin B and Binns I 1985. J Electrostatics 16, 259-266
2. Makin B and W Wisniewski 1987. Oxford Inst. Phys. Ser. No. 85
3. Dastoori K PhD Thesis (in Prep), University of Dundee
4. Elder O BEng 1997 Thesis, University of Dundee
5. Stroud, K.A. 1987, "Engineering Maths", The MacMillan Press Ltd, ISBN 0-333-4487-1

Inst. Phys. Conf. Ser. No 163
Paper presented at the 10th Int. Conf., Cambridge, 28–31 March 1999
© *1999 IOP Publishing Ltd*

Surface charge distribution mapping of insulating materials

J L Davidson and A G Bailey

Department of Electrical Engineering, University of Southampton,
Southampton, SO17 1BJ, UK

Abstract. Electrostatic charge may build up on the surface of insulators in many industries due to the movement of insulating products over surfaces. This paper reports on the design of a computer controlled system which is able to measure the distribution of charge on the surface of an insulator. The capability of the system is demonstrated using the example of the well understood charge decay behaviour of polyethylene film on a ground plane. The experimental arrangement accurately monitors charge into the interior of polyethylene with no evidence of lateral spreading.

1. Introduction

Landers [1] has investigated the distribution of charge due to discharges from insulating materials. A similar study has been undertaken by Matsui *et al* [2]. Both studies have scanned the surface of a charged insulator using an induction probe prior to and after a provoked discharge. However, any charge migration or neuralization by the discharge not on the scan line could not be detected directly. In a related area, several studies have investigated the decay characteristics of the surface potential of corona charged polyethylene and PET films in some detail [3-5]. These studies have obtained complete two dimensional raster-mode scanning of charged dielectric surfaces and via computer logging have monitored the changes with time of the surface charge distribution. In this present study, a better understanding of electrostatic discharge from insulating materials is sought. To achieve this, it is necessary to map the surface charge density in a two dimensional way and to calculate the total charge present. To study this a computer based experimental arrangement is currently being developed.

2. Measurement of charge distribution across the surface of an insulator

The basis of the experimental design is a measurement of the electric field across a charged surface using a perpendicular movement of an electrostatic fieldmeter. The fieldmeter is fully guarded and positioned by means of a micrometer. The main consideration has been that a single scan should be performed quickly to prevent charge migration delay errors in the associated fieldmeter measurements. In the arrangement, the measuring probe can be stepped automatically over a spinning grounded turntable measuring 200 mm in diameter that supports the sample to be studied. The electric field generated by the surface charge is measured using a JCI 140F static monitor via a sensing aperture.

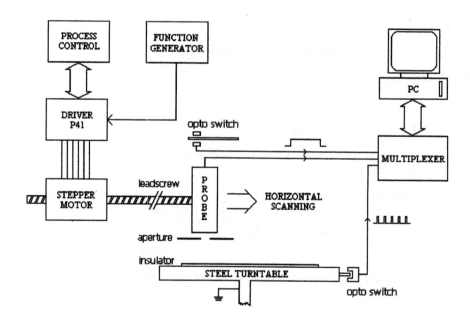

Fig. 1 Schematic of the experimental arrangement used for charge mapping.

Consideration has been given to the height of fieldmeter above the insulator. Since if the distance between the insulator and probe is large, charge accumulating on the edge of the insulator may effect the result. However, if the insulator to fieldmeter distance is too small, corona discharge may occur at high surface potentials. Therefore, the distance is chosen to be 3 mm in our experiment. The diameter of the sensing aperture measures approximately 3 mm. This arrangement should provide good averaging of any local variations in surface potential but sufficient resolution to study the behaviour of surface charge in detail.

The electrostatic field measurements used to calculate the surface charge density σ are recorded using a JCI 168 multiplexer via a Hewlett-Packard Brio PC. JCILOG4 software provides a simultaneous graphical display and storage of the input channels at a total of up to 400 readings a second. The stored fieldmeter readings can be converted into a surface charge density distribution by an analysis of the equivalent capacitor circuit and considering the probe to insulator distance z_1, the thickness of the insulator z_2 and the associated relative permitivity ε_2. In the case of an insulator on a grounded plane, the capacitance of the system is essentially composed of the insulator to ground capacitor which is in parallel with the fieldmeter to insulator capacitor. Analysis of this situation yields the result for the one dimensional case:

$$\sigma = Ez_1\left(\frac{\varepsilon_1}{z_1} + \frac{\varepsilon_2}{z_2}\right) \tag{1}$$

where ε_1 and ε_2 are the permitivities of the regions above and below the charged surface respectively.

3. Experimental

Initial experiments have concentrated on the decay characteristics of low-density polyethylene by recording simple line scans. The polyethylene sheet of thickness 30 μm and measuring approximately 10cm square was tightly bound to a grounded backing and initially charged by corona. The corona charging was performed using a single needle electrode suspended 6 mm above the surface and held at –4 kV for 45 seconds. The corona point was removed in a direction perpendicular to subsequent line scans with the HV still applied. This ensured no loss of deposited charge by inverse corona. Immediately after the cessation of charging the polyethylene film was scanned and the distribution of the surface potential determined at various times after charging. During scanning the velocity of the fieldmeter remained constant at 22 mm s^{-1} and the fieldstrength readings were recorded at 7 ms intervals. Calibration of the fieldmeter was provided by a separate experiment, replacing the polymer with a metal surface located at the same fieldmeter to sample distance and held at a known potential. All the measurements were carried out over normal laboratory conditions with no humidity control. However, as the purpose of these measurements was to evaluate the experimental arrangement and not to identify the exact nature of charge migration, the lack of close control over the environmental conditions was considered justified.

4. Results and Discussion

Figure 2 shows the surface charge density decay calculated using Eq. (1) for a polyethylene sheet initially corona charged as described. The initial corresponding maximum surface potential V_s is in the region of 800 V and falls exponentially to a value of 150 V over a period of 48 hrs. The decay constant of the central region can be measured to be approximately 1.5×10^4 s. This is in good agreement with the calculated value by assuming a volume resistivity of approximately 10^{15} Ωm and a relative permittivity of 2.4. Thus the decay constant can be estimated from $\varepsilon_0 \varepsilon_r \rho$ and is in the order of 10^4 s. The exponential decay is to be expected since the initial normal E field for the central region is given by V_s/d and is approximately 25 MV m^{-1}. Studies by Ieda et al [6,7] have experimentally shown the decay of surface voltage for E fields less than 30 MV m^{-1} occurs exponentially in accordance with Ohms Law.

Due to the tight coupling between the insulating layer and the underlying ground plane the recorded surface charge densities are high. Furthermore, by monitoring the decay curves it is seen that while the initial decay is relatively rapid, the material retains almost 25 % of the initial charge after 48 hrs. In fact even after significant decay, the remaining charge density is a factor of four times greater than the Gausssian surface charge limit ($\approx$25 μC m^{-2}) predicted for an insulator away from any ground influences.

It may be observed that the peak values of the surface charge densities decay without any significant lateral spreading of charge i.e. the width of the profiles remain unaltered over the entire measured time scale. This suggests a high normal component of the electric field with respect to the polyethylene surface compared with the associated tangential E field, resulting in an exponential bulk migration of charge to the grounded electrode. The absence of the lateral spreading of charge deposited on low-density polyethylene has also been observed by Das-Gupta [8].

442

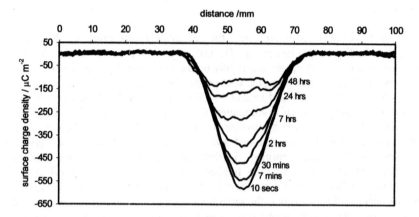

Fig. 2 Linescans of charged polyethylene surface as described in the text.

5. Summary

A purpose-built scanning system has been designed in which a fieldmeter is stepped automatically over an insulator such that a surface charge distribution results. The arrangement can be used to study the decay of surface charge over extended periods of time. Results obtained to date indicate that the system introduces a considerable degree of sophistication into the measurement of surface charge. The level of demonstrated resolution will prove useful in determining the partial relief of charge during a discharge process. Future work will assess the amount of charge relieved from various materials and relate this to the energy of the discharge and therefore to potential electrostatic hazards.

5. Acknowledgements

The authors wish to acknowledge the valuable contribution made by Mr D White in the construction of the experimental arrangement. We would also like to thank Dr J Chubb of John Chubb Instrumentation for his helpful comments regarding the electrostatic fieldmeter and the JCI software. The work is part of a programme supported by the Defence Evaluation Research Agency of the United Kingdom.

References

[1] E. U. Landers, *J. Electrostatics* **17** (1985) 59-68.
[2] M. Matsui, N. Murasaki, K. Fujibayashi and X. Wei, *J. Electrostatics* **35** (1995) 73-82.
[3] E. A. Baum, T. J. Lewis and R. Toomer, *J. Phys. D: Appl. Phys.* **10** (1977) 287-297.
[4] E. A. Baum, T. J. Lewis and R. Toomer, *J. Phys. D: Appl. Phys.* **10** (1977) 2525-2531.
[5] E. A. Baum, T. J. Lewis and R. Toomer, *J. Phys. D: Appl. Phys.* **11** (1978) 963-977.
[6] D. K. Das-Gupta, *J. Electrostatics* **23** (1989) 331-340.
[7] M. Ieda, G. Sawa and U. Shinohara, *Elect. Eng. Jap.* **88** (1968) 67-73.
[8] M. Ieda, G. Sawa and U. Shinohara, *Japan. J. Appl. Phys.* **6** (1967) 793-794.

Inst. Phys. Conf. Ser. No 163
Paper presented at the 10th Int. Conf., Cambridge, 28–31 March 1999

Experience with electrostatic fieldmeter instruments with no earthing of the rotating chopper

John Chubb

John Chubb Instrumentation, Unit 30, Landown Industrial Estate, Gloucester Road, Cheltenham, GL51 8PL, UK.
Tel: +44 (0)1242 573347 Fax: +44 (0)1242 251388 email: jchubb@jci.co.uk

Abstract In 1990 two novel designs of 'field mill' fieldmeter were described [1] which do not use the traditional arrangement with an earthed rotating chopper. The instruments described were based on 'constant capacitance' and 'back to back' fieldmeter operation. This paper reports the experience gained since 1990 in the operation of commercial instruments based on both the above principles. Also included are comments on features required for high performance and reliable long term continuous operation.
Experience has been that although the 'back to back' fieldmeter approach appears more complex, both mechanically and electronically the complexities can be handled without difficulty so that the significant benefits can be realised. The main benefits from avoiding the need to earth the rotating chopper are:
1) avoidance of wear of earthing brushes: which means long and quiet operational life, limited only by the motor drive and bearings
2) avoidance of drag of earthing brushes: which means easy operation at high rotational speeds so fast response fieldmeters are feasible
3) avoidance of the need for special precious metal alloys for the earthing brushes, the careful mechanical setting required and the progressive degradation of zero noise as brushes wear.
   To minimise the influence of contamination all surfaces within the sensing region need to be gold plated and gaps between surfaces need to be large to minimise the influence of dust and surface contamination and to avoid the risk of bridging by fibres. For operation in wet environments (for example for measurement of atmospheric electric fields) critical gaps need to be at least 6mm to avoid water bridging. Long surface insulation paths are needed.

1. Introduction

Traditionally electrostatic fieldmeter instruments have been based on the use of an earthed rotating chopper to modulate the observed electric field at a sensing surface [1,2]. This approach works well, but has a number of limitations for practical and commercial instruments. Making a good low noise earthing contact to a rotating shaft is not easy. No lubrication can be used and the contact wears. Wear can be minimised by using a smooth shaft of small diameter, by keeping the rotational speed and the brush contact pressure down. The simple approach is a thin spring wire pressing lightly on the side of the rotating shaft. For low noise a precious metal earthing brush contact is needed. The earthing problems become significant for instruments needed for long continuous operation (greater than several months) and for fast response (below say 10ms). At the high rotational speeds for fast response instruments a higher contact pressure is needed to avoid contact bounce – and this both exacerbates wear and increases the motor power required. Also, it may not be easy to mount and adjust the spring pressure of earthing brushes in small scale instruments.

In 1990 two novel designs of 'field mill' fieldmeter were described [2] which did not use the traditional arrangement with an earthed rotating chopper. The two instruments described were based on 'constant capacitance' and 'back to back' fieldmeter operation. This paper reports the experience gained since 1990 in the operation of commercial instruments based on both the above principles. Also included are comments on features required for high performance and reliable long term operation.

2. General fieldmeter instruments

The 'constant capacitance' fieldmeter approach is an arrangement of the rotor system that ensures the potential of the rotor assembly remains constant during rotation and is not affected by any nett charge on it. With this condition the signals observed at the sensing surface relate only to modulation of the observed external electric field by rotation of the rotor assembly. Achievement of a constant total capacitance during rotation of the rotor requires very accurate setting of the value of the sectored compensating capacitor to provide exact counterbalancing against the variation of rotor capacitance as it rotates behind the sectored sensing aperture. Critical to this is the absence of end float in the motor drive. Good mechanical stability of the whole sensing region is also important. Although over 100 compact handheld instruments were made with this mode of operation, it became clear that the approach was not well suited to easy and reliable manufacture. A change was hence made to the 'back to back fieldmeter' approach.

The 'back to back' fieldmeter approach is in principle two fieldmeters driven by the same motor with the two rotor assemblies electrically connected together. The 'secondary' fieldmeter is in a fully shielded enclosure. The secondary sensing surface signal arises from variation of the voltage of the rotor assembly by the variation of its capacitance as the primary chopper rotates behind the sectored sensing aperture. By backing off the signal of the primary fieldmeter by an appropriate fraction of the signal observed by the secondary fieldmeter signal it is possible to fully compensate for any nett charge held on the rotor assembly. A useful simplification for practical design was realisation that the function of the secondary fieldmeter is, in fact, just to observe the variation in voltage of the rotor assembly arising from variations in capacitance of the rotor as it rotates. The absolute level of the rotor voltage is not relevant. Figure 1 shows the basic arrangement for the rotor, sensing surfaces, motor drive and phase sensitive detection. In practical instruments the phase sensitive detection signal may be obtained by sensing rotation of a sectored secondary rotor surface or from a logic signal for some electronically commutated motors.

An original arrangement for a 'back to back' fieldmeter had the two sensing regions at either end of a motor with a double ended shaft. This worked, but the capacitance coupling of motor commutation noise signals to the shaft gave large common mode signals to the fieldmeter sensing circuits. These were difficult to null. Common mode commutation noise can be radically reduced by mounting the two rotors together on one side of the motor and by insulating the rotor assembly from the motor shaft. Having the two rotor assemblies together is mechanically much simpler. If arrangements are made to operate phase sensitive detection either from the motor or from the secondary chopper then single ended motors may be used – and this is convenient for compact instruments.

Infra red detectors are often, and conveniently, used for obtaining rotational information to drive phase sensitive detection circuits. Care needs to be taken, particularly with reflective opto detectors, to avoid interaction of such detectors with ambient radiation. This is only really a problem with strong illumination by tungsten lamps or from sunlight. Such interaction can upset operation of phase sensitive detection. To avoid the possibility of such problems a change was made to use of a reluctance magnetic pick-up.

The 'back to back' fieldmeter approach also requires absence of end float in the rotor assembly drive. The arrangement has proved easier to set up and more stable than the 'constant capacitance' approach. Immunity to charge on the rotor assembly is easily tested by adding charge to, and then earthing, the rotor assembly. This is used in both fieldmeter approaches for setting up instruments.

To minimise the influence of corrosion, and different electrochemical potentials, it is best to gold plate all surfaces in and around the sensing region. It is also advantageous to keep gaps between surfaces as large as compatible with other design requirements to minimise the influence of dust and surface contamination - and to avoid the risk of bridging by debris and fibres.

Experience has been that although the 'back to back' fieldmeter approach appears more complex, both mechanically and electronically, than the traditional field mill with an earthed rotor the complexities can be handled without difficulty and significant benefits can be realised. The main benefits from avoiding the need to earth the rotating chopper are:

1) avoidance of wear of earthing brushes: which means long and quiet operational life, limited only by the motor drive and bearings
2) avoidance of drag of earthing brushes: which means easy operation at high rotational speeds - so fast response fieldmeters are feasible
3) avoidance of the need for special precious metal alloys for the earthing brushes, the careful mechanical setting required and the progressive degradation of zero noise as brushes wear.

In indication of the performance achieved, it is noted that a fieldmeter with a sensing aperture diameter of 25mm has a noise level and ability to measure electric fields to within $20V\ m^{-1}$ pk-pk with a response time around 50ms. This capability allows measurement of surface potentials to better than 1V at 100mm. Response times down to below 3ms can be achieved, and these enable observations to follow 50/60Hz fields.

3. Fieldmeters for adverse environments

For operation in wet environments (for example for measurement of atmospheric electric fields) critical gaps need to be at least 6mm to avoid water bridging between horizontal plane surfaces. This requires an appropriately large sensing aperture to achieve sensible coupling of the external electric field to the primary sensing surface. Insulation for the sensing surfaces needs to be provided with suitably long surface tracking paths and, of course, the signal processing circuits need to be well protected from the environment [3].

A fieldmeter for long term continuous measurement of atmospheric electric fields (for a lightning warning system on St Kilda) was built using a sensing aperture diameter of 95mm (in an overall diameter of 101mm) and a four sector chopper with the arrangement shown in Figure 1. For long operational life the chopper assembly was driven by an electronically commutated motor. A logic signal from this motor provided the basic reference to synchronise operation of the phase sensitive detection circuits.

This instrument provided the opportunity to measure electric fields up to $2000kV\ m^{-1}$ with noise around $1\ Vm^{-1}$ on the most sensitive range. Calibration [4] remained very stable.

To justify confidence for observations in adverse conditions, such as long term monitoring of atmospheric electric fields, a system was developed for continuous operational health monitoring. This involved modulating the potential of the whole fieldmeter assembly, relative to the local earth, by a low amplitude square wave signal of known voltage excursion. The modulation of observations was compared with expectations, taking account of any general local slope of the variation of observations. Immunity to environmental conditions was shown by the ability to make continuous measurements through periods of heavy rain

446

with an upward facing instrument. The utility of the operational health monitoring facility was demonstrated by detection of a fine spider's web across the sensing aperture during atmospheric electric field measurements!

4. Conclusions

Experience has shown that the 'back to back' design of fieldmeter, devised in 1990 and developed since then, is suitable for use in practical commercially manufactured instruments. It has proved an approach that offers good electric field measurement capability - in terms of low noise, stable zero, accurate measurement of electric field over a wide range and opportunity for fast response.

References:
[1] P. E. Secker *"The design of simple instruments for measurement of charge on insulating surfaces"* J. Electrostatics 1 1975 p27

[2] J. N. Chubb *"Two new designs of 'field mill' fieldmeter not requiring earthing of rotating chopper"* IEEE Trans Ind. Appl. 26 (6) Nov/Dec 1990

[3] I. E. Pollard, J. N. Chubb *"An instrument to measure electric fields under adverse conditions"* 'Static Electrification 1975' Inst Phys Confr Series 27 p182

[4] *"Methods for measurements in electrostatics"* BS 7506: Part 2: 1996

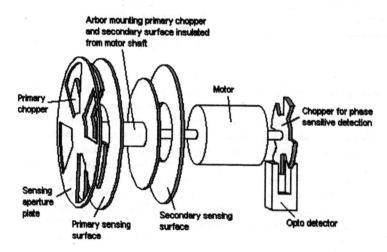

Figure 1: Basic arrangement for 'back to back' fieldmeter

Inst. Phys. Conf. Ser. No 163
Paper presented at the 10th Int. Conf., Cambridge, 28–31 March 1999

Second Generation of Electrostatic Powder Coating Diagnostic Instrument

B. Makin and K. Dastoori

Department APEME, University of Dundee, Dundee, DD1 4HN

Abstract: A second generation of EPCDI (Electrostatic Powder Coating Diagnostic Instrument) based on PC control has been designed at the University of Dundee to analyse the electrostatic powder coating deposition efficiency by measuring the powder adhesion properties in the pre cured state. The first generation instrument operates with a BBC computer using an 8 bit ADC. Interfacing this system to a PC will have several shortcomings in the hardware and software. The second generation runs under the Pentium PC control, with 12 bit ADC card mounted within the PC. There are several advantages in the hardware and software. These include: -

- The instrument incorporates an improved pressure regulator that is used to control the air pressure.
- A constant distance is kept between the IR-LED and the photodiode.
- The interface electronics has also been improved to give more accurate results. The light signal is very carefully screened from noise due to other mechanical components and careful use of amplifiers has greatly reduced the drift that nullified much of the previous results.

1. Introduction

A metal coated glass substrate is electrostatically powder coated and small amounts of powder are selectively removed by a pulsed air jet under computer control. The thickness of the remaining powder layer is determined by measuring the transmission of an infra red light beam. Results from the instrument are presented for different types of electrostatic guns which are used with industrial coating installations. The data is analysed to optimise the coating conditions to ensure maximum deposition efficiency. The EPCDI (Electrostatic powder coating diagnostic instrument) is thus split into modules.

i. The mechatronic module - This contains the mechanical devices responsible for changing and monitoring the sample position, varying the air pressure and sealing the probe against the sample surface.

ii. The electronic interface module - This is split into two parts. It contains the circuitry used to convert the small computer signals to voltages able to move the mechanical components. It also houses the circuitry used to create the light signal and detect the transmitted light signal.

iii. The computer module - The computer has an interface card mounted inside it. This card can be controlled using software to measure incoming signals and send out signals to the interface circuitry. The software used is a mixture of C language and that written specifically for the diagnostic and some commercially available packages.

448

Figure 1. shows the two halves transmitter and receiver sections of the probe assembly as shown in figure 2. These contain the IR led, the photodiode and the air supply from the valve.

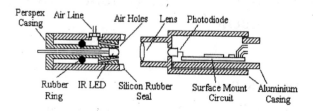

Figure 1 General Assembly of Optical / Pneumatic Module

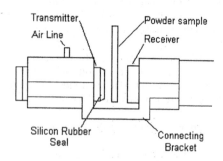

Figure 2 - Probe Assembly

2. Testing and Results

The sample slides are checked to make sure they are clean before being placed in the jig and coated evenly using either a corona or tribo gun. The samples are immediately taken to the powder coating diagnostic instrument for testing. Before carrying out any tests on the powder samples the diagnostic instrument was checked for the mechanical components using option 1 on the software main menu and check the light transmission reading for an empty sample holder is around 100%. The software tests the thickness of the powder coating at 43 sites over the sample. The limits of the sample's movement are defined by the positioning of the two optoschmitt devices that act as end stops. When the run scan option is chosen:
The sample position is reset to the right end stop. A calibration light reading is taken in the first position. The calibration value is taken as the maximum light that could be transmitted through the slide. The first light reading is taken 6 steps to the left where the powdered area of the slide begins; the value is displayed on the screen. The stepper motor moves to the left one step and the next light reading is obtained and displayed. This process is repeated until the sample position reaches the left limit of travel. When the sample reaches the left limit it is returned to the right end stop. The screen displays the 43 light readings, along with the average and the range of values, graph is then plotted of step number v. light reading and returning to the main menu. The adhesion test is run under software control. This tests the adhesion of the powder to the sample at 4 sites on the slide. This is done by sealing the probe against the surface of the sample and measuring the changing powder thickness as a series of short air puffs removes the powder coating. The pressure of these air puffs has been previously set. There are two types of adhesion tests, the fixed and the variable pulse width [1,2]. These two tests will give quite different results.

2.1 Scanner Accuracy

It was important to check that the changes in the interface circuitry had produced a more accurate light transmission reading. Previous versions of the diagnostic instrument had an accuracy of round 3-10%. This had meant that the scan readings were quite meaningless, because large variations in light transmission levels were not due to a slightly inconsistent powder coat thickness, but to poor readings. The scanner was checked by running the scan function with no sample in place. A perfect measurement would give a consistent result of 100%. The plot given in figure 7, shows the result of five such scans. The figure is scaled between 99.8% and 100% to highlight the errors. The average reading over the five scans was 99.94% light transmission. The average range of results of the five readings at every point was 0.06%. This was a massive improvement over earlier results and showed the diagnostic was quite capable of producing accurate powder thickness comparisons.

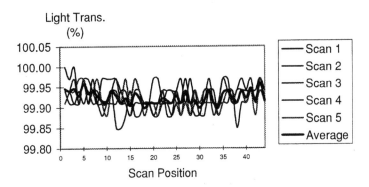

Figure 7. - Scan results with no sample in place.

2.2 Tribo Gun Testing

(a) Variations in thickness

The decreasing powder thickness approximately follows the order in which they were sprayed see figure 8. The EPCDI confirms that the problem experienced with most tribo guns. The inside surface of the gun barrel gradually becomes coated in powder, reducing the ability of the gun to charge the powder, resulting in thinner coats. It is unclear when the charging ability of the gun will stabilise. The results do show the limitations of the tribo gun and also show the need to let the gun spray for at least a few minutes before any calibration tests are made on the powder layer.

(b) Adhesion Measurements with Varying Air Pressure

The adhesion of the tribo gun powder layer was tested. Figure 9 shows the change in adhesion value [5][6] with varying air pressure. A constant pulse width of 50ms was used for the air pulse. As expected, the adhesion of the powder layer dropped linearly with increasing air pressure. Each air pressure value was repeated at least twice to prevent the result being greatly influenced by the order the samples were sprayed. The adhesion results for each pressure value were quite consistent.

450

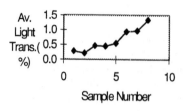

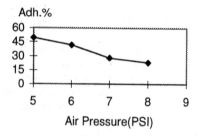

Figure 8 - Variation in powder thickness v. order sprayed

Figure 9 - Tribo powder layer adhesion with varying air pressure

2.3 Powder Coat Thickness with Varying Corona Gun Voltage

The powder coat thickness is expected to increase with increasing powder charge and hence with increasing gun voltage. While inconsistencies in the test results make firm conclusions difficult, experiment results did indicate that the thickest powder samples were usually made using a gun supply of 50kV. Figure 11 show the results from two experiments that seem to support this conclusion.

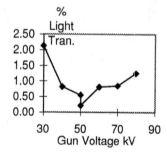

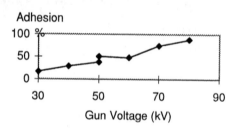

Figure 11- Powder coat thickness with varying corona gun voltage

4. Conclusions

A second generation of EPCDI(Electrostatic Powder Coating Diagnostic Instrument) has been developed and tested successfully. There are many variables that can affect the powder coat thickness and adhesion. The difficulty in obtaining consistent results could be attributed to a number of environmental factors. The accuracy of EPDCI (Electrostatic Powder Coating Diagnostic Instrument) was demonstrated using Tribo, Corona and Super Corona guns.

References

1. Makin B and Binns I 1985. J Electrostatics 16, 259-266
2. Makin B and W Wisniewski 1987. Oxford Inst. Phys. Ser. No. 85
3. Dastoori K PhD Thesis (in Prep), University of Dundee
4. Elder O BEng 1997 Thesis, University of Dundee
5 Makin B and Dastoori K 1999 Cambridge Inst. Phys "Using Electrostatic corona gun aand EPCDI to measure the thickness & adhesion of powder coated substrate"
6. Stroud K A 1987, "Engineering Maths", The MacMillan press ltd, ISBN 0-333-4487-1

Inst. Phys. Conf. Ser. No 163
Paper presented at the 10th Int. Conf., Cambridge, 28–31 March 1999

A System For Obtaining High Resolution Macroscopic Surface Charge Density Distributions On Contoured Axi-Symmetric Insulator Specimens

D.C. Faircloth N.L. Allen

Ferranti Building, Department of Electrical and Electronic Engineering, UMIST, Manchester, M60 1QD.

Abstract. This paper describes the method of operation of a system for measuring high-resolution surface charge density maps on practical insulator specimens using an electrostatic probe. Apparatus is described that can scan the electrostatic probe over the surface of the insulator and record the probe signal. The process of obtaining a charge density map from the probe signal measurements is explained. An example charge distribution is shown.

1. Introduction

Atmospheric air provides the basic insulation for many practical high voltage installations such as transmission lines, switchgear etc. However, the high voltage conductor in any such system must be mechanically supported, and the only practical solution is to use solid insulation. This introduces an insulating surface between the high voltage conductor and ground. An insulators surface is intrinsically the weakest part of the solid-gas insulation system. Thus, physical knowledge of the insulating properties of solid insulation surfaces is very important. A major factor which influences surface discharges is the charge deposited on the surface of the solid insulation. This has led to the development of the surface charge scanning apparatus described in this paper. The apparatus is capable of measuring high-resolution charge density maps on all or part of almost any contoured axi-symmetric insulator specimen.

2. Surface Charge Density Measurement

2.1. Electrostatic Probe

The electrostatic probe principle [1] is used, where charge on the surface induces a voltage on the centre conductor of a coaxial probe positioned above the surface. The outer conductor is grounded and the voltage induced on the centre conductor is measured via a very high input impedance ($>10^{15}\Omega$) op-amp. The outer diameter of the probe is 2.9mm and the diameter of centre conductor is 0.6mm. A multi-point measuring technique is employed, where the probe voltage is recorded at many points above the insulator surface.

2.2. Calibration Problem

The probe voltage measurements must be converted into surface charge density measurements. Early investigators [1] used a simple capacitive method in which each value was linearly related to a charge density. This technique yields limited accuracy and makes no account for the probe's response to neighbouring charges on the surface. In recent years techniques have evolved to model the probe response to distant charges [2]. The technique used here is an adaptation of Pedersen's λ-function [3]. He related the Poisionian charge (q) induced on the probe to the surface charge density (σ) on a surface element: $q = \lambda\sigma$.

The technique employed here relates contribution to the total probe voltage (v) directly to the surface charge density (σ) on a surface element: $v = \phi\sigma$, where ϕ is the constant of proportionality measured in $Cm^{-2}V^{-1}$. Each element has a different associated ϕ-value depending on its distance from the probe. The total probe voltage (V) is the sum of the contributions from all the elements of surface charge: $V = \Sigma v = \Sigma\phi\sigma$. For the probe in one particular position the ϕ-values for all the elements on the surface make up the probe response function. The probe has a different response function for each voltage measurement position.

2.3 The Φ-Matrix Technique

The surface area is divided into elements as shown in figure 1. The elements do not have to be square and there does not have to be an equal number of horizontal and vertical divisions.

The probe voltage in position (i,j) is given by:
$$v_{ij} = \sum_{n_y}^{y=1}\left[\sum_{n_x}^{x=1}\phi_{ij}(xy)\,\sigma_{xy}\right]$$

where, $\phi_{ij}(xy)$ is the value of the probe's response function to charge at position (x,y) for the probe at position (i,j) and, σ_{xy} is the surface charge density on the surface element at position (x,y). This is a first order function of the $n_x n_y$ surface charge densities.

There are $n_x n_y$ probe voltage measurements in total and each of these voltages is a function of $n_x n_y$ surface charge densities. The problem is reduced to the solution of $n_x n_y$ simultaneous equations, which is solved using the matrix inversion technique.

The measured probe voltages and the unknown charge densities are grouped to two vectors $\overline{V}$ and $\overline{\sigma}$. They are related by the matrix equation: $\overline{V} = \overline{\sigma}\,\Phi$
where, Φ is a matrix containing all the ϕ-function values that are coefficients of the simultaneous equations. Hence the unknown charge density's can be found from: $\overline{\sigma} = \overline{V}\Phi^{-1}$

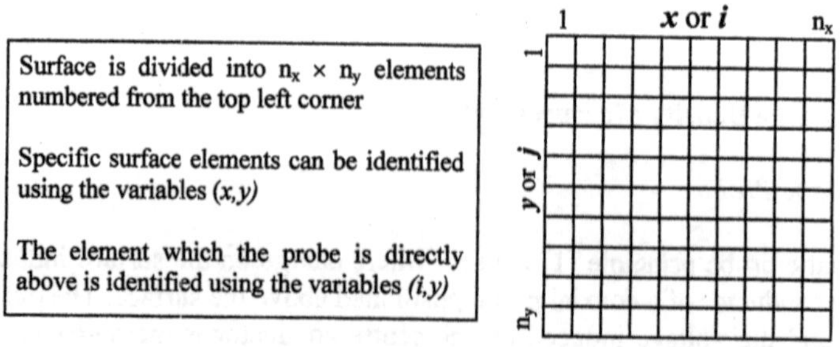

Surface is divided into $n_x \times n_y$ elements numbered from the top left corner

Specific surface elements can be identified using the variables (x,y)

The element which the probe is directly above is identified using the variables (i,y)

Figure 1: The division of the surface

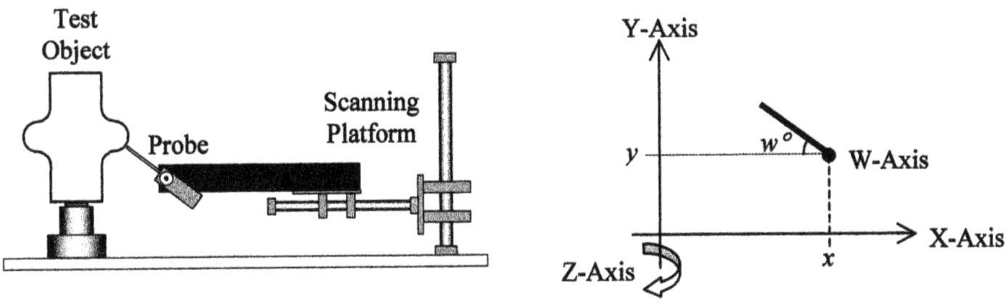

Figure 2: The scanning apparatus and the four axes that define the position of the probe above the surface.

3. The Scanning Apparatus

The probe is positioned using 4-stepper motors, as shown in figure2. The scanning platform was designed and constructed at UMIST. A computer controls the whole scanning system using specially written software. The contour of the test object can be entered into the computer manually or the system can measure the shape of the test object automatically using a spring-loaded sensor. The computer then calculates the probe positions to scan the surface. The probe must be kept perpendicular to, and a constant distance from, the surface. The surface is divided up into measurement points at which the probe voltage is recorded.

The surface is scanned in layers by rotating the test object. After each rotation the probe moves to the start position of the next layer until the whole surface has been scanned.

As the probe moves over the surface the probe voltage is measured at each point and stored in a file on the computer. All the settings for the scanning system are also stored in the same file; this allows the user to recall measurement parameters at a later date. The software has the facility to graphically display the scanned in voltage measurements and monitor each layer as it is scanned in. The scanning system has been developed as a versatile, easy to use piece of equipment with many useful facilities to help the user.

4. Implementing The Calibration Technique

The probe's response functions are found using a 3-D electrostatic field solver, key values are found, then a computer generates the Φ-matrix by quadraticaly interpolating any missing ϕ-values.

The matrix is inverted in Matlab running under UNIX on a Fujitsu AP3000. The solution of the charge distributions is automated using script files; files generated by the scanning system are read in, the required Φ-matrix is automatically generated and the charge distributions are solved and saved, all without any user intervention.

The solution time increases with the resolution of the surface. As the number of surface divisions increases so does the size of the matrix required to solve the charge distribution. The amount of computer memory required increases as n^4 where n is the 1-dimensional division of the surface (i.e. a surface is divided into $n \times n$ elements). For example a surface scanned to a 100×100 resolution requires a 10,000×10,000 Φ-matrix which needs 800MB of RAM and takes just under 24 hours to solve. The amount of RAM required increases very quickly with resolution; for a 150×150 surface division 4GB of RAM is required.

454

5. Example Surface Scans

To demonstrate the resolution of the system, a scan is made of the corona produced by a single -10kV 1.2/50μs impulse applied to a point electrode 1mm above the surface of a cylindrical PTFE insulator specimen. The effect of calibration using the Φ-Matrix technique is clearly visible by the differences between figures 3 and 4. After the surface was scanned a dust figure was obtained (figure 6) using black photocopier toner which adheres to positive charge. The dust figure compares favourably with the charge density contour map (figure 5).

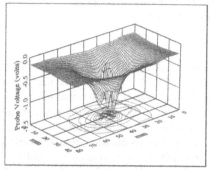

Figure 3: The measured probe voltage

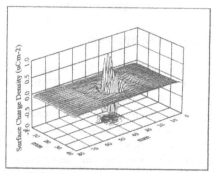

Figure 4: The calibrated surface charge density

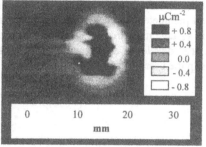

Figure 5: Surface charge contour map

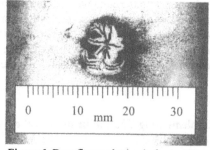

Figure 6: Dust figure obtained after surface had been scanned.

6. References

[1] D.K. DAVES, "The Examination of the Electrical Properties of Insulators by Surface Charge Measurement", Journal of Scientific Instrumentation, Vol. 44, pp. 521-524, 1967.

[2] T.O. Rerup, et al, "Using the λ Function to Evaluate Probe Measurements of Charged Dielectric Surfaces", IEEE Trans. Dielectrics EI, Vol. 3, pp. 770-777, 1996.

[3] A. Pedersen, 'On the Electrostatics of Probe Measurements of Surface Charge Densities', Gaseous Dielectrics V, Pergamon Press, pp. 235-240, 1987.

Inst. Phys. Conf. Ser. No 163
Paper presented at the 10th Int. Conf., Cambridge, 28–31 March 1999

Electrostatic mass air flow meter

W.Balachandran, W. Machowski, W. Horner

Department of Manufacturing and Engineering Systems
Brunel University, Uxbridge UBX 3PH, U.K.

Abstract. The paper presents the designing and constructing of an electrostatic air flow meter capable of operating in an environment with wide range of humidity levels. A revised model of operation for the electrostatic air flow meter is proposed. The new model takes into account the effects of the space charge on the electric field inside the meter, resulting in a measurement that is insensitive to the air humidity. The results of the tests carried out in humidity controlled environment (40-70%) support these findings.

1. Introduction

The electrostatic air flow meter is an instrument that uses the displacement of ions in an air flow to measure the mass flow rate of the air. The meter could be used in aerospace, automobile industries as well as for air conditioning applications.

There are several methods presently being used for air mass flow measurement. The most common method is the use of a hot wire (sometimes known as hot flume). This method works measuring the rate of cooling of a hot wire placed in the air flow. This is achieved by measuring the current required to keep the wire at a constant temperature. Other older methods include the measurement of mass flow by measuring the rotation of a mechanical flap placed in the flow. The main disadvantages of all the current methods is that none can measure the reverse flows that can be present and none are responsive enough to changes in flow rate.

There have been several previous attempts to produce a meter based on mobility of ions [1,2,3]. These attempts appear to have encountered measurement in particular the sensitivity of the meter to humidity variation.

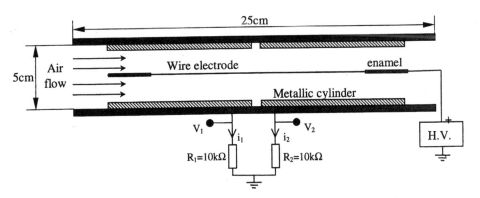

Fig.1 Schematic diagram of the electrostatic mass flow meter used in the study

2. The Flow meter

A schematic diagram of the meter is shown in Fig.1. This meter is based on a 0.125mm diameter Platinum/Rhodium alloy (ratio 9/1) wire as the corona source. This wire was maintained in the geometrical centre of the cylindrical electrodes using a PTFE frame. The production of ions was prevented at the ends of the wire by the use of an anti corona enamel. This enabled the electric field to be constant along the whole length of the meter so that the ions always travel through a uniform electric field. The ions are collected by means of two cylindrical electrodes around the inside wall of the pipe. The measurement can then be made in terms of the current difference between the two electrodes using the following equation:

$$\dot{M} = m\frac{i_2 - i_1}{i_2 + i_1} = m\frac{\Delta i}{i_{total}}$$
1

where: $\dot{M}$ - is mass flow rate of air, m - is calibration gain constant, i_1 and i_2 are measured ionic currents captured by the first and the second electrode respectively.

3. Modelling the Operation

Research to date has tended to model the operation of the meter based on the assumption of a single ion travelling through free space. Since the operation of the meter relies on the ionisation of the large volume of air, the cumulative field effect of all the ions present (the space charge) will add to the electric field from the corona source. Hence, the ions will be accelerated in a different manner to how the original model of operation predicts.

The space charge creates an electric field in radial direction from the wire (r) as well as in the direction of air flow (z). In order to model the electric field due to the ions generated, it is therefore necessary to consider how the ions are distributed as they travel through coaxial tube arrangement in the non uniform electric field.

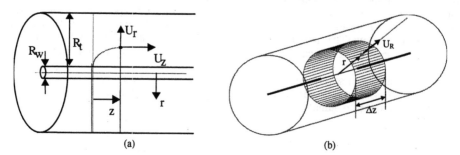

(a)   (b)

Fig.2 Diagram of (a) single ion trajectories, and (b) space charge distribution inside the tube

Firstly, the trajectory of the single ion with known mobility in (r, z) plane, as shown in Fig.2a was calculated. Well known equations describing the axial velocity U_Z of a laminar air flow in the coaxial pipe and radial velocity U_R due to the influence of electrical field were used for this purpose. In the electrostatic air flow meter, the cumulative electric field of the space charge adds to the electric field from the corona wire. The distribution of space charge is visualised by replacing the single ion with a cylinder of ions centred around the wire. This cylinder of charge travels in a radial direction out from the wire, as shown in Fig.2b, with a radial velocity U_R dependent on the electric field. It is assumed that the wire is of sufficient length for the effects of the electric field at the end of the wire to be negligible.

Since the imaginary cylinder represents a number of charged ions, travelling at a certain velocity U_R then the total corona current, i_{total} may be calculated using the following equation:

$$i_{total} = 2\pi r \, \Delta z \, \rho_I U_R \qquad\qquad\qquad\qquad 2$$

Where: ρ_I – is ion density, Δz - is the length of the cylinder. This equation assumes that there will be a constant density of ions at the cylinder surface. Using the ion mobility relationship $U_R=kE$, where k is the ion mobility, the corona current can be used as a measure of both the electric field present and the ion mobility. This implies that the total current, i_{total} is dependent on the ion mobility, which varies with humidity. This leads to the conclusion that it is possible to develop a model of operation which can compensate for humidity variations.

A numerical model of the meter was developed where the velocity of an ion through the air gap is used to determine the displacement of the ion due to the air flow. The velocity of each ion can be separated into two components – the radial velocity due to the electric field and the velocity in the direction of air flow due to the ions being carried by the air. The radial velocity U_R can be expressed in terms of the total collected current from equation 2. The current difference Δi measured between the two electrodes can become a direct measurement of the air mass flow rate, regardless of the total corona current or the electric field. The measurement can be made directly for any corona voltage using the following equation:

$$\dot{M} = m(\Delta i + c) \qquad\qquad\qquad\qquad 3$$

where: c – is calibration offset constant. The gain constant m should remain constant for all electric field strengths.

The electric field related to ions space charge also affect ion trajectory by repelling them in the direction of air flow. Hence, there will be an error in the measurement because the current difference measurement will contain both a displacement component due to the air flow and a drift component due to the other ions. However, at higher wire potentials the ion drift velocity will become less significant in comparison to the radial velocity. It can therefore be expected that ion drift will tend to become insignificant at the higher electric field strengths.

4. Experimental set-up and results

A small flow rig capable of delivering air flow up to 33g/s located in an enclosed space was used in the experimental study. The level of relative humidity was varied from 40% to 70%. The measurement was obtained by taking the current measurement directly from each electrode into a computer based data acquisition system.

In order to establish the values of gain constant m and calibration offset constant c, the meter was calibrated at various operational conditions. The potential applied to the wire electrode ranging from 6kV to 10kV while air mass flowrate was maintained within 8g/s to 33g/s range. The results indicate that the measurements can be performed with 1% accuracy with respect to the maximum flow rate.

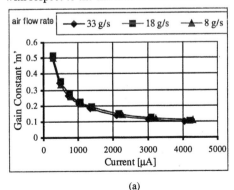
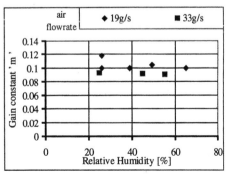

(a) (b)

Fig.3. Gain constant as a function of total current

a) effect of the space charge; and b) effect of air humidity; on the gain constant 'm'

458

The effects of varying the electric field applied to the wire and hence the influence of space charge density on the gain constant m, are shown in Fig.3a. As the electric field strength increases, hence the total current measurement also increases, the calibration gain constant tends to converge to a constant value. Therefore the ion drift becomes less significant at the higher electric field strengths. The results shown in Fig.3b are taken at a constant electric field strength (potential applied to the wire electrode – 7kV) and constant flowrate of 19g/s and 33g/s, while varying the air humidity. It would appear from these results that the meter is still sensitive to humidity variation, despite the revised model.

However, the humidity effects were measured at varying electric field while keeping the mass flow rate constant. These results are shown Fig.4a and Fig.4b, where the calibration gain constant is shown plotted against the total measurement current for two air flow rates: 19g/s and 33g/s.

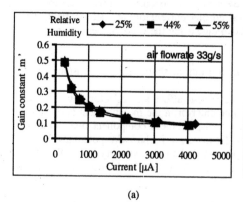
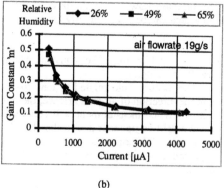

(a)   (b)

Fig.4. The effect of relative humidity on the gain constant m measured at potential varied from 7 to 12kV and
a) 33g/s, b) 19g/s of air flow

The results obtained indicate that the gain constant is only marginally influenced by the humidity variations, especially for higher flowrates.

This discrepancy in measurement results can be explained as follows. An increase in relative humidity should produce a decrease in the ion mobility which directly affect the current. As the ion mobility decreases, the total current measurement also decreases. In an ideal situation where the ion drift is negligible (i.e. at high wire potentials), variations in the total current should automatically compensate for ion mobility variations. Hence the measurement will become independent of humidity.

The properties of the ion drift discussed earlier are a major feature of the results. At the lower electric field strengths the ion drift becomes more significant and the calibration gain constant increases. However, the characteristic curve of the corona current against the gain constant will remain identical. This explains why the measurements from the varying electric field investigation appear to be independent of humidity. At the lower electric field strengths, variations in current have the effect of producing a small variation in the gain constant. Therefore, the calibration of the meter will be sensitive to small variations in ion mobility. At higher electric field strengths the measurement will become insensitive to humidity variations because the ion drift tends to become insignificant. Therefore, the calibration gain constant will converge towards a constant value.

5. Conclusions

A novel method of operation of the electrostatic air flow meter was developed which enable to operate this within a wide range of relative humidity. This was achieved by producing a revised model of operation that incorporates the additional electric field from the space charge created by ions.

Making the measurement independent of humidity represents a significant improvement in performance and may make the meter viable for various industrial applications. The wide dynamic flow range of the meter and the ease of varying the flow range by varying only the high voltage supply output may open other potential applications.

6. References

[1] Asano K., Higashima Y., Katsuzuka K., Urayama K.. *Ion-Flow Anemometer Using High Voltage Pulse.* Conference Record, IEEE/IAS Annual Meeting, 1990, pp: 802-808

[2] Cockshott C.P., Vernon J.P., Chambers P. *An Air Mass Flowmeter For Test Cell Instrumentation.* IEE Conference Publication, 1983. No. 229, pp: 20-26.

[3] Cops M.H., Moore J.H.. *Electronic Fuel Injection system utilising corona discharge air mass flow transducers,* Lucas Engineering Review. Apr. 1978 Vol. 7, No 2, pp 30-36.

[4] Malaczynski G.W., Schroeder T. *An ion-drag air mass-flow sensor for automotive applications,* Conference record - IAS Annual Meeting. 1989, pp: 2196-2202.

[5] Miller C.G., Loeb L.B. *Positive Coaxial Cylindrical Corona Discharges in Pure N_2O_2 and Mixtures Thereof.* Journal of Applied Physics, 1951. Vol.22, No. 4. pp: 494.

[6] Miller C.G., Loeb L.B.. *Negative Coaxial Cylindrical Corona Discharges in pure N_2, O_2 and mixtures thereof.* Journal of Applied Physics, 1951, Vol. 22, No. 5, pp 614.

Inst. Phys. Conf. Ser. No 163
Paper presented at the 10th Int. Conf., Cambridge, 28–31 March 1999

In-line Continuous Flow Measurement of Pneumatically Conveyed Solids Using Electrostatic Sensors

Y Yan   J Ma

Advanced Instrumentation and Control Research Centre, School of Engineering, University of Greenwich, Medway Campus, Chatham Maritime, Kent ME4 4TB, UK

Abstract. Electrostatic sensing technology in conjunction with advanced signal processing techniques offers a promising solution to the in-line continuous measurement of the mass flow rate of pneumatically conveyed solids. This paper describes the fundamental principle, advantages and limitations of the technology and examines the effects of variations in solids velocity, particle size and material type on the measurement of mass concentration of solids. Key aspects of the optimum design of electrostatic sensors and associated signal processing elements are discussed.

1. Introduction

In-line continuous measurement of mass flow rate of pneumatically conveyed solids plays an increasingly important role to achieve increased productivity and improved product quality and process efficiency. It has been recognised that the mass flow metering of solids in pneumatic pipelines is a technically challenging task. Amongst all the technologies being proposed, the electrostatic sensors combined with advanced signal processing algorithms offer the most promising solution to this long-standing industrial problem [1-3]. It has long been known that particulate solids in a gas stream carry a certain amount of net electrostatic charge due to particle-particle collision, particle-wall impact and particle-gas friction. The charge on the particles can be detected by an insulated electrode in conjunction with a suitable electronic circuit. This apparently simple approach has attracted many academics and engineers all over the world to develop electrostatic sensors for solids flow metering. Although substantial time and effort have been spent, real advances in the subject area are very limited and many issues remain to be examined. This paper describes the fundamentals, advantages and limitations of the technology and examines the effects of variations in solids velocity, particle size and material type on the measurement of mass concentration of solids.

2. Electrostatic Sensing Mechanism

The movement of a particulate material in a pipeline or other elements of process plant generates electrostatic charge. It is understood that the amount of charge on the particles depends on many factors, including physical and chemical properties of the particles (particle size, moisture content, chemical compositions, precharge, etc.) and surrounding environment (humidity, pipeline roughness, gas velocity, etc.). Although the amount of charge carried on the particles is usually unpredictable, the charge can be detected by an insulated electrode with a signal processing circuit, which derives an 'electrostatic' signal from the fluctuations in the electric field caused by the passage of the charged particles.

462

If the electrode is embedded in the pipeline or coated with an electrically non-conductive material resulting in no direct contact between the particles and the electrode, the sensing interaction will be pure electrostatic induction. In contrast, if the electrode such as a rod is exposed directly to the particulate fluid, direct charge transfer due to the contact between the particles and the electrode can take place. The signal derived in such a way is often referred to as *triboelectric* signal. In practice, whether an electrostatic sensor produces an inductive signal, a triboelectric signal or a combination of both depends dominantly upon the structure, geometry and dimensions of the sensor. In general, if the axial dimension of an exposed electrode is small compared to the pipe size, electrostatic *induction* will be the dominant interaction. Since the electrostatic signal derived from the sensor consists of AC and DC components, the signal processing electronics can be designed to provide either AC and/or DC components of the signal. Since the electrostatic signal depends on a wide range of parameters which are often ill-defined and varied irregularly, it is therefore very difficult to obtain an absolute measurement of the mass concentration of solids. However, the technology is highly suited for the indicative monitoring, trending or balancing of pneumatically motivated solids and providing useful signals for the solids velocity measurement using advanced signal processing algorithms.

3. Measurement Strategy

Fig.1 shows a schematic diagram outlining the overall measurement strategy using an electrostatic sensor with ring-shaped electrodes. The solids velocity is obtained by determining the time difference between the two electrostatic signals derived from a pair of axially spaced electrodes, whilst the relative mass concentration is inferred from the RMS magnitude of one of the electrostatic signals. The mass flow rate of solids ($\dot{M}$) is then calculated from the following equation:

$$\dot{M} = A \rho_m \upsilon \tag{1}$$

where A is the cross-sectional area of the pipe (m^2), ρ_m is the mass concentration of solids (kg/m^3) and υ is the solids velocity (m/s). It should be noted that, due to the relative nature of the electrostatic sensing technology, the sensor could only provide relative mass concentration and hence relative mass flow rate of solids [1].

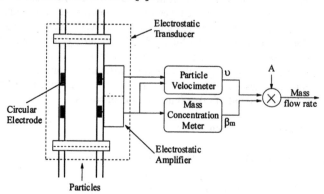

Fig.1. Structure of an electrostatic mass flow monitor

4. Sensor Design and Signal Processing

Many different geometrical shapes of electrodes can be used as a sensing element. Fig.2 shows typical examples of the sensing arrangement. The first three types are non-restrictive in nature and can be implemented either in an exposed or embedded arrangement as required.

From a practical point of view, the shape and dimensions of the electrode can have a significant effect on the overall cost, installation and maintenance of the sensor. A ring electrode has certain advantages over other forms in that it is less susceptible to vibrations and able to average the flow streams circumferentially and has a higher overall sensitivity than other non-restrictive types of electrodes of the same axial width. However, the cost of a large-scale electrostatic sensor with ring electrodes and its associated installation cost would be exceedingly high, making it unattractive in many applications such as coal-fired power stations. In contrast, a rod or a bar electrode has the advantages of low cost and easy installation, making them highly suited for applications in large-scale processes.

| stud | quarter-ring | ring | rod | bar |

Fig.2. Different geometrical shapes of electrodes

For a given pipe size, the shape and the axial dimension of the electrode are two crucial factors affecting the fundamental characteristics of the sensor, including *spatial sensitivity*, *sensing volume* and *spatial filtering effects*. For example, the electrostatic sensor with a ring electrode of a small axial width exhibits: (a) an axially symmetrical sensing field with higher sensitivity to particles closer to the pipe wall; (b) a sensing volume stretched axially well beyond the physical dimension of the electrode along the central axial lines permitting more particles around the centre of the pipe to be detected than those moving closer to the pipe wall.

In general, any electrostatic sensor exhibits some form of low-pass filtering effects and can therefore be regarded as a low-pass filter. The bandwidth of the filter should be maintained such that the cross-correlation velocity measurement can achieve steady and repeatable readings [1]. The bandwidth (B) of an electrostatic sensor with a narrow non-restrictive electrode can be estimated by the following equation [4, 5]:

$$B = \frac{\upsilon}{W} \tag{2}$$

where W is the axial width of the electrode. Equation (2) indicates that the bandwidth of the signal derived from a non-restrictive sensor varies with the velocity of solids. However, if a restrictive electrode is used, the bandwidth of the signal is expected much wider than the above estimation as the collisions between the electrode and particles produce higher frequency components depending upon the velocity and size of the particles. The original wide-band electrostatic flow noise due to the two-phase flow turbulence is transformed into an electrical signal via the electrostatic sensor before reaching an amplifier. Due to the low-pass nature of the sensor, some high frequency components are inevitably lost in the sensing process. The amplifier must be designed such that it has a suitably wide bandwidth, permitting all the information contained in the sensor signal to be conveyed to the next stage for further processing.

5. Experimental Investigation
One of the major problems in applying the electrostatic technology for mass concentration measurement is that the magnitude of the electrostatic signal depends not only on mass concentration but also on solids velocity. Experiments with a commercial electrostatic sensor

464

have been conducted to quantify the effect of variations in solids velocity on the electrostatic signal for different materials and different particle sizes. The sensor tested uses a pair of axially spaced ring electrodes (Fig.1). The velocity of solids was measured by a digital purposed built digital signal processor, whilst the mass concentration data was represented by the RMS magnitude of the AC signal. Typical results obtained are plotted in Fig.3. It can be seen that the signal magnitude (V_{rms}) is a function of both the mass concentration (ρ_m) and the solids velocity (υ). A number of empirical relationships between V_{rms}, ρ_m and υ have been proposed in the past [4, 5], but the authors believe that the relationship can be generalised in the following form:

$$V_{rms} = k\,\beta_m^a\,\upsilon^b \tag{3}$$

where k, a and b are constants, depending upon the properties of the solids, their surrounding environment and the sensor design. It must be stressed that an inappropriately designed electrostatic sensor can result in a more complex relationship than Equation (3).

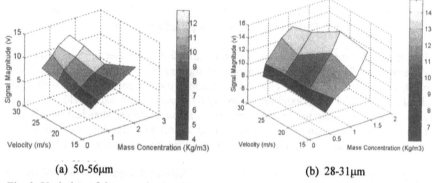

(a) 50-56μm   (b) 28-31μm

Fig. 3. Variation of the magnitude of the electrostatic signal with mass concentration and velocity of solids for different sizes of silcaride black silicon carbide

6. Concluding Remarks
The electrostatic sensors provide a promising solution to the in-line flow monitoring and measurement of pneumatically conveyed solids. The measurement of solids velocity by combining electrostatic sensors and advanced signal processing techniques is superior to any other known techniques. It is a challenging task to develop electrostatic sensors suitable for the absolute measurement of mass concentration of pneumatically conveyed solids due to the dependence of the signal on many other parameters. Such dependence can be minimised by optimising the design of the electrostatic sensors, in particular, the shape and dimensions of the electrodes and the dynamic properties of signal processing electronics.

Acknowledgements
The authors wish to acknowledge the support of PCME Ltd who provided the equipment as required in the experimental investigation.

References
[1]   Yan Y and Johnson P A, A Report for NMSPU, DTI, 1998.
[2]   Yan Y, Measurement Science and Technology, vol.7, no.12, pp.1687-1706, 1996.
[3]   O'Neill B C, Proceedings of the 8th International Conference on Electrostatics, no.118, pp.135-140, 1991
[4]   Gajewski J B, Measurement Science and Technology, vol.7, pp.1766-1775, 1996.
[5]   Yan Y, 5th World Congress for Particle Technology, San Diego, USA, July 1996.

Author Index